Reimar Spohr

# Ion Tracks and Microtechnology

Reimar Spohr

# Ion Tracks
# and Microtechnology
## Principles and Applications

Edited by
Klaus Bethge

**vieweg**

**Author:** Dr. Reimar Spohr
GSI Darmstadt
P.O. Box 11 05 52
D-6100 Darmstadt

**Editor:** Prof. Dr. Klaus Bethge
Institut für Kernphysik der Johann Wolfgang Goethe-Universität
August-Euler-Str. 6
D-6000 Frankfurt 90

Vieweg is a subsidiary company of the Bertelsmann Publishing Group International.

Produced by W. Langelüddecke, Braunschweig

ISBN-13: 978-3-322-83104-0      e-ISBN-13: 978-3-322-83103-3
DOI: 10.1007/978-3-322-83103-3

# Contents

# Preface

The penetration of heavy charged particles through matter has been the subject of investigations since the early days of Bohr's atomic model. Much later it was found that the resulting traces have dimensions close to the atomic scale and can be revealed in the form of fine patterns. Quite recently, this characteristic attracts applications in micro-electronics and -mechanics, biology and medicine, surface and membrane technology, magneto-optics and low temperature physics — applications which require a high subtlety of geometric control on a microscopic scale.

Progress in advanced technologies depends crucially on the refinement of the available tools. On the road into the submicron regime, customary lithographies using visible and ultraviolet light, x rays, and electrons are steadily nearing their physical limits. A central point in the search for better tools is the improvement of irradiation technology.

Ions have a well-defined range of penetration, a high local confinement of the deposited energy and can be generated conveniently in great quantity. The generated damage zones can be stored indefinitely in many insulators and be used to initiate a phase transformation process that changes, removes, or collects material along the latent tracks. Up to now the most common development process is track etching, which acts as a chemical amplifier that dissolves the damaged zone of the latent tracks preferentially and creates etch pits or channels that can be extremely fine, starting around 10 nm and increasing linearly with the etching time.

The most outstanding distinction of the ion track technique is that already one single particle suffices to produce a developable damage of defined lateral width, angle, and depth. In contrast to conventional lithography, the resulting two-step technique is applicable even to mechanically stable, chemically inert, and radiation-resistant dielectrics. This unique combination of recording and revealing properties results in a wide array of potential applications and may earn the ion track technique a place among the structural tools of the future.

This comprehensive introduction to the field of ion tracks represents the first attempt to give a coherent picture of ion tracks in the context of microstructure technology. It describes the physical processes involved in track creation and their development, provides an overview of the main features of the technique and establishes some guidelines of relevant properties which can be the starting point for a new approach to microstructure technology.

The book is intended to introduce the field to students, engineers, and scientists interested in microtechnology. It is the objective of the author to convey insight in the basic mechanisms of the ion track technique and help the reader to acquire the necessary know-how to handle this microtool successfully.

Frankfurt, January 1990

K. Bethge

# Dedication

This work is dedicated to

## P. Armbruster

## B.E. Fischer

and

## Ch. Schmelzer

as a token of sympathy and appreciation

# Acknowledgement

This work is due to the criticism, encouragement, stimulation, and tolerance of the following persons:

| | | | | | |
|---|---|---|---|---|---|
| N. | Angert | F. | Granzer | T.D. | Märk |
| P. | Armbruster | K.O. | Groeneveld | W.D. | Myers |
| O. | Bernaola | S.L. | Guo | Z. | Pengji |
| J.P. | Biersack | J.W. | Hansen | P.B. | Price |
| E. | Bonderup | P. | Hansen | C. | Riedel |
| R. | Brandt | W. | Heinrich | W. | Roesch |
| H. | Bruchertseifer | N. | Itoh | W. | Scharf |
| L.T. | Chadderton | J. | Lindhard | Ch. | Schmelzer |
| M. | Chlupka-Spohr | H.B. | Lück | E. | Schopper |
| H.H. | Cui | R. | Katz | H. | Seidel |
| H. | Delfs | P. | Kienle | G. | Siegert |
| J. | Dietrich | H. | Klein | P. | Sigmund |
| W. | Ehrfeld | H.A. | Khan | G. | Somogyi |
| W. | Enge | O. | Klepper | F. | Studer |
| G. | Eska | J. | Knoll | K. | Thiel |
| G. | Fiedler | P. | Koczon | L. | Tommasino |
| K. | Finlayson | G. | Kraft | C. | Trautmann |
| B.E. | Fischer | A. | Kukoc | E. | Varoquaux |
| R.L. | Fleischer | W. | Kulcke | P. | Vater |
| H. | Geissel | Y. | Langevin | J. | Vetter |
| M. | Gering | D. | Marx | H.S. | Virk |
| B. | Gondesen | W. | Meckbach | K. | Wien |

This work was realized on Apple Macintosh, using Claris MacDraw, Microsoft Basic, Microsoft Excel, Microsoft Word, Wolfram Mathematica.

# Introduction

### Search for improved lithographies

The fabrication of microelectronic circuits depends decisively on the use of lithography. The main steps of this technique are the irradiation and the development of a thin radiation sensitive film — the resist — deposited onto a silicon substrate. The irradiation by visible light, uv, x rays or electrons changes the solubility of the resist film in the developing medium and enables in this way the transfer of information from the mask onto the resist film. The structured resist film itself serves in turn as a stencil for transferring the resist film pattern onto the silicon substrate.

Since the invention of the transistor in 1948 by Bardeen, Brattain, and Shockley, the packing density of transistors — combined to integrated circuits by Kilby about 1959 — has roughly doubled every two years, starting at about 1 bit per $cm^2$ and reaching approximately 4 Mbit per $cm^2$ at present. The steadily rising need for higher and higher storage capacities and shorter switching times has led to resolution requirements that have almost reached the physical limits of the conventional techniques using light.

Deep ultraviolet light permits to generate patterns with line-widths down to approximately 0.4 µm. This may be adequate to manufacture a 64 Mbit chip on a surface of about 2 $cm^2$. For the then following generations of semiconductor memories with structures down to 0.25 µm electron or x ray lithography can be used. On the way to still finer structures and higher packaging densities, alternative structuring technologies have to be found.

Besides miniaturization in the field of microelectronics, another trend becomes visible in the steady refinement of microtechnology for which ion tracks can offer new perspectives. In a quiet revolution, micromechanical parts are starting to be manufactured and combined with microelectronic circuits to autonomous microsystems. The mechanical parts comprise membranes, valves, nozzles, tuning forks, accelerometers, and even mechanical gears, and crank shafts. The conferences taking place in this field [1] are a lively expression of something qualitatively new which may change our technological perspectives decisively. The goal of micromechanics is to create miniaturized multi-functional devices — autonomous microsystems containing mechanical as well as electronic

---

[1]     Gabriel, K.J., W.S.N. Trimmer (editors): "**Proceedings of the 1987 IEEE Micro Robots and Teleoperators Workshop. An Investigation of Micromechanical Structures, Actuators, and Sensors.**" IEEE Catalog Number 87TH0204-8, Library of Congress Number 87-82657, (1987).

components on the same piece of solid material — such as microsensors and their counterpart, microactors. This combination of microequipment requires deep-cutting lithographic tools with high lateral control and depth-resolution. Todays technologies are just overcoming the still dominating planar technique, in which the structured patterns have a negligible depth in comparison to their lateral dimensions.

The limitations of the conventional lithographies have led to an intensive search for alternative microtools. Since diffraction is the main limitation in visible light lithography, one trend is to use photon radiations of shorter wavelength, such as uv radiation and x rays. Another technique to achieve three-dimensional structures is "reactive ion etching". In this technique ion milling is enhanced by chemical reaction in statu nascendi. Depending on the pressure of reactive gases, etching can be performed unidirectionally or isotropically.

When considering lithography, there are various practical reasons for using charged particles in future lithographies instead of photon radiations. In contrast to photons, charged particles — such as electrons and ions — can be conveniently generated, deflected, and focussed electromagnetically without any interaction with matter. The energy, intensity, and angle of charged particles can be adapted to the specific needs simply by adjustment of electromagnetic fields that accelerate, focus, or deflect the charged particles. In this way the penetration-depth and direction of charged particles can be conveniently accommodated to the specific needs. Charged particles can thus be handled much more easily than photons. Finally, charged particles have a strong interaction with matter, in contrast to photons, which are absorbed much less strongly in solids and follow an exponential law with rather large penetration depths.

**Figure 1** *Stochastically distributed etched ion tracks* in polyethylene terephtalate (Hostaphan, Hoechst AG, D-6200 Wiesbaden), etched in NaOH solution with methanol and a detergent.

A promising step in the search for improved irradiation techniques is the ion track technique which enables linear structures of considerable depth (Figure 1) and may become an important microtool below 0.1 μm line-width. There exist principally two modes of operation of such a microtool. Either — using a proper masking or aiming technique — regular track distributions can be employed to change local properties, or, stochastical track distributions can be employed to change global properties.

### Single-particle recording

In comparison with the much lighter electrons, ions create a much higher local concentration of the deposited radiation damage. This leads to a quantitatively new behavior — known in microelectronics as "single event upset" — , the possibility to record and develop discrete tracks, corresponding to one single particle. In other words, already one ion suffices to induce — physically or chemically — a submicroscopic change in the recording material (resist or other dielectric material) and thereby can render it susceptible to a development process. Until the advent of the ion track technique, the radiation-sensitive film could only be sensitized through the accumulated interaction of a large number of particles in the same critical volume. Single-particle recording therefore appears to be a most exciting aspect for future uses of the ion track technique (Figure 2).

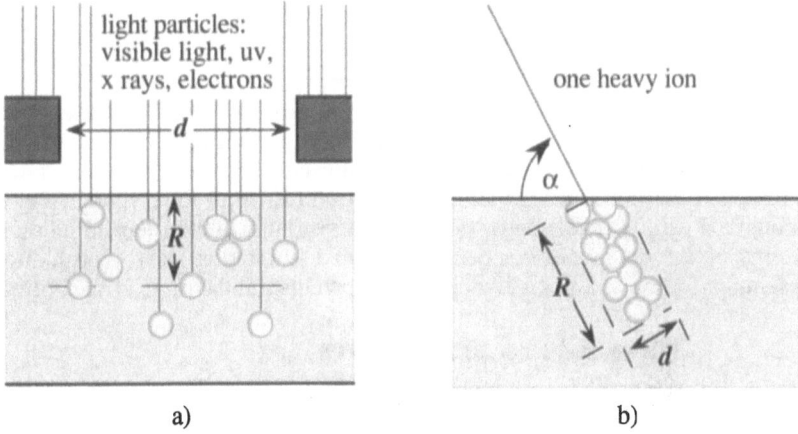

a)                                                                    b)

**Figure 2   Comparison of conventional lithographies with the ion track technique.**

(a) *Conventional lithography*. Light particles deposit their energy in few interactions within the radiation sensitive polymer. The radiation damage caused by one single photon or electron remains in most cases undetectable. Only the joint effect of many particles within a given volume renders a contingent zone of the resist developable. The absorption events occur at various depths and the radiation effect decreases — often exponentially — with increasing penetration-depth. A mask is necessary to restrict the exposed zone laterally.

(b) *Ion track technique*. One energetic ion deposits its energy quasi-continuously in many interactions within the insulating solid, whereby the strength of the interaction is determined by its velocity and effective charge. During the slowing-down process the direction of the penetrating ion is maintained with a high degree of precision. Since the radiation effect of a single ion is large and highly concentrated, already one individual ion is capable to register as a developable damage. A mask is not necessary to create an elementary microstructure.

Another aspect of ion tracks is that they cover a much wider class of materials and are not constrained to specifically designed radiation sensitive resists. Most insulating materials belong to the class of track recording materials. Many radiation resistant polymers, glasses, and crystals have been already structured,  for example quartz glass which in the form of a silicon dioxide layer is a frequently used protective and insulating film in semiconductor technology. In this way the class of lithographically accessible materials widens dramatically and covers most homogeneous insulators, materials which usually cannot be structured directly by light particle radiations, such as light, uv, x rays or electrons.

Inaccessible to a direct track process are still some crystals with high binding energies, such as diamond and sapphire. Furthermore, materials with high electronic or atomic mobilities, such as metals and most semiconductors which cannot be structured directly by ion tracks — at least at present. However, in some cases — giving up the feature of single-particle recording as in conventional lithographies — the combined damage of many tracks concentrated within a small volume can still render some materials susceptible to a development in which the collectively damaged volume is removed preferentially. As is common practice in conventional lithography, many of the materials inaccessible to a direct structuring process — for example metals — can be shaped indirectly by transferring a structure from some suitable track recording insulator to the desired material. Thus ion lithography might be used ultimately to create structures close to the natural limits of digital electronics [1], [2], as given by the local fluctuations due to the finite size of the atoms.

## Perspectives of ion microbeams

Ion microbeams offer several new features which are absent in the present lithographies using light particles. Such an "ultimate" microtool could provide the versatility of a computer-controlled mechanical lathe enabling the generation of microstructures with individual ions, at high lateral resolution (down to the size of latent ion tracks of about 0.01 μm), variable angle of incidence, variable depth (depending on the available energy of the ion accelerator between about 1 μm and at least 1 mm), and high depth-resolution (of the order of a few percent, depending on the mass of the projectile) [3].

## Three-dimensional structures

While the present microelectronics is mainly based on the lithographic shaping of thin photoresist films roughly of the order of 1 μm in thickness — known as the planar technique — micromechanical devices such as sensors or actors require often tools which are capable to cut much deeper into the solid.

There exists an advantage of heavy ions in comparison with light ions since the scattering of ions in a solid decreases with increasing ion mass. If the employed ions are much heavier than the matrix atoms, the incident ions follow almost a straight path (Figure 3) and come to rest at a nicely defined depth. Thus the "machining" process can be stopped at a well-defined depth. This is a decisive advantage of heavy ions over lighter particles, even lighter ions, such as protons or helium ions.

Due to its well-defined direction and depth, the ion track technique can be compared to a lathe for which high aspect ratios — the ratio between depth and width of the removed zone— can be obtained at arbitrary angle of incidence [4].

[1]    Keyes, R.W.: "Physical Limits of Digital Electronics." Proc. IEEE 63, 740, (1975).
[2]    Wallmark, J.T.: "Fundamental Physical Limitations in Integrated Electronic Circuits." Inst. Phys. Conf. Ser. 25, 133, (1975).
[3]    Fischer, B.E., Gesellschaft für Schwerionenforschung (GSI), Planckstr. 1, D-6100 Darmstadt.
[4]    Reactive ion etching is another such "microlathe", employed to cut vertical structures of 1 μm width and 7 μm depth into silicon for manufacturing condensors in the new 4 Mbit memories.

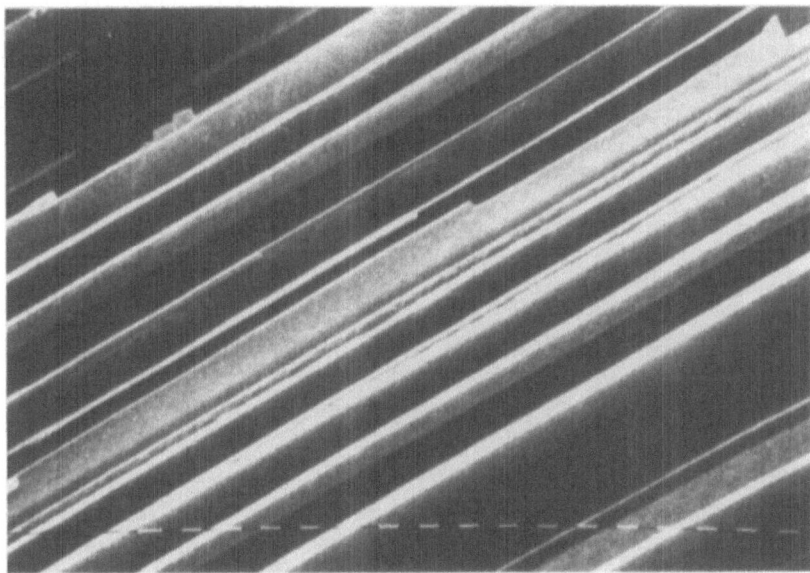

**Figure 3** *Ion tracks of high length and parallelism* in mica, $KAl_2Si_3AlO_{10}(OH,F)_2$, etched in HF solution [1]. Bar unit = 1 μm.

## Comparison with conventional lithographies

In comparing different lithographic techniques, the most important term up to now was "line-width" (or its inverse "lateral resolution") [2]. Line-width is given by the smallest attainable width of a trench or ridge of approximately rectangular cross section cut into the resist film. In planar lithographies, in which the depth of the generated structure plays a minor role, this characterization is sufficient for most applications. However, if deep trenches or high ridges are required, simultaneously with the line-width the depth of the structure has to be specified. Alternatively the term "aspect ratio", defined by the ratio between the depth (or height) of the structure and its lateral width, is used.

In Figure 4 the possibility of the ion track technique to generate fine structures of great depth is compared with conventional lithographies. Thereby visible light lithography is used as the basis of comparison. Visible-light lithography, capable of structures approximately 1 μm wide and 1 μm deep, is diffraction-limited, corresponding to the wavelength of visible light. Electron lithography provides an improved resolution and aspect ratio by about one order of magnitude. Its present line-width is about 0.1 μm at a penetration range of about 1 μm, yielding an aspect ratio (depth divided by width) of about 10:1. The line-width of x ray lithography corresponds roughly to the fineness of the the employed electron lithographic mask of about 0.1 μm. It achieves a much higher aspect ratio of about $10^3$:1. Both, x ray and electron lithography are not limited by diffraction but by scattering during the energy transfer process. The present limit of electron and x ray lithography — restricted to thin films — is around 0.05 μm.

[1] Vetter, J., Gesellschaft für Schwerionenforschung (GSI), Planckstr. 1, D-6100 Darmstadt.
[2] In practice, "overlay" is as important as line-width. Overlay relates the local placement of different manufacturing steps mutually. The presently achieved overlay precision is about $1 / 10^5$.

Ion tracks enable the generation of deep and fine channels. For certain glasses aspect ratios up to $10^2{:}1$ and slightly above have been obtained [1], [2], [3], [4], [5]. In allyl diglycol carbonate aspect ratios of $10^3{:}1$ have been observed [6]. For polycarbonate aspect ratios between $10^2{:}1$ and $2{\cdot}10^4{:}1$ have been observed at diameters slightly below $0.015$ μm [7], [8]. The maximum aspect ratio in ionic crystals (mica) should be of the order of $10^5{:}1$ at diameters of the order of $0.007$ μm [9].

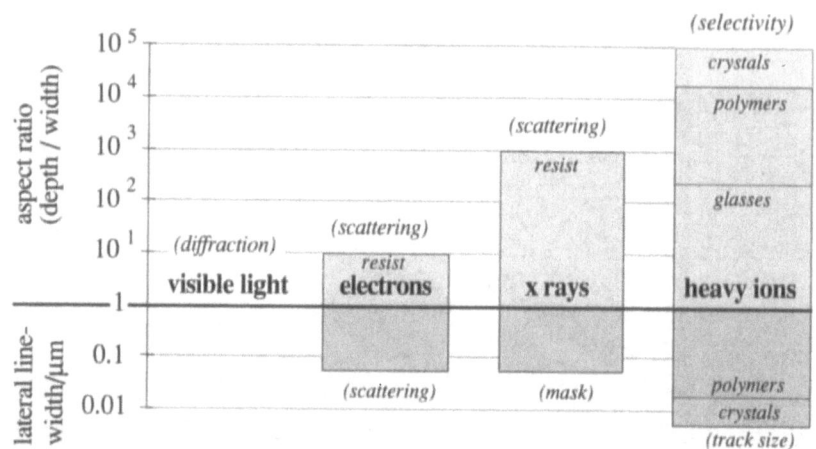

**Figure 4**  *Aspect-ratio and lateral line-width of various microtools.* The technological limits are indicated in brackets.

In contrast to the conventional techniques using light particles, such as photons or electrons, the ion track technique is limited not by the electronic but by the much more concentrated atomic damage around the ion trajectory in the solid which has a diameter of the order of $0.01$ μm. The penetration-depth depends on the available energy of the ion and can be $1000$ μm and more. The aspect ratio is limited by the selectivity of the track-etch process. This is the ratio between the dissolution rate of the track zone and the dissolution rate of the undisturbed matter.

[1]   Aschenbach, J., G. Fiedler, H. Schreck-Köllner, G. Siegert: "Special Glasses as Energy Detectors for Fission Fragments." Nuclear Instruments and Methods (NIM), **116**, 389-395, (1974).

[2]   Fiedler, G., J. Aschenbach, W. Otto, T. Rautenberg, U. Steinhauser, G. Siegert: "Further Developments with Glass Detectors for Heavy Ions and Fission Fragments." NIM, **147**, 35-39, (1977).

[3]   Fiedler, G., U. Steinhauser, T. Rautenberg, R. Haag, P.A. Gottschalk: "Glass Detectors for Heavy Ions." NIM, **173**, 85-92, (1980).

[4]   Dürolf, H., B. Genswürger, R. Spohr: "Eignungsuntersuchung von photoätzbarem Glas als Kernspurmaterial." GSI-Jahresbericht 1977, p. 131, (1977).

[5]   Wang, S., S.W. Barwick, D. Ifft, P.B. Price, A.J. Westphal: "Phosphate Glass Detectors with High Sensitivity to Nuclear Particles." Nuclear Instruments and Methods in Physics Research, **B35**, pp. 43-49, (1988).

[6]   Heyna, U., W. Enge, G. Sermund, R. Beaujean: "Measurements of Transversal Etching Rates of Uranium Tracks in CR-39." Nuclear Tracks and Radiation Measurements, **12**, No. 1-6, p. 33, (1986).

[7]   Schnoor, G., H. Schütt, R. Beaujean, W. Enge: "Electrolytical Studies of Submicroscopic Nuclear Tracks in Plastic Detectors." Nuclear Tracks, **3**, p. 51, (1982).

[8]   Guillot, G., F. Rondelez: "Characteristics of Submicron Pores Obtained by Chemical Etching of Nuclear Tracks in Polycarbonate film." J. Appl. Phys. **52**, 7155, (1981).

[9]   Bean, C.P., M.V. Doyle, G. Entine: "Etching of Submicron Pores in Irradiated Mica." J. Appl. Phys. **41**, pp. 1454-1459, (1970).

## Materials technology

Besides the structural aspect of conventional lithographies, a new property emerges in the ion track technique, namely the possibility to change properties of materials by ion tracks in a controlled way on a microscopic scale. As the magnitude of the generated structures approaches certain minimum dimensions such as domain size, wavelength, or mean-free-path, it becomes possible to access new phenomena and manipulate materials in new and unexpected ways [1].

The track-induced effects increase with increasing nominal porosity — the product of the number of ion tracks per unit area and the cross section of the individual tracks. Depending on the track density, three regimes of use can be distinguished (Figure 5). With increasing track density the irradiated solid undergoes a gradual transition from a practically unchanged solid (single tracks) over a modified solid (track arrays) to a completely transformed solid (overlapping tracks).

First, the single track regime enables extremely localized changes of the solid properties, defined by the size and direction of the track. Second, the regime of more or less discrete ion tracks enables to tailor globally distributed properties which gradually dominate the behavior of the solid with increasing track density. By the collective interaction of many ion tracks, new directionally dependent properties can be created, whereby the track angle controls the direction of the anisotropy. Third, the regime of overlapping tracks leads ultimately to a complete transformation, replacement, or dissolution of the original solid. Thereby the original properties of the solid vanish gradually. This enables an ion lithography with high depth of structuring and steep edges.

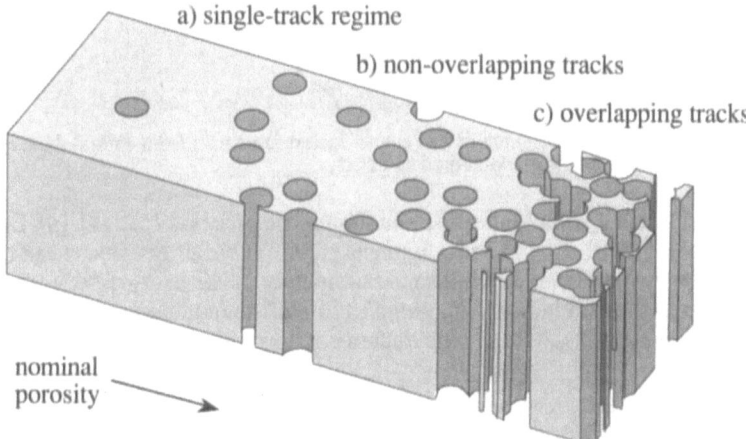

**Figure 5 Track density regimes in materials technology.** (a) *Single-tracks*. Every track represents a discrete, highly localized physical change of the material properties at the site of the track. (b) *Non-overlapping tracks* induce global changes of surface and volume properties. (c) *Overlapping tracks*. The original bulk properties of the solid vanish gradually until ultimately the solid is completely removed.

[1]  Smith, H.T.: "The Impact of Submicrometer Structures in Research and Applications." Proceedings of the International Conference on Microlithography, Amsterdam, 30 Sept. - 2 Oct., 1980, edited by R.P. Kramer, Delft University, Delft, (1981).

### Historic origin of ion tracks

Wilson's cloud chamber [1] enabled in 1898 the observation of individual alpha particles in supersaturated moist air. Thereby a string of microscopic water droplets is formed, triggered by the secondary ions created along the path of an alpha particle. The rapidly fading phenomenon reveals a direct glimpse into the atomic realm.

Sixty years later, in 1958, Young observed the first persistent "nuclear tracks" in the solid phase [2]. While tracks in fluids are transient phenomena of short duration, wiped-out after fractions of a second, tracks in solids can endure millions of years in the solid, often without perceivable change (Figure 6). The "frozen" damage of fossil tracks in solids is thus capable to record and reveal events from the far back history of our world. This discovery established the basis for the ion track technology outlined here.

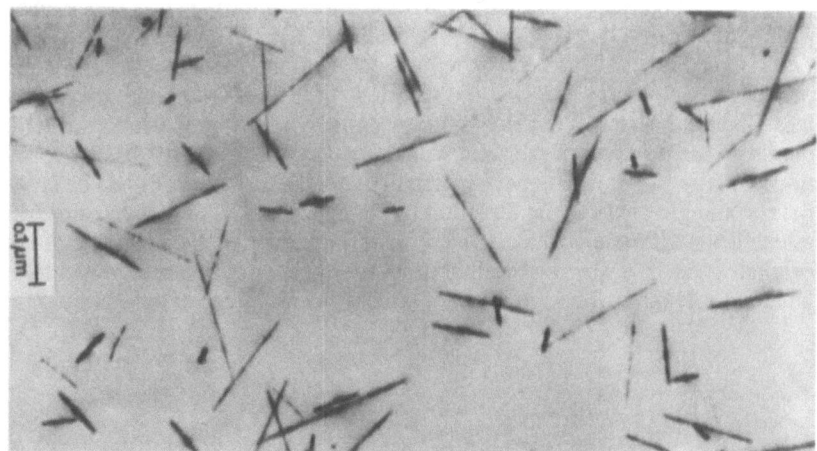

**Figure 6** *Fossil latent tracks* due to fission fragments from natural uranium in mica [3], the entree into the field of particle tracks.

The close relation between photographic processes [4], [5], [6], [7] and etched ion tracks is exemplified by the development of the "nuclear" emulsions used for registration of fast heavy particles, which resulted ultimately in the development of single-crystal silver-halide track recorders [8], [9], representing (as homogeneous-phase systems) an intermediate step between photographic track recording and track etching.

[1]     Wilson, C.T.R. Phil.Trans. **A189**, 265, (1897).
[2]     Young, D.A.: "Etching of Radiation Damage in Lithium Fluoride." Nature **182**, 375-377 (1958)
[3]     Fleischer, R.L., P.B. Price, R.M. Walker: **"Nuclear Tracks in Solids: Principles and Applications."** University of California, Berkeley, 1-605, (1975).
[4]     James, T.H. (editor): **"The Theory of the Photographic Process."** 4th edition, MacMillan Publ. Co. Inc., New York, 714 pp. (1977)
[5]     Frieser, H., G. Haase, E. Klein: **"Die Grundlage der photographischen Prozesse mit Silberhalogeniden."** Vols. **1** - **3**, Akademische Verlagsgesellschaft, Frankfurt am Main, (1968).
[6]     Granzer, F., E. Moisar: "Der photographische Elementarprozess in Silberhalogeniden." Physik in unserer Zeit, **12.2**, p. 36, (1981).
[7]     Böttcher, H., J. Epperlein: **"Moderne photographische Systeme. Wirkprinzipien, Prozesse, Materialien. "** VEB Deutscher Verlg für Grundstoffindustrie, Leipzig, 328 pp., (1983).
[8]     Childs, C., L. Slifkin, Bull. Amer. Phys. Soc. **6**, 52, (1961).
[9]     Haase, G., E. Schopper, F. Granzer: "Solid State Nuclear Track Detectors: Track Forming, Stabilizing and Development Processes." Radiat. Eff. **34**, 25, (1977).

The observation of "historically" recorded events developed rapidly. Several mysteries in geologic, solar and stellar history were tackled. The principal aspects of the diagnostic technique and its applications are laid down by Fleischer, Price and Walker in 1975 [1] who recognized the practical uses of the technique. The recent developments in track diagnostics [2] are reported in a conference series [3], [4], [5], [6], [7], [8], [9], and in the journals Radiation Effects and Defects in Solids, Nuclear Tracks and Radiation Measurements, Nuclear Instruments and Methods, and in Applied Physics.

In the literature, covering the observation of recorded ion tracks, most frequently the term "Solid State Nuclear Track Detector" with the acronym "SSNTD" is applied. This term stresses the diagnostic perspective but, does not apply to the tool aspect of the technique. In the context of structural applications therefore term "ion track" is preferred, pointing directly to the particle type and the resulting effect.

Few years after the first observation of etchable tracks in solids, it was recognized that particle tracks can be created at will in the sense of a structural tool, using heavy fragments from nuclear decay or heavy ions from accelerators. Slowly the concept of a passive diagnostic tool was replaced by that of an active structural tool for imprinting structure onto solids [10]. The first practical application that found wide acceptance was the "nuclear track filter", consisting of a track-perforated thin polymer membrane [11].

While photography became the most important recording medium in the world of observation, lithography became the foremost tool in microstructure technology. In the same way as photography and lithography have undergone a fruitful co-evolution, track observation and track structure technology can profit from each other. A mutual response has just begun as well as revisiting atomic physics in the context of ion-matter interaction [12]. In one direction, the techniques for the observation of ancient tracks can be applied to improve the precision of the artificially generated structures. In the opposite direction, artificially created ion tracks from accelerators can be used to calibrate naturally occurring tracks and thus establishing their charge and energy. The following section gives an overview on the observational roots of the technique.

[1]     Fleischer, R.L., P.B. Price, R.M. Walker: "**Nuclear Tracks in Solids: Principles and Applications.**" University of California, Berkeley, 1-605, (1975).

[2]     Durrani, S.A., R.K. Bull: "Solid State Nuclear Track Detection." Pergamon Press, Oxford, 304 pp., (1987).

[3]     Granzer, F. H. Paretzke, E. Schopper (editors): "**Solid State Nuclear Track Detectors.**" Proceedings of the 9th International Conference, Munich, 30 Sept. - 6 Oct. 1976, Pergamon Press, (1978).

[4]     Khan, H.A., K. Kristiansson (editors): "**Solid State Nuclear Track Detectors.**" Proceedings of the International Spring School, Rawalpindi, 5-14 March, 1979, Nucl. Instrum. & Meth., **173**, (1980).

[5]     François, H., N. Kurtz, J.P. Massue, M. Monnin, R. Schmitt, S.A. Durrani (editors): "**Solid State Nuclear Track Detectors.**" 10th Int. Conf., Lyon, 2-6 July 1979, Pergamon Press, Oxford, (1980).

[6]     Fowler, P.H., V.M. Clapham (editors): "**Solid State Nuclear Track Detectors.**" Proceedings of the 11th International Conference, Bristol, 7-12 Sept. 1981, Pergamon Press, Oxford, (1982).

[7]     Espinosa, G. R.V. Griffith, L. Tommasino, S.A. Durrani, E.V. Benton (editors): "**Solid State Nuclear Track Detectors.**" Proc. 12th Int. Conf., Acapulco, Mexico, 4-10 Sept. 1983, Pergamon Press, Oxford, (1984).

[8]     Tommasino, L., G. Baroni, G. Campos-Venuti (editors): "**Solid State Nuclear Track Detectors.**" Proc. of the 13th Internat. Conference, Rome, 23-27 Sept. 1985, Pergamon Press, 1-988, (1986).

[9]     Khan, H.A., I.E. Qureshi, I. Ahmad, S.A. Durrani, E.V. Benton (editors): "**Solid State Nuclear Track Detectors.**" Proc. 14th Int. Conf., Lahore, Pakistan, 2-6 April 1988, Pergamon Press, 1-802, (1988).

[10]    Fischer, B.E., R. Spohr,: "Production and Use of Nuclear Tracks: Imprinting Structure on Solids." Rev.Mod.Phys., **55**, No.4, 907-948, (1983).

[11]    Price, P.B., R.M.Walker: "Molecular Sieves and Methods for Producing Same." US Patent 3,303,085, Feb. 7 1-6 (1967).

[12]    Armbruster, P., J.C. Jousset, J. Remillieux (chairmen): "**Symposium on Swift Heavy Ions in Matter.**" Caen, May 16-19, 1989, Radiation Effects and Defects in Solids, **110**, Nos. 1-2, (1989).

## Diagnostic applications of particle tracks [1]

The physical parameters of particle tracks — such as length, diameter, direction, areal- and volume density, and thermal stability — contain a wealth of information on the particle source, the emitted particles, the medium between source and target, and the target material itself.

The development of the particle track technique and its application in various fields of science and technology is firmly based on the successful deciphering of this information. Particle tracks turned out to be a powerful analytical and diagnostic tool in such different disciplines as nuclear physics, elementary particle physics, cosmic ray physics, geo- and cosmochemistry, geophysics, astrophysics, cosmology, dosimetry, archeology, biology, medicine, and health physics. The large diversity of these applications may be illustrated by some highlights:

- Tracks of spontaneous-fission products of $^{238}U$ allow the dating of sample ages between a few years and more than $10^9$ years [2], [3]. On the other extreme — applying the "blocking effect" to heavy ions in regular crystal matrices — it is possible to measure life-times of compound nuclei between $10^{-16}$ and $10^{-19}$ seconds [4].

- Autoradiography of chemical elements by means of neutron, proton, or $\alpha$ particle induced tracks is often the only accessible way to determine the spatial distribution of spurious elements in microscopic samples [5], [6], [7]. In the macro-world, particle tracks provide an economic alternative for prospecting uranium ore deposits by the determination of large-scale (1 to 100 $km^2$) uranium distributions in rocks. This technique is based on alpha tracks originating from the decay of $^{222}Rn$, recorded in polymer films, thus serving as uranium indicators [8].

- The selective sensitivity of track recorders for heavy particles — even at high background levels of $\beta$ and $\gamma$ radiation — gives the opportunity to determine fission cross sections in the picobarn ($10^{-36}$ $cm^2$) range [9] and to push down the detection limits in the search for super heavy elements in nature [10], [11]. One of the most spectacular efforts in the search for magnetic monopoles was an experiment using a balloon stack of Cerenkov films, nuclear emulsions, and organic polymer films [12].

[1]    This section is due to K. Thiel, Department for Nuclear Chemistry, Institute for Biochemistry, University of Cologne, D-5000 Köln.
[2]    Fleischer, R.L., P.B. Price, R.M. Walker: **"Nuclear Tracks in Solids: Principles and Applications."** University of California, Berkeley, 1-605, (1975).
[3]    MacDougall, J.D.: Scientific American, **235**, 114 (1976).
[4]    Brown, F., D.A. Marsden, R.D. Werner: Phys. Rev. Letters **20**, 1449, (1968).
[5]    Fleischer, R.L., O.G. Raabe: Health Phys. **32**, 253, (1977).
[6]    Thiel, K., W. Herr, J. Becker: Earth Planet. Sci. Lett. **16**, 31, (1972).
[7]    Thiel, K., R. Saager, H.D. Stupp: "Some Applications of Uranium Tracing in Geological Samples Using Thermal Neutron Induced Fission." Proc. 14th Int. Conference on Solid State Nuclear Track Detectors, Lahore, Pakistan, 2-7 April 1988, Pergamon Press, N.Y., p. 775, (1989).
[8]    Gingrich, J.E.: Power Eng. no **8** (Aug.), 48, (1973).
[9]    Burnett, D.S., R.C. Gatti, F. Plasil, P.B. Price, W.J. Swiatecki, S.G. Thompson: Phys. Rev. **134 B**, 952, (1964).
[10]   Brandt, R.: "Search for Superheavy Elements in Nature and in Heavy Ion Reactions." Superheavy Elements, Proc. Int. Symp. on Superheavy Elements, Lubbock, Tex. March 9-11, (1978), Editor: M.A.K. Lodhi, Pergamon Press, p. 103, (1978).
[11]   Brandt, R., T. Lund, D. Molzahn, P. Vater, H. Jungclas, A. Marionov: Nucl. Instr. Meth. **173**, 121, (1980).
[12]   Price, P.B., E.K. Shirk, W.Z. Osborne, L.S. Pinsky: Phys. Rev. Lett. **35**, 487, (1975).

&#x25E6;     The existence of a track-etch threshold can be exploited to observe rare emission events in a high background of $\alpha$ particles which is many orders of magnitude higher than the observed heavy-particle flux. This selective detection technique enabled the discovery of new decay modes of radioactive nuclei associated with the emission of carbon, neon, magnesium, and silicon [1], [2], [3], [4], [5].

&#x25E6;     By inverting the technique of fission track dating, track recorders — especially glasses — of known age containing uranium allow the determination of the $^{238}U$ spontaneous-fission decay constant. A review of the various attempts to establish this basic physical constant is given in [6].

&#x25E6;     A unique application of heavy-ion tracks in mica and glasses is the investigation of ion-induced multiple fission of very heavy nuclides [7], [8]. Since the trajectories of the reacting particles are recorded in three dimensions, details of the reaction mechanism in extremely rare events can be deduced by measuring angular and range correlations of the observed tracks.

&#x25E6;     Using a special holographic technique — zone plate coded imaging — it was possible to record the spatial distribution of the thermonuclear $\alpha$ emission in a laser-induced fusion experiment of very short duration [9].

&#x25E6;     A completely different localizing technique is based on the registration of high energy ions in polymer films after passing through organic tissue of inhomogeneous mass density [10]. This heavy-ion radiography allows cancer diagnosis in mammography at radiation levels far below conventional x ray diagnostics. On a microscopic scale, heavy-ion radiography has been applied to reveal inner details of small insects and microorganisms, a technique which has direct consequences for ion lithography [11].

&#x25E6;     Particle tracks turned out to be a very powerful tool for measuring extremely small ion sputtering rates of solid surfaces [12], for unraveling quite complex uranium-mobilization processes in geological samples [13], and for studying the redox character of the early Precambrian atmosphere [14].

&#x25E6;     The availability of lunar samples for laboratory analysis after 1969 — in the

[1]    Price, P.B., J.D. Stevenson, S.W. Barwick, H.L. Ravin: "Discovery of Radioactive Decay of $^{222}Ra$ and $^{224}Ra$ by $^{14}C$ Emission.". Phys. Rev. Letters, **54**, 297, (1985).

[2]    Barwick, W.W., P.B. Price, J.D. Stevenson: "Radioactive Decay of $^{232}U$ by $^{24}Ne$ Emission." Phys.Rev. C, **31**, 1984, (1985).

[3]    Wang, S., P.B. Price, S.W. Barwick: "Radioactive decay of $^{234}U$ via Ne and Mg Emission." Phys. Rev. C, **36**, 2717, (1987).

[4]    Wang, S., D. Snowden-Ifft, P.B. Price: Heavy-Fragment Radioactivity of $^{238}Pu$: Si and Mg Emission." Phys. Rev. C, **39**, 1647, (1989).

[5]    Price, P.: "Heavy-Particle Radioactivity." Annual Rev. of Nuclear and Particle Sci. **39**, 19, (1989).

[6]    Thiel, K. W. Herr: Earth Planet. Sci. Letters **30**, 50, (1976).

[7]    Vater, P., H.J. Becker, R. Brandt, H. Freiesleben: Phys. Rev. Lett. **39**, 594, (1977).

[8]    Khan, H.A., P. Vater, R. Brandt: Proc. 10th Int. Conf. on Solid State Nuclear Track Detectors, Lyon, 2-6 July 1979, Editors: H. François, N. Kurtz, J.P. Massue, M. Monnin, R. Schmitt, S.A. Durrani, Pergamon Press, p. 915, (1980).

[9]    Ceglio, N.M., E.V. Benton: 10th International Conf. on Solid State Nuclear Track Detectors, Lyon, 2-6 July 1979, Paper No 3-6-2, (1979).

[10]   Benton, E.V., C.A. Tobias, R.P. Henke, M.R. Cruty: Proc. 9th Int. Conf. on Solid State Nuclear Track Detectors, Neuherberg/München, Sept. 30-Oct. 6, 1976, Editors: F. Granzer, H. Paretzke, E. Schopper, Pergamon Press, Vol. **2**, p. 739, (1978).

[11]   Fischer, B.E., B. Genswürger, R. Spohr: 10th Int. Conf. on Solid State Nuclear Track Detectors, Lyon, 2-6 July 1979, Paper No 3-5-3, (1979).

[12]   Thiel, K., H. Külzer, W. Herr: Nucl. Tracks, **4**, 19, (1980).

[13]   Thiel, K. R. Vorwerk, R. Saager, H.D. Stupp: Earth Planet. Sci. Letters, **65**, 249, (1983).

[14]   Thiel, K., R. Saager, R. Muff: Nucl Instr. and Methods **173**, 223, (1980).

course of the sample return missions of the USA and the USSR — became a tremendous challenge for the particle track technique. The valuable sample material prompted the refinement of existing and the development of new, ultra-sensitive track techniques [1]. A similar incitement is due to the recovery of an increasing number of Antarctic meteorites [2].

° A crucial in-situ track experiment on the lunar surface was the so-called Lunar Neutron Probe Experiment (LNPE) [3]. The resulting depth profile of the thermal neutron-flux was the basis for a better understanding of cosmic ray energy spectra and lunar surface dynamics [4].

° The selective sensitivity of track recorders for heavy particles at high background levels of lower energy-loss particles enabled to investigate the effect of cosmic ray nuclei on biological objects in space experiments [5].

° Plastic track-recorders have been widely used in heavy-ion cosmic-ray research (Z>2), using high altitude balloons, rockets, and satellites [6], [7]. In 1971 plastic track recorders [8] and independently electronic instruments were the first to detect the isotopes of nitrogen and boron in galactic cosmic rays, representing the only source of matter from outside our solar system [9], [10]. Recent studies refined the observation of latent tracks in single cosmic dust grains down to sub-$\mu$m sizes [11], [12].

° Plastic track-recorders are well suited to measure the rare heavy cosmic ray ions (Z> 26) because of their large collecting areas, as compared with electronic detectors. A rich harvest of such rare events is expected from a long duration satellite flight [13], carrying 20 $m^2$ of plastic recorders, which should be recovered by the end of 1989, after about 6 years of exposure.

° The investigation of a possible fission track excess due to the spontaneous fission of extinct $^{244}$Pu was a first step to obtain reliable fission track ages of

[1]   Proc. Apollo 11 Lunar Sci. Conf., Geochim. Cosmochim. Acta, Suppl. 1, (1970) and the following LPSC Proceedings.

[2]   Proceedings of the Annual Meetings of the Meteoritical Society, published in Meteoritics (Journal of the Meteoritical Society).

[3]   Woolum, D.S., D.S. Burnett: Earth Planet. Sci. Lett. **21**, 153, (1974).

[4]   Proc. Apollo 11 Lunar Sci. Conf., Geochim. Cosmochim. Acta, Suppl. 1, (1970) and ff.

[5]   Bücker, H., G. Horneck, R. Facius, G. Reitz, M. Schäfer, J.U. Schott, R. Beaujean, W. Enge, E. Schopper, W. Heinrich, J. Beer, B. Wiegel, R. Pfohl, H. Francois, G. Portal, S.L. Bonting, E.H. Graul, W. Rüther, A.R. Kranz, U. Bork, K. Koller-Lampert, B. Kirchheim, M.E. Starke, H. Planel, M. Delpoux: "Radiobiological Advanced Biostack Experiment." Science, **225**, pp. 222-224, (1984).

[6]   Benton, E.V.: "Summary of Radiation Dosimetry Results on US and Soviet Manned Spacecraft." Adv. Space Res. **6**, pp. 315-328, (1986).

[7]   Heinrich, W., B. Wiegel, T. Ohrndorf, H. Bücker, G. Reitz, J.U. Schott: "LET Spectra of Cosmic-Ray Nuclei for Near-Earth Orbits." Rad. Res. **118**, pp. 63-82, (1989).

[8]   Beaujean, R. W. Enge: "Isotopenanalyse an niederenergetischen Teilchen der Elemente Bor, Kohlenstoff, Stickstoff und Sauerstoff aus der kosmischen Strahlung in Plastik Detektoren." Z. Physik, **256**, 416-440, (1972).

[9]   Enge, W.: "Isotopic Composition of Cosmic Ray Nuclei." Nucl. Instr. Meth. **147**, pp. 211-220, (1977).

[10]  Fukui, K., W. Enge, R. Beaujean: "Beryllium Isotopes in Cosmic Radiation Measured with Plastic Detectors." Z. Physik, **A 277**, pp. 99-105, (1976).

[11]  Bradley, J.P. D.E. Brownle, P. Fraundorf: Science **226**, 1432, (1984).

[12]  Thiel, K. J.P. Bradley, R. Spohr: "On the nature of latent nuclear tracks in cosmic dust particles". Proceedings of the 14th International Conference on Solid State Nuclear Track Detectors, Lahore, Pakistan, 2-7 April 1988, Pergamon Press, N.Y., p. 685, (1989).

[13]  Thompson, A., D. O'Sullivan, C. Domingo, K.-P. Wenzel, V. Domingo: "Extended Exposure for Ultra Heavy Cosmic Ray Experiment on the LDEF Space Craft." 20th Int. Cosmic Ray Conf., Moscow, Vol.2, pp. 402-405, (1987).

extraterrestrial samples [1]. Simulation experiments with protons revealed a more complex situation and confirmed the importance of cosmic-ray induced heavy-element fission for fission track dating of lunar surface samples [2].

° The combination of the simple technique of track recording with advanced image processing [3] has opened new possibilities in the field of heavy ion physics [4], enabling relativistic heavy-ion fragmentation experiments in which typically $10^6$ etched tracks were identified and precisely measured.

## Ion tracks and accelerator technology

The practical aspects of the ion track technique were mainly developed at high energy heavy-ion accelerators above 1 MeV/u since about 1970 which enable ion ranges adequate for convenient observation. The developed techniques can be transferred now to lower energy machines, especially if high lateral resolution and high beam intensities are more important than large penetration depth. Such accelerators can provide the breeding ground for practical applications [5] as basic research gradually vanishes at the low-energy machines.

The rapid progress in ion accelerator technology has led to a dramatic reduction of the total expenses of ion accelerators during the last decade. This change commands a fresh evaluation of the technological feasibility of ion track technology at accelerators. Recently, several commercial applications of ion accelerators in the field of ion tracks have been started. This reminds of the history of implantation, a well-known pioneer in the transition from a scientific instrument to a commercial tool.

The history of ion accelerators shows a fruitful interaction between science and technology. Ion accelerators were originally developed for obtaining a better understanding of the interaction between ions and matter. Directly related with the trend of basic science to higher energies was the search for practical application of the lower energy machines which were left behind. In the wake of nuclear physics and its associated technologies a variety of practical applications developed which now belong to our common technology pool. Already since about 1930 energetic x rays and electrons — end products of the acceleration process — were used for practical purposes, for example in radiation therapy. This interaction between basic research and applied work has remained quite natural until today. Always, the motivation to raise the energy of the accelerators originated in basic research and almost predictably some time later practical applications emerged.

The applications of ion beams nowadays range from the use of low energy ions in the field of surface technology to the application of relativistic heavy ions in radiation therapy. Commercially most important are low energies, based on the earliest developments with a strong tendency to higher and higher energies as is evident by the emergence of the ion track technique described in this book.

[1]   Crozaz, G., R. Drozd, H. Graf, D.M. Hohenberg, M. Monnin, D. Ragan, C. Ralston, M. Seitz, J. Shirk, R.M. Walker, J. Zimmermann: Proc. Third Lunar Sci. Conf., Suppl. 3, Geochim. Cosmochim. Acta, Vol 2, 1623, (1972).
[2]   Damm, G., K. Thiel, W. Herr: Earth Planet. Sci. Lett. 40, 439, (1978).
[3]   Trakowski, W., B. Schöfer, J. Dreute, S. Sonntag, C. Brechtmann, J. Beer, H. Drechsel, W. Heinrich: "An Automatic Measuring System for Particle Tracks in  Plastic Detectors." Nuclear Inst. Meth. 225, pp. 92-100, (1984).
[4]   Heinrich, W., C. Brechtmann, J. Dreute, D. Weidmann: "Applications of Plastic Nuclear Track Detectors in Heavy-Ion Physics." Nucl. Tracks Radiat. Meas. 15, pp. 393-402, (1988).
[5]   Scharf, W.: "Particle Accelerators. Applications in Technology and Research." Research Studies Press Ltd, John Wiley & Sons, New York, 663 pp., (1989)

# I Principles of track creation

## 1    Irradiation technology

The generation of nuclear tracks requires the availability of a source of heavy ions of sufficiently high nuclear charge and energy. There exist in principle four different ways to create artificial ion tracks in solids: Nuclear reactors, radioactive sources, ion accelerators, and scanning ion microbeams (Figure 1-1).

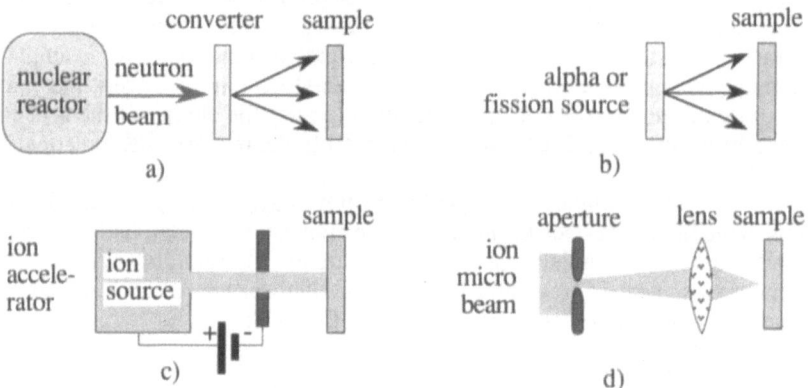

**Figure 1-1  Irradiation techniques for generating ion tracks**: (a) *Fission-fragments* created by irradiating fissionable material with neutrons from a nuclear reactor. (b) *Alpha particles or fission-fragments* created by a spontaneously decaying nuclide in a thin-film radioactive source. (c) *Ion beam* of well-defined nuclear charge, atomic weight, energy, and angle from ion accelerator. Broad-beam irradiation concept. (d) *Ion microbeam* utilizing ions from an accelerator. Fine-beam irradiation concept.

The irradiation of a solid sample is characterized by the following projectile parameters, physical units, and typical parameter ranges (Table 1-1):

**Table 1-1** Irradiation parameters, units, and typical parameter ranges for creating ion tracks in solids.

| nuclear charge number $Z$ | $1 \leq Z \leq 92$ |
|---|---|
| kinetic energy of ion $T$ (MeV) or | $10^0 \leq T \leq 10^4$ |
| specific energy of ion $T_s$ (MeV / $u$) | $10^{-2} \leq T_s \, 10^2$ |
| impact angle of ion $\alpha$ (degrees) | $0° \leq \alpha \leq 90°$ |
| angular spread of ion beam $\Delta\alpha$ (degrees) | $0° \leq \Delta\alpha \leq 180°$ |
| particle-beam intensity $I$ (particles / s) | $10^0 \leq I \leq 10^{14}$ |
| particle flux $F$ (particles / (cm$^2$ s$^1$)) | $10^0 \leq F \leq 10^{15}$ |
| areal dose or fluence $N$ (particles / cm$^2$) | $10^0 \leq N \leq 10^{12}$ |

# 1.1    Radioactive sources

Radioactive sources provide ion beams of limited energy. While $\alpha$ sources produce exactly one isotope — $^4$He — at very well defined energies, sources of heavy ions such as fission fragment sources have a broad energy and mass distribution. Due to their limited intensity, radioactive sources are used normally with relatively broad angular distribution. These apparently crude irradiation conditions, however, are very useful in many technologically relevant applications. Low-intensity alpha and fission sources are still the most convenient method to generate ion-tracks in solids. They are valuable in many investigations of the principal properties of ion tracks, in searching for improved track-recording materials and refined developing procedures, and in low-dose irradiations.

## 1.1.1    Nuclear reactors

Reactors enable large-scale commercial applications such as the production of "nuclear" track filters with areal densities up to $10^{10}$ fission-fragments per cm$^2$. The irradiations require access to a neutron outlet port of a research reactor. They also require a neutron converter foil that transforms the neutron flux coming from the reactor core into fission-fragments. For this purpose a $^{235}$U foil, inserted between the reactor and the track recorder, can be used. Nuclear reactors yield quite broad energy- and mass-distributions of fission-fragments. However, due to their higher intensity, a better beam collimation and a large throughput of sample material is possible.

## 1.1.2    Alpha and fission sources

Radioisotopes offer the possibility to irradiate samples with wide-angle low-intensity beams of alpha particles and fission-fragments of about 1 MeV/$u$ specific energy, corresponding to ion ranges of the order of 10 to 20 μm. They enable to study track-etching phenomena satisfactorily in a dose range up to about $10^6$ to $10^8$ ion tracks per cm$^2$ and — due to their convenience at low intensities— offer the fastest entrance into the field of ion-track technology. Weak sources (0.1 μCi [1] or less) pose a small radiation hazard

---

[1]    Curie (Ci), a radiation unit named after Marie Curie (1867-1934) who discovered radium; 1 Ci = $3.7 \cdot 10^{10}$ nuclear transformations per second = $3.7 \cdot 10^{10}$ Bq ≈ activity of 1 g of $^{226}$Ra.

and can be handled even at schools. Because of their convenience, such sources should be part of any newcomer's "starter kit" [1]. Radioisotope sources can be obtained from various suppliers of radioactive calibration sources and have typically the size of a coin. While there exist many nuclides emitting alpha particles ($^4$He ions), there exists practically only one convenient nuclide for small-scale fission-fragment irradiations, the $^{252}$Cf alpha and fission-fragment source. Information on existing nuclides, emitted particles, emission energies, and half-lifes can be obtained from nuclear data tables [2], [3], [4].

### General nature of radioisotopes

Very useful information on the production and properties of alpha and fission sources can be found in the manuals of radiation chemistry facilities [5], the essence of which is resumed in the following.

The nuclei of radioisotopes contain either too many or too few neutrons as compared with the naturally occurring stable isotopes of the same elements. Such nuclei will ultimately change into stable configurations by radioactive decay. The primary decay is always either the emission of a charged particle ($\alpha$, $\beta^+$ or $\beta^-$) or the capture of an orbital electron by the nucleus. These processes change the electric charge of the nucleus thus leading to a nucleus which belongs to a chemically different element. The product nucleus always has a lower energy content than the parent nucleus, and the difference in energy appears as the energy of the emitted radiations.

The emitted radiations consist of charged particles, such as alpha particles — designated by the symbol $\alpha$ and equivalent to $^4$He ions —, electrons ($\beta^-$) or positrons ($\beta^+$), accompanied in many cases by electromagnetic radiation in the form of $\gamma$ rays or x rays. Electron and positron emission is also accompanied by the emission of uncharged particles of small mass — neutrinos — which carry away part of the energy.

It is the properties of the charged particles and the electromagnetic radiations and the fact that radioisotopes are chemically indistinguishable from stable isotopes of the same element which give rise to the varied applications of radioisotopes. For example, all the therapeutic uses of radioisotopes depend on the ability of the radiations to ionize atoms of the substances through which they pass. This ionization in turn leads to chemical and biological changes in the material. Industrial uses such as the dispersal of static electricity and the sterilization of medical products also depend on this power of ionization. Radiography and the multitude of industrial thickness gauging applications are made possible because the radiations are absorbed by matter to an extent which is dependent both on the nature and the density of the material, and on the type of radiation and its energy. Finally, the facts that the radiations can be detected easily and measured quantitatively are responsible for the uses of radioisotopes in medical diagnosis and for the enormous variety of tracer applications in research and technology.

[1]    Enge, W.: "Introduction to Plastic Nuclear Track Detectors." Nuclear Tracks 4, 283, (1980).
[2]    Yoshihara, K. , H. Kudo, T. Sekine: **Periodic Table with Nuclides and Reference Data**. Springer Verlag, Berlin, (1985).
[3]    Lederer, C.M , V.S. Shirley (editors): **Table of Isotopes**, 7th Edition, John Wiley & Sons, Inc., New York, (1978).
[4]    Brown, E., R.B. Firestone: "**Table of Radioactive Isotopes**." V. S. Shirley (Editor), John Wiley, New York, (1986).
[5]    "**The Radiochemical Manual**." Amersham Radiochemical Centre, White Lion Road, Buckinghamshire, HP79LL, UK, (1966).

### Half-life

The number $dN$ of nuclei disintegrating per unit time interval $dt$ is proportional to the number $N_0$ of nuclei present at the time $t = 0$:

$$dN = -\lambda N_0 dt ,\tag{1.1}$$

where the "decay constant" of the particular radioisotope $\lambda$ corresponds to the fraction $dN / N_0$ of the nuclei that decay per unit time interval and the minus sign takes care of the fact that the number of nuclei decreases due to the decay process. Integration of (1.1) yields the familiar form of a decay function:

$$N = N_0 e^{-\lambda t} ,\tag{1.2}$$

where $N$ is the number of remaining nuclei at time $t$. From this equation the time $\tau$ required for the activity to fall to one half of its initial value is given by

$$\tau = -\frac{\ln (0.5)}{\lambda} = \frac{0.693}{\lambda} .\tag{1.3}$$

This time $\tau$ is the characteristic half-life of the particular radioisotope. Half-lives of radioisotopes vary over many orders of magnitude. Nuclides with half-lives down to a few minutes and up to $3 \cdot 10^5$ years corresponding to $^{36}$Cl are in use. Within this range there exist about 200 radioisotopes with useful half-lives.

Using eq. (1.3) for the half-life $\tau$, the decay equation (1.1) can be written

$$N = N_0 e^{-0.693\frac{t}{\tau}} \approx N_0 e^{-0.7\frac{t}{\tau}} , \text{ leading to}\tag{1.4}$$

$$\ln \left(\frac{N}{N_0}\right) = -0.693\frac{t}{\tau} .\tag{1.5}$$

Thus by plotting $\ln (N/N_0)$ against time expressed in units of half-lives, a straight line graph is obtained which can be used to determine graphically the proportion of a radioisotope remaining after a given time.

The half-life of a radioisotope defines the initial specific activity $I_0$ of that isotope. One gram of a pure radioisotope will contain $6.02 \cdot 10^{23} / W$ atoms, where $W \approx A u$ is the atomic weight of the isotope, $A$ its atomic mass number, and $u$ the atomic mass unit. Of these, a fraction $\lambda$ will disintegrate per time unit. Since 1 Curie $= 3.7 \cdot 10^{10}$ disintegrations per second, the initial specific activity $I_0$ of one gram of the radioisotope is

$$I_0 = \frac{6.02 \cdot 10^{23}}{W} \frac{\lambda}{3.7 \cdot 10^{10}} \text{ Curies} .\tag{1.6}$$

If $\lambda = 0.693 / \tau$ is inserted from (1.3), we obtain:

$$I_0 = \frac{3.56 \cdot 10^5}{W \tau} \text{ Curies per g, where } \tau \text{ is given in years} .\tag{1.7}$$

### Production of radioisotopes

Radioisotopes can be obtained either by bombardment of suitable target materials with neutrons or charged particles, or by extraction from naturally occurring radioactive substances or from fission products. Neutron bombardment, most frequently performed at nuclear reactors, leads to "neutron-rich" isotopes, in which the number of neutrons in the nucleus is greater than in the stable isotopes of the same element. Ion bombardment at accelerators leads by contrast usually to "proton-rich" — neutron deficient — isotopes. The naturally occurring radioisotopes, with some exceptions, decay by the emission of $\alpha$ particles.

### Production rates

For short irradiations, and irradiations in low fluxes, the decreasing number of target nuclei due to their transformation can be neglected. In this case the specific activity $I$ of the product nuclide is building up with time $t$ according to:

$$I = I_\infty \left( 1 - e^{-\frac{t}{\tau}} \right) ,$$

(1.8)

where $I_\infty$ is the activity for $t \to \infty$, and $\tau$ is a characteristic time.

### Design of radioactive sources

Fission- and $\alpha$ sources must be sufficiently stable to prevent radioactive contamination. The sources should be permanently sealed. Due to the large energy-loss of charged particles in matter, the radioisotope film as well as the sealing radiation window must be very thin. It is difficult to make alpha sources which emit a high fraction of the generated particles, and at the same time, to prevent the leakage of the highly toxic radioisotope. The maximum alpha energies from the commonly used radioisotopes are between 4.5 and 7.6 MeV, which allows an equivalent window thickness of no more than 1 mg per $cm^2$. Even then there is considerable loss and broadening of the alpha energy. Because of the very high radiation dose to which the protective coating is subjected, organic materials cannot be used. Even mica, aluminum, and gold become perforated within a few weeks if the intensity of the source is more than a few m Ci per square centimeter. For example, a mica window of 1 mg per $cm^2$ is exposed to $>10^7$rad / h from a $^{210}$Po source of 0.1 Curie per $cm^2$. This limits the use of radioactive sources to irradiations in which relatively low particle fluences (ions per $cm^2$) are required.

Mainly two types of alpha sources exist:

1. *Uncovered sources*: For calibration sources up to 1 $\mu$ Curie, emitting some $10^6$ $\alpha$ particles per minute, a thin vacuum-evaporated deposit of the radioisotope can be used uncovered. This method provides relatively small energy spread. For example the mean width at half maximum of the alpha-particle energy-peak may be as small as 10-20 keV.

2. *Covered sources*: For industrial applications up to 100 $\mu$ Curie foil sources are used. They consist of an inner active layer of a dispersion of the radioisotope compound in a metal matrix protected by a thin covering of corrosion-resistant metal (gold, platinum or palladium) and are backed by a 0.25 mm thick disk of silver or nickel. $^{241}$Am is the preferred isotope. The thickness of the front cover is about 3 $\mu$m. However, such sources must be safeguarded against abrasion and corrosive atmospheres and the spectral resolution of the radiation from these well-protected sources is relatively poor.

---

[1]    "The Radiochemical Manual." Amersham Radiochemical Centre, White Lion Road, Buckinghamshire, HP79LL, UK, (1966).

### Alpha emitters

There exist several useful nuclides emitting alpha particles ($^4$He ions) with energies up to 6 MeV or 1.5 MeV/$u$. (Table 1-2) which can be employed for investigating track-etch properties and for applications at track densities up to about $10^8$ tracks per cm$^2$. Thereby one has to consider, that alpha particles are yielding relatively low ionization densities along the ion trajectory in solids, compared with heavier ions.

**Table 1-2**  Some useful alpha emitters [1], [2], [3].

| nuclide | half-life (years) | α energies (MeV) | % | associated β- and γ-radiation (MeV) |
|---|---|---|---|---|
| $^{241}$Am | 433 | 5.440 | 13 | $\gamma_{max}$ 0.060 |
| | | 5.490 | 85 | |
| | | diverse energies | 3 | |
| $^{210}$Pb (+daughters) | 22 | 5.305 | 100 | $\beta_{max}$ 1.17 $\gamma_{max}$ 0.8 (very weak) |
| $^{238}$Pu | 87 | 5.456 | | $\gamma$ 0.0435 (very weak) |
| | | 5.499 | | |
| $^{239}$Pu | 24100 | 5.105 | 10.7 | $\gamma_{max}$ 0.051 (weak) |
| | | 5.143 | 16.8 | |
| | | 5.155 | 72.5 | |
| $^{210}$Po | 0.38 | 5.305 | 100 | $\gamma$ 0.8 (very weak) |
| $^{226}$Ra (+daughters) | 1620 | 4.589-7.68 | | $\beta_{max}$ 3.26 $\gamma_{max}$ 2.43 |
| $^{228}$Th (+daughters) | 1.91 | ≈5.2 | 1 | |
| | | 5.338 | 28 | |
| | | 5.421 | 71 | |
| $^{244}$Cm | 18.1 | 5.669 | | |
| | | 5.766 | | |
| | | 5.808 | | |

### Fission-fragment emitters

There exists only one nuclide with a half-life of the order of one year, convenient for generating fission-fragments, namely $^{252}$Cf. Such sources are produced by vacuum evaporation onto a stainless steel disc of 25 mm diameter and 0.5 mm thickness [4]. The active spot has a diameter of 4 or 7 mm. The initial source strengths range between 0.4 KBq up to 370 KBq [5], or 0.01 µCi up to 10.0 µCi. The sources can be manufactured with a protective gold coating of ca. 0.1 mg cm$^{-2}$ equivalent thickness, corresponding to a thickness of about 0.05 µm which somewhat broadens the energy distribution but reduces the radiation hazard in the long-term handling.

[1]    "**The Radiochemical Manual**." Amersham Radiochemical Centre, White Lion Road, Buckinghamshire, HP79LL, UK, (1966).

[2]    Office des Rayonnements Ionisants (ORIS), Laboratoire de Metrologie des Rayonnements Ionisants, Boite Postale No 21, F-91190 Gif-sur-Yvette, France: **Standards of Radioactivity**, p. 61 ff., (1986).

[3]    Brown, E., R.B. Firestone: "Table of Radioactive Isotopes." V. S. Shirley (Editor), John Wiley, New York, (1986).

[4]    Atomic Energy Research Establishment (AERE), Chemistry Division, Harwell, Oxfordshire, OX11 0RA, England.

[5]    Becquerel (Bq), a radiation unit named after Henry Becquerel who found the radioactive decay of natural uranium in 1896; 1 Bq = 1 radioactive decay per second.

The nuclide $^{252}$Cf (Figure 1-2 and Table 1-3) disintegrates with a probability of 96.91% by emission of an $\alpha$ particle and with a probability of 3.09% by spontaneous fission. The fission-fragments correspond to two broad peaks around 80 MeV and around 104 MeV. Its half-life is 2.64 years. The $\alpha$ emission corresponds to mainly three groups of energies: 84.0% of the $\alpha$ particles are emitted at 6.1184 MeV, 15.8% at 6.076 MeV, and 0.2% at 5.977 MeV. The fission rate is about $10^3$ fissions per s per $\mu$Ci, emitted into the full solid angle. Besides fission-fragments and alpha particles such a source emits about $4 \cdot 10^3$ neutrons per s per $\mu$Ci, or a total of 4 neutrons per fission.

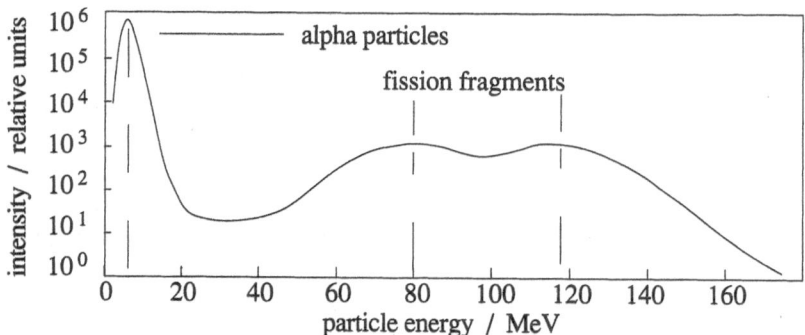

**Figure 1-2**  *Energy spectrum of* $^{252}$*Cf source* — the most useful "mini"-source for track generation.

**Table 1-3**  Data of $^{252}$Cf fission and alpha source [1].

| nuclide | half-life (years) | fission energies (MeV) | $\alpha$ energies (MeV) | $\alpha$ percentage % | fission-to-$\alpha$ ratio |
|---------|-------------------|------------------------|-------------------------|------------------------|---------------------------|
| $^{252}$Cf | 2.64 | $\approx 80 \pm 14$ $\approx 104 \pm 9$ | 6.1184 6.076 5.977 | 84.0 15.8 0.2 | 0.032 |

## 1.2    Ion accelerators

In contrast to radioactive sources, ion accelerators are much larger and much more expensive. However they provide ion beams of well-defined nuclear charge and mass, energy, impact angle and angular spread. In addition, the ion energy can be adapted to the specific requirements of the intended application and the accelerator can be safely switched-off after completing the irradiation. Below the Coulomb threshold for nuclear excitation — which is of the order of 1 MeV/$u$ — the radiation hazard is restricted to the period of the irradiation. Above the Coulomb threshold the irradiated material can be activated and has to be subjected to a radiation safety control.

[1]    Office des Rayonnements Ionisants (ORIS), Laboratoire de Metrologie des Rayonnements Ionisants, Boite Postale No 21, F-91190 Gif-sur-Yvette, France.

## 1.2.1    Characteristic parameters of accelerators

Ion accelerators are characterized by the following parameters (Table 1-4):

**Table 1-4**  Characteristic parameters of accelerators. For a given atomic number $Z$ usually all its natural isotopes can be accelerated.

| | |
|---|---|
| atomic number $Z$ | $1 \leq Z_{min} \leq Z \leq Z_{max} \leq 92$ |
| accelerating voltage $U$ or | MV (Mega Volt) |
| ion energy $T$ or | MeV (Mega electron Volt) |
| specific ion energy $T_s$ | MeV / $u$ (MeV per atomic mass unit) |
| accelerator frequency | number of accelerated beam pulses per second |
| pulse intensity | intensity during one ion bunch |
| beam intensity $I$ | time-averaged intensity of ion beam |
| beam emittance | $r \cdot \alpha$ (waist radius times waist angle at focus) |
| beam brightness | $I / (r \cdot \alpha)^2$ (intensity / emittance$^2$) |

**Energy classes of ion beam devices**

Ion beam devices can be classified according to their maximum attainable energy, ion range, and overall costs which are closely related with each other. According to the historic development, the devices of lowest energy became first available, have already found the most widespread application, and attained technological perfection by now. With increasing energy — and overall costs — the maximum attainable ion range increases and new applications are possible. Accordingly, with increasing energy the number of scientific uses increases and that of practical uses decreases. This fact reflects the delay between scientific exploration and technological exploitation. Starting from very low acceleration energies, five classes can be distinguished (Figure 1-3).

1.  *Ion beam sputtering and deposition devices* provide ion energies up to a few keV, corresponding to ion ranges up to the order of 0.01 μm. The electronic industry relies heavily on the use of very low energy ion beams as a key tool to structure silicon.

2.  *Ion implanters* with energies between 1 and a few 100 keV are widely used to modify electric properties of silicon and surface properties [1]. The corresponding ion ranges — between a few atomic layers and about 1 μm — are well suited to structure solids. Among the applications are short-range ion-tracks, such as used for the modification of solid surfaces. Besides structuring applications, ion beams can simultaneously be used to analyze the elemental composition of solids.

3.  *Low energy ion accelerators* with energies between 10 keV and a few 10 MeV were originally built for nuclear physics. They are now frequently used to analyze the elemental composition of solids in larger depth and for modifying solid properties by deep implantation. Ion ranges between about 1 and about 100 μm provide good opportunities for applying the ion-track technique. Large-scale applications of ion tracks to imprint structure onto solids are possible. Random track arrays can be employed to induce global property

[1]    Feldman, L.C., J.W. Mayer: "**Fundamentals of Surface and Thin Film Analysis.**" North Holland, New York/Amsterdam/London, pp. 1-352, (1966).

changes of the solid volume or its surface. Below the Coulomb threshold for nuclear activation which is of the order of a few MeV/$u$ even very high ion doses — necessary for implantation — will not lead to a radioactivation of the irradiated samples. Beyond the Coulomb threshold care has to be taken not to exceed certain radiation levels. Most ion-track applications associated with track etching, however, are far below any radiation hazard, as long as single tracks or track arrays of low areal track-densities are used. Another benefit for the creation of ion tracks between 1 and 10 MeV/$u$ is their high energy deposition along the ion track, corresponding to the so-called "Bragg peak" of the energy-loss function. This leads to a high damage-density in the track recording solid and thus to tracks which can be easily developed. Besides wide-beam irradiations locally confined ion beams can be used. Scanning ion microbeams are being developed at present and may soon reach the verge of becoming commercially feasible tools for the generation of well-defined ion-track patterns.

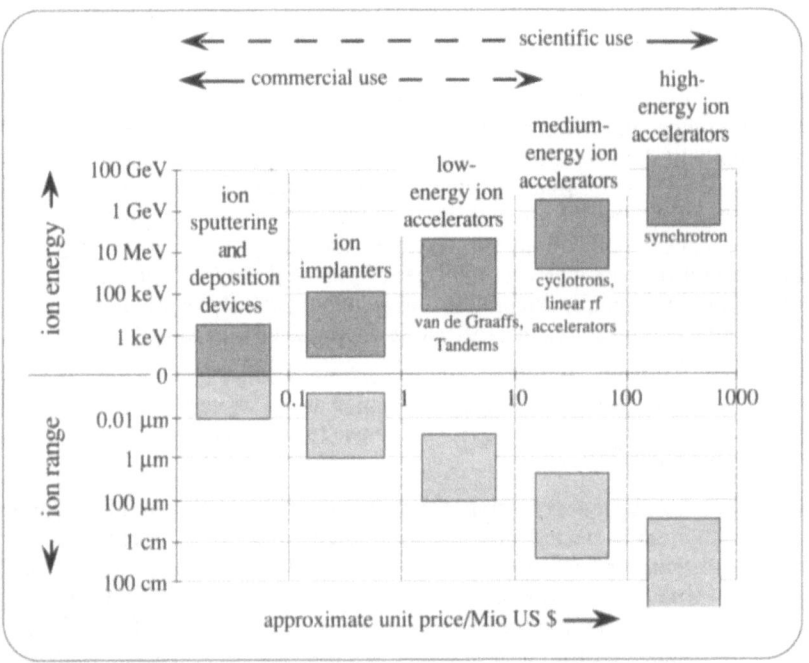

**Figure 1-3** *Ion energies and ion ranges for five classes of ion accelerators.* The upper energy and range limits belong to lighter ions and the lower energy and range limits to heavier ions.

4.    *Medium energy ion accelerators* with energies between 10 MeV and a few GeV provide ion ranges between 100 μm and a few cm. They are at the verge of being commercially used in the field of the ion-track technology, for example for the production of ion-track filters. Due to the large ion ranges, beam windows can be employed, greatly facilitating irradiations under normal atmospheric conditions outside the vacuum of the accelerator. Medium energies are especially suited for the generation of long ion tracks at not too high areal

track densities. Due to their high costs at present, medium energy ion accelerators are still mostly used scientifically, whereby radiation biology is close to a practical use.

5.    *High energy ion accelerators* with energies above 1 GeV are still the exclusive domain of scientific research and still far out of reach for practical applications. Also, at present there exist very few such machines to explore their potential fields of practical uses. The probability of fragmentation increases with the length of the track, which will lead to an activation of the irradiated material. Very interesting, however, is the production of long ion tracks at low areal densities or the generation of track patterns of many tracks as templates for a multiple reproduction. In this way the irradiation costs could be reduced to reasonable values. At the same time, this is a way to circumvent potential problems of radioactivation.

## 1.2.2    Production of highly charged heavy ions

### Ion sources

A necessary precondition for the acceleration of particles is their electrical charge. The ion charge should be as high as possible in order to facilitate the acceleration in electrical fields of discharge-limited strength. Ion accelerators therefore depend on the availability of high-intensity ion sources with high charge states of the generated ions.

The development of ion sources started with the generation of ions from low pressure gas discharges [1], [2] (Figure 1-4 a). At that stage of development, the only accessible tuning-parameters were the gas pressure and the current through the discharge tube. In order to produce intense beams of heavy, highly charged positive ions, a series of improvements had to be invented. Heated filaments increased the number of electrons available for the electron-impact ionization process of the originally neutral gas atoms. Magnetic confinement was introduced in the form of a magnetic bottle together with an electrostatic extraction to release the ions into the accelerator. Sputtering was introduced to generate ions of solid elements in a rare-gas discharge-plasma in connection with extraction and mass-separation of the sputtered ions. The "operational" or "drive" gas thereby serves to maintain a stable discharge plasma. In the case of a highly developed ion source, the art of creating a stable beam of ions thus depends on a successful multi-parameter optimization.

Figure 1-4 b shows the principle of an ion source for the production of highly charged heavy ions of gaseous or solid elements at high intensities. Electrons are released from a indirectly heated cathode, accelerated and injected into an inert drive gas like argon. They are confined in an axial magnetic field which increases the efficiency of ionizing the drive gas. For generating ions from solid elements, a sputter electrode made of this element is in contact with the drive gas plasma. Neutral sputtered particles are incorporated into the plasma, ionized and extracted together with the ions of the drive gas. The sputtered ions are separated from the ions of the drive gas using a magnetic mass separator.

[1]    Goldstein, E.: "Über eine noch nicht untersuchte Strahlungsform an der Kathode inducirter Entladungen." Sitzungsberichte der königlichen preussischen Akademie der Wissenschaften zu Berlin, **39**, pp. 691-699, (1886)
[2]    Dietrich, J.: "Stand und Entwicklung von Ionenquellen für Beschleuniger." Akademie der Wissenschaften, Zentralinstitut für Kernforschung, Rossendorf bei Dresden, Bericht ZfK-516, 34 pp., (1984).

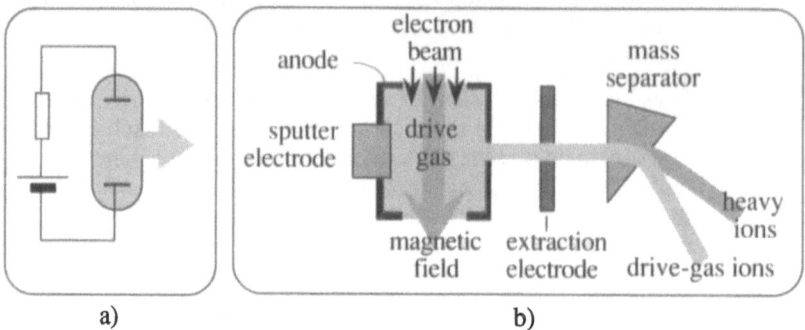

a)                                           b)

**Figure 1-4   Ion sources. (a)** *Simple gas discharge*, the origin of ion source development.
**(b)** *Penning ion source* for production of highly charged heavy ion beams of gaseous or
solid elements. An electron beam from a heated cathode bombards the operational gas —
most frequently argon — leading to an ion plasma. Sputtered atoms of the solid are incorporated into this plasma, ionized, extracted and mass-separated before being injected into the
accelerator.

## 1.2.3     Ion deflection and focussing

### Ion deflection

Deflection of energetic heavy ions is most frequently performed in electric or
magnetic fields. If the ion is subject to a constant and uniform magnetic field of induction
$B$ at 90° to its trajectory, it is subject to a Lorentz force, $F_L = |F_L| = |Q (v \times B)| = Q\, v\, B$,
where $Q$ is its charge, $v$ its velocity and $B$ the magnetic induction. For ions of constant
velocity $v$, this leads to a constant radial acceleration of the ion $b_r = dv / dt =$
$v\, d\alpha / dt = v\, (v / R) = v^2 / R$ at 90° to its trajectory (Figure 1-5).

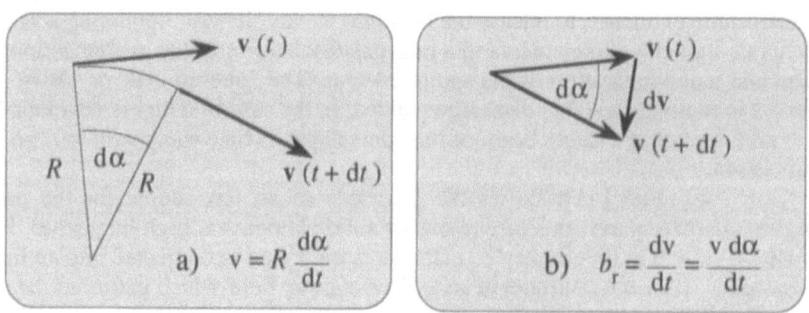

**Figure 1-5**   Definition of tangential velocity $v$ **(a)** and radial acceleration $b_r$ **(b)**.

The fundamental equation of motion $F_L = m\, b_r$ leads therefore to $Q\, v\, B =$
$m\, v^2 / R$ and the ion describes a circle of radius

$$R = \frac{m\, v}{Q\, B} = \frac{P}{Q\, B} \quad , \quad \text{where } P \equiv m\, v \quad . \tag{1.9}$$

Numerically the orbit radius of a 1 eV electron in a magnetic field of 1 Gauss
(0.0001 V s / m²) is 3.4 cm. The corresponding radius of a proton is 62 m. The orbit

radius of a $^{238}$U ion in a charge state of 20$^+$ and with a specific energy of 10 MeV/$u$ in a field of 1 Tesla (10$^4$ Gauss) is about 5 m. In accelerator technology the particles are often characterized by their rigidity, their power to resist a deflection in a magnetic field:

$$\text{rigidity} \equiv B\,R = \frac{m\,\text{v}}{Q} = \frac{P}{Q}\,. \qquad (1.10)$$

For a deflection of an ion in a cyclic accelerator or for beam scanning by an angle $\alpha$, a corresponding magnetic field sector is required (Figure 1-6). For small deflections one obtains $\alpha \approx L/R$, where $L$ is the length of the magnetic field sector, therefore the deflection angle is

$$\alpha \approx \frac{L}{R} = \frac{Q\,B}{m\,\text{v}}\,L\,. \qquad (1.11)$$

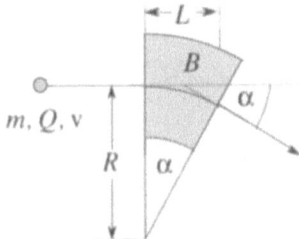

**Figure 1-6** *Ion deflection by a short magnetic field sector*, where $\alpha$ = sector angle = deflection angle of the ion of mass $m$, charge $Q$, and velocity v in a magnetic field of induction $B$, and $R$ is the orbit radius of the ion within the magnetic field.

### Ion focussing

Without focussing only a small fraction of the ion beam injected into the accelerator from the ion source would be available at the exit of the accelerator. Ion focussing at high energies and masses is usually achieved by magnetic quadrupole lenses. They offer sufficient focussing power to keep the ion beam on its prescribed path and adapt the beam width ultimately to the desired size at the position of the target. The cross-section through a magnetic quadrupole is shown in Figure 1-7.

For a travelling ion the magnetic quadrupole field represents a dynamic potential in the form of a saddle. For a positive ion travelling parallel to the axis of the quadrupole, the Lorentz force results in a focussing force in the vertical plane and a defocussing force in the horizontal plane, as indicated by the larger arrows. This corresponds approximately to the action of a cylinder lens. Therefore the combined action of two such quadrupoles rotated about the beam axis by 90°, a so-called quadrupole doublet, represents approximately a spherical lens, able to focus in the vertical as well as in the horizontal plane.

However, a quadrupole doublet does not produce symmetrical effects in both planes. Often a third quadrupole is added for reducing its distortions, forming a quadrupole triplet. The middle singlet has twice the length of the end singlets. Quadrupole triplets are normally operated symmetrically, the field strengths of the first and third singlet being identical.

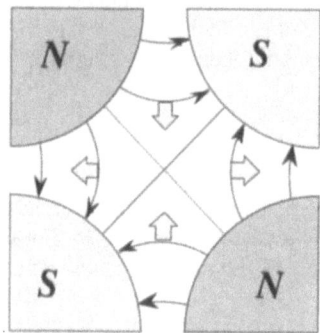

**Figure 1-7** *Cross section through a magnetic quadrupole lens.* Black arrows: magnetic field lines. Shaded arrows: Lorentz force acting on a positively charged particle passing parallel to the beam axis into the drawing plane, pulling it towards the beam axis in a vertical plane through the beam axis and away from the beam axis in a horizontal plane through the beam axis. This corresponds to a focussing action in the vertical plane through the beam axis and to a defocussing action in the horizontal plane through the beam axis.

## 1.2.4   Acceleration techniques

### Principles of ion acceleration

In practice the only existing method for accelerating charged particles is to use an electric field in the proper direction. The different types of accelerators differ essentially only in the way the electric field is produced and acts on the particles.

The operation of an accelerator consists of the following sequence of steps. The charged particles from the ion source are collimated into a beam and injected with a defined initial energy into the accelerator. They are forced to follow certain trajectories under the action of an accelerating electric field until they have reached the required energy. The electric field can be continuous or pulsed. Accordingly one distinguishes between electrostatic and high frequency (h.f.) accelerators — for which alternatively the terms alternating current (a.c.) or radio frequency (r.f) accelerators are used.

### Accelerator types

1. *Electrostatic accelerators.* The high static voltage is traversed by the ion beam only once or twice. Van de Graaff accelerators and tandem accelerators belong to this group.

2. *High frequency accelerators* in which the ion beam encounters the same acceleration voltage several times. Depending if the trajectories of the particles are straight or curved paths, two subgroups exist:

   a) *Linear accelerators* use the successive passage of the ions through a linear array of gap electrodes subject to a high frequency voltage. Wideroe and Alvarez accelerators belong to this group.

   b) *Cyclic accelerators* use magnetic deflection for "recycling" the particles again and again through the same accelerating gap subject to a high frequency voltage. Thereby energy is accumulated in each successive passage through the same acceleration field. Cyclotrons and synchrotrons belong to this group.

**Historic development.** In the beginning only electrostatic accelerators were used. Later on the principles of high frequency and cyclic acceleration were found and applied to multiply the attainable energy of electrostatic accelerators by many orders of magnitude. Thereby the electrostatic (or d.c) accelerators are often used as injectors for higher energy a.c. (or r.f.) accelerators. Detailed information on accelerator technology can be found in the books by R. Kollath [1] and by E.Persico and collaborators [2].

First attempts to accelerate particles to high energies required the generation of a high static voltage that was traversed by the accelerated particles only once. This class of accelerators comprises the van-de-Graaff accelerators. Their upper limit of acceleration of about 25 million electron volts (MeV) is given by electrical discharges occurring between the high voltage terminal and ground.

A completely new approach was necessary to overcome the discharge limit of electrostatic accelerators. The new idea was to pass the accelerated particles not only once but several times through the same accelerating field. In this way the particle accumulates more and more kinetic energy at each field passage and can ultimately reach velocities approaching the velocity of light. In high frequency accelerators the accelerating voltage therefore corresponds only to a small fraction of the final energy attained by the particles.

### Electrostatic accelerators

Electrostatic accelerators provide high beam intensity, good beam collimation, and energy stability. They are capable to produce a continuous current of ions.

In electrostatic accelerators the final energy of the ions is defined by the constant voltage difference between the acceleration electrodes. Whatever the trajectory of the particles in these fields, the kinetic energy gained depends only on the point of departure and on the point of arrival and hence it cannot be larger than the potential energy difference corresponding to the maximum voltage drop existing in the machine.

The kinetic energy $T$ gained by the particle in an electrostatic accelerator is equal to the product of the voltage drop $U$ between the electrodes and the charge $Q$ of the particle:

$$T = Q \, U \ . \tag{1.12}$$

$Q$ is the charge of the ion, with $0 \le Q / |e| \le Z$, where $Z$ is the nuclear charge number of the ion, and $e$ is the electron charge. According to (1.12) the kinetic energy increases linearly with the charge $Q$ of the ion. Therefore the availability of highly charged ions from the ion source is most important. The kinetic energy is usually given in "electron volts" (eV), which is the kinetic energy gain of an arbitrary particle carrying the elementary charge of one electron when accelerated through a potential difference of 1 volt.

Electrostatic accelerators essentially consist of a vacuum system with two electrodes between which the ions are accelerated (Figure 1-8). One electrode at the potential of the ion source is at high positive voltage. The other electrode is on ground potential.

Tandem accelerators use the voltage drop of the high voltage generator twice. First, single charged negative ions are accelerated from an ion source on ground potential toward a stripper electrode on positive high voltage. By passing through the stripper foil, the negative ions throw off some of their electrons and the resulting positive ions are accelerated toward a second electrode which is again on ground potential.

[1]   R. Kollath: "Teilchenbeschleuniger", F. Vieweg & Sohn, Braunschweig, pp.1 - 335, (1962).
[2]   E. Persico, E. Ferrari, S.E. Segre: "Principles of Particle Accelerators", W.A. Benjamin, Inc., New York, pp. 1 - 301, (1968).

The ultimate voltage drop which can be attained in accelerators is mainly limited by surface discharges along insulating surfaces within and outside the vacuum system. In practice, the limiting voltage is of the order of 1 MV. This limit can be increased by more than one order of magnitude, if the insulating surfaces outside of the vacuum system are enclosed in a tank containing a gas of high electro-negativity, such as sulfur hexafluoride, ($SF_6$), at a pressure of the order of several bar.

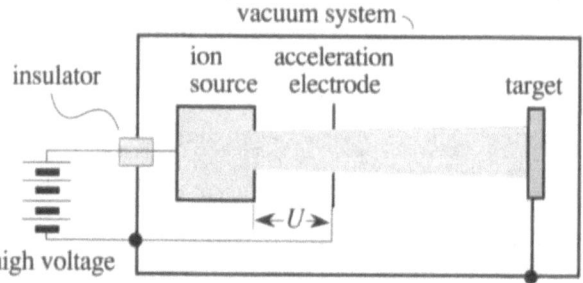

**Figure 1-8**  *Principle of electrostatic accelerator*. The kinetic energy gain $T$ of the ion corresponds to the product of the ion charge $Q$ and the acceleration voltage $U$ and is limited to roughly 25 MeV, the discharge limit of insulator surfaces and sparking gaps.

### High frequency accelerators

In order to overcome the discharge limit of electrostatic accelerators one must use fields which vary in time and space. In linear high frequency accelerators the particles are accelerated step by step in several acceleration gaps. In cyclic high frequency accelerators they are passed in repeatedly through the same acceleration gap. In both cases the acceleration is not limited by the maximum voltage of the individual gap.

The particle receives during each gap transit the same energy $Q\,U$, where $Q$ is the particle charge and $U$ is the voltage drop across the gap at the instant of the transit. The total kinetic energy gain $T$ of the ion corresponds to the number $n$ of gap passages times the energy gain per gap passage, $Q\,U$,

$$T = n\,Q\,U\ . \tag{1.13}$$

This expression can be split-up in two factors, the first representing the number of nucleons $A$ — or atomic mass number — and the second the kinetic energy of one nucleon:

$$T = \frac{m_0\,v^2}{2} = A\,\frac{u\,v^2}{2}\ ,\ \text{with}\ m_0 = A\,u\ , \tag{1.14}$$

where $m_0$ is the mass of the ion, $v$ its velocity, $A$ its (fractional) atomic mass number, and $u = 1.6606 \cdot 10^{-24}$ g the atomic mass unit. The term $u\,v^2/2$ characterizes the used accelerator structure and operation voltage and does not depend on the mass $A$ of the ion. Therefore it is possible to split the kinetic energy $T = A\,T_s$ acquired in a high frequency accelerator into a term $A$, characterizing the mass of the accelerated isotope, and the specific energy $T_s$ characterizing the accelerator setup:

$$T_s = \frac{T}{A} = \frac{u\,v^2}{2}\ ,\ \text{where}\ u\ =\ 1.6606 \cdot 10^{-24}\ \text{g}\ . \tag{1.15}$$

From eq. (1.15) and eq. (1.13) we obtain $v = (2\,n\,U\,(Q\,/\,m_0))^{1/2}$ or $v = (2\,n\,U\,[Q\,/\,(A\,u)])^{1/2}$. Thus, ions of the same specific charge $Q\,/\,m_0$ or $Q\,/\,(A\,u)$ will be accelerated to the same velocity $v$ if the gap voltage $U$ is the same. Thus for a given acceleration voltage $U$, heavier ions need to be higher charged or heavier ions of the same charge require a higher acceleration voltage.

According to the acceleration principle, ions of different mass have to pass at the same time through the same gaps. In order to keep them in phase with the accelerating high frequency, therefore the accelerating voltage has to be increased with increasing ion mass. As a consequence, linear accelerators will accelerate ions of different masses $m_0 = A\,u$ to the same velocity $v$. By adjusting the accelerating voltage to the required value of the chosen ion — using the same high frequency accelerator structure and operating frequency — any type of ion will be accelerated to the same specific energy $T_s$. This is the justification for factorizing the acquired kinetic energy according to eq. (1.14) and eq. (1.15) into a factor characterizing the atomic weight of the ion and a factor characterizing the accelerator configuration.

In practical applications of ion beams, such as the generation of ion tracks, the use of the term "specific energy" has the additional benefit that ions of the same specific energy $T_s$ and correspondingly of the same velocity $v$ will have ranges of roughly the same magnitude in matter.

### Principle of phase stability

For an efficient acceleration of ions in high frequency accelerators, ions of slightly deviating injection phase still have to be accelerated to roughly the same final velocities. This requirement of phase stability is fortunately an inherent feature of high frequency accelerators, increasing the accepted range of injected phase, mass, and energy and thus the available beam intensity.

The principle of phase stability can best be visualized for a travelling wave in which the wave velocity increases smoothly with increasing distance from a reference point (Figure 1-9). Only particles within a given phase-range are accelerated, similar to the technique used in wave surfing to keep the board on the slope of a wave. This principle of acceleration is most closely realized in the travelling wave accelerators used for electron acceleration.

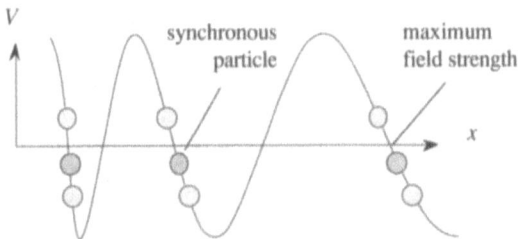

**Figure 1-9** *Principle of travelling-wave acceleration.* In analogy with a wave surfer on a water wave, the potential $V$ along the $x$ - axis is shown. Only particles within a certain phase-range close to $V = 0$ are accelerated — comparable to the strategy used in wave surfing. Since the accelerating field $E$ is obtained from the potential $V$ by differentiation, the maximum field strength $E_{max}$ corresponds to $x$-values, for which the potential $V = 0$.

In high frequency accelerators the acceleration is provided by an electric field localized at the gap. The gap field changes periodically in time. Particles which satisfy proper injection conditions and gain the correct amount of energy at their first transit across the first accelerating gap will, on successive transits, find always a field of the same value, directed so as to accelerate it. Any particle satisfying such conditions is called a "synchronous particle". It is accelerated by the desired amount at each transit and reaches the required final energy at the end of the acceleration process. Only a negligible number of particles fulfill this condition exactly. For the practical operation of high frequency accelerators it is therefore necessary that not only the particles in exact synchronism with the accelerating electric field but also those which have slightly different phases are accelerated by the same amount on the average over a large number of turns, so that all these particle reach the desired final energy. The synchronism between the particle motion and the accelerating electric field is of essential importance in the design of the machine when the particles must cross the accelerating gap a very large number of times, as in the case of the cyclotron and synchrotron. Phase stability essentially is equivalent to an elastic restoring force encountered by particles deviating slightly from the phase of the synchronous particle.

The acceleration of the particle depends on the phase $\varphi$ of the high frequency voltage at the time of transit through the acceleration gap (Figure 1-10) [1]. Thereby the phase-change during the gap-transit can be neglected.

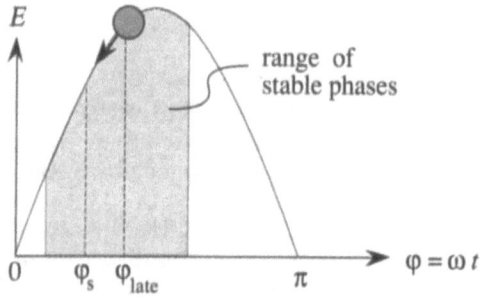

**Figure 1-10** *Principle of phase-stability*. Shown is the accelerating field $E(t)$ for a given acceleration gap. Only the synchronous particle (index s) will pass the next acceleration gap at the same phase $\varphi_s$. A "late" particle crossing the gap at phase $\varphi_{late}$ will receive the bonus of a higher accelerating field, but — in compensation — will be subject to a lower acceleration field in the next gap, since it passes too early through it. In this way a harmonic oscillation around the phase of the synchronous particle is established.

The synchronous particle passes the acceleration gap at the phase $\varphi_s$ and will by definition be accelerated by just the right amount to pass the next acceleration gap at the same phase $\varphi_s + 2\pi$, corresponding to the same electric field strength $E$. If another particle passes through the acceleration gap slightly later than the synchronous particle at the phase $\varphi_{late}$ it encounters a slightly larger electric field and thus will obtain a slightly larger velocity as the synchronous particle. It will therefore pass the next acceleration gap slightly earlier. That corresponds to a restoring force driving the late particle back to the phase of the synchronous particle. This process reverses as soon as the particle has arrived at some following acceleration gap slightly earlier than the synchronous particle. In this way an oscillation around the phase of the synchronous particle is established.

---

[1]    according to the usual notation $\varphi \equiv 2\pi \, \nu \, t \equiv 2\pi \, t / T \equiv \omega \, t$

### Linear accelerators

The linear accelerator accelerates charged particles along approximately straight trajectories by means of high frequency electric fields (Figure 1-11). For a fixed high frequency the velocity of the accelerated ions will be independent of the ion mass as long as the acceleration voltage is within a certain acceptance range — as long as the charge to mass ratio of the ions, $Q / m$, is sufficient for maintaining phase stability — and the ion is pulled along in synchronism with the travelling wave. This enables the simultaneous acceleration of several charge states within a given charge spectrum in the accelerator.

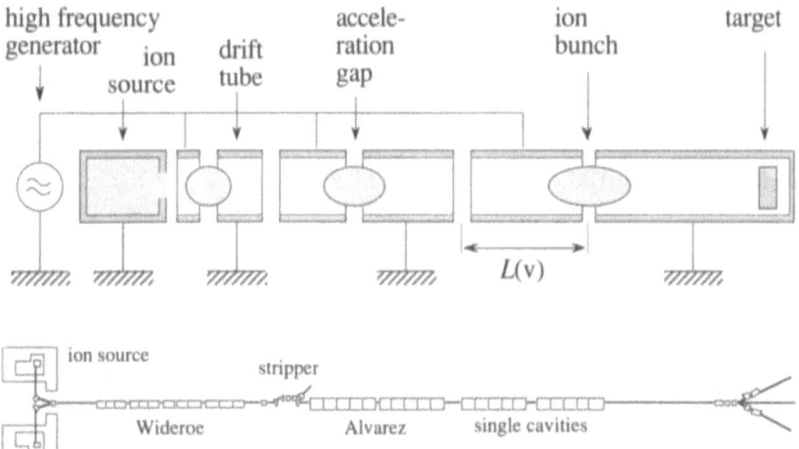

**Figure 1-11** (**top**) *Principle of linear high frequency accelerator.* Ions within a certain phase range are accelerated, forming a "bunch". (**bottom**) *Layout of the UNILAC accelerator*, the heavy ion accelerator of GSI, D-6100 Darmstadt, which accelerates heavy ions — for example $^{238}U$ — to specific energies of roughly 15 MeV/$u$. Approximate length 100 m.

The simplest structure was proposed in 1928 by R. Wideroe. The ion beam is injected along the axis of the accelerator. Successive electrodes in the form of "drift tubes" have alternating polarity, turning to the opposite voltage during the passage of the ion bunch through the drift tubes. The length $L$ of the drift tubes increases along the accelerator with the increasing velocity v of the accelerated ion and is chosen in such a way that a particle which encountered the accelerating field in the first gap and is accelerated thereby to a velocity v will also be accelerated in the second gap, and so on. Only the second, forth, and so on gap will accelerate an ion bunch at a given time. This corresponds to the relation $L / v = \tau / 2$, where $\tau$ is the cycle time of the oscillator, or $L(v) = v \tau / 2$. Thereby the synchronous particle always encounters the field in each successive gap in the same phase as in the first gap and receives during each transit the same energy $Q U$, where $Q$ is the particle charge and $U$ is the voltage drop across the gap at the instant of the transit. The total kinetic energy gain $T$ of an accelerated ion corresponds according to eq. (1.13) to the total number $n$ of gaps times the energy gain per gap, $Q U$.

With increasing velocity of the particle, the length $L$ of the drift tubes increases, thus enlarging the overall size and the total cost of the accelerator prohibitively for high final energies. In order to avoid enormous machine lengths, the frequency must be as high as possible. But then the transmission of energy becomes difficult in practice because of large energy losses in the conductors between the h.f. generator and the acceleration gaps. This problem is circumvented by using resonating structures — for example resonating cavities — in which very high voltages occur, whereby the energy consumption in the transmission lines is minimized. In this way the power requirements of the feeding high frequency generator can be drastically reduced.

### Cyclotrons

The principle of the classical cyclotron was discovered by E.O. Lawrence in 1930. Conceptually it is similar to a linear accelerator, in which the trajectory is wound up into a spiral by a magnetic field (Figure 1-12). More precisely, the ion trajectory consists of a series of semicircles increasing stepwise during the passage of the ion through the acceleration gap. The invention is based on the fact that the orbital time period $\tau$ of a charged particle in a constant uniform magnetic field is independent of the particle velocity $v$, as long as the particle behaves classically and $m \approx m_0$. According to eq. (1.9) the time $\tau$ for completing one orbit is:

$$\tau = \frac{2\pi R}{v} = \frac{2\pi}{v} \frac{m v}{Q B} = \frac{2\pi m}{Q B} \quad , \tag{1.16}$$

which is independent of the ion velocity $v$ and consequently independent of the instantaneous radius $R$ of the ion trajectory.

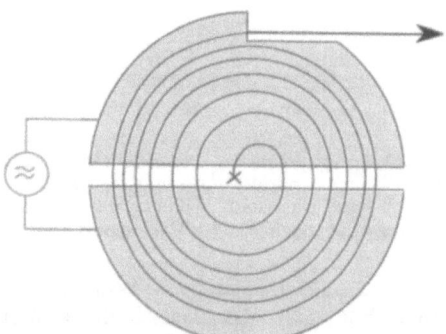

**Figure 1-12** *Principle of cyclotron.* The ion beam starts at the center, is accelerated in the gap between the high frequency electrodes, bent by a constant and homogeneous magnetic field and ejected after reaching the outer magnetic field boundary. Since the time for completing a semicircle is independent of the particle velocity, as long as the ion velocity is small in comparison with the velocity of light, the frequency as well as the magnetic field can be kept constant.

The particles move inside a vacuum chamber in the form of a flat cylinder. This encloses two hollow electrodes whose shape has the form of a letter "D", commonly called a "D". The high frequency voltage is applied between the electrodes, so that there is an alternating electric field in the gap, but no electric field inside the hollow electrodes. The vacuum chamber is placed between the pole pieces of a large magnet which produces a constant uniform magnetic field perpendicular to the plane of the electrodes. The ion source is placed at the center of the vacuum chamber so that the ions, leaving the source at low speed, are accelerated in the gap and enter the inner space of one of the electrodes. Here the magnetic field forces them to follow a semicircular orbit until they cross the gap again. If in the meantime the electric field has reversed its direction, the particles are again accelerated and follow a semicircle of larger radius in the other electrode, and so on. This operation of the machine is possible since the time required for the particles to describe each semicircle is constant. The angular frequency ω of the high frequency generator is chosen such that the transit time inside the high frequency electrodes is equal to the half period of the accelerating field. Thus the particles always encounter the electric field in the gap with the same phase and they gain the same amount of energy at each transit, similar as in the case of the linear high frequency accelerator. Accordingly, the total energy gain $T$ of the accelerator corresponds to the number $n$ of gap passages times the energy gain $Q\,U$ in the acceleration gap, similar as in the case of the linear accelerator.

According to eq. (1.16), the required angular frequency ω of the high frequency generator is

$$\omega \equiv \frac{2\,\pi}{\tau} = \frac{Q\,B}{m} \quad , \tag{1.17}$$

which is independent of the particle velocity as long as the particle behaves classically and $m \approx m_0$. The angular frequency ω is called the cyclotron frequency or (nonrelativistic) magnetic resonance frequency. The condition $m \approx m_0$, for the validity of the classical approximation, limits the maximum kinetic energy in a cyclotron to a few percent of the particle rest mass $E_0 = m_0\,c^2$. If the kinetic energy increases to values not negligible with respect to $E_0$, as the mass increases, so does the time interval between succeeding transits according to eq. (1.16) until eventually it encounters a decelerating field. For this reason the practical specific energy limit of the classical cyclotron is of the order of $T_s \approx$ 10 MeV / $u$, corresponding to a particle velocity of about 15 % of the speed of light.

This limitation can be overcome by applying the magnetic field only within specific sectors, resulting in the so-called split-field cyclotrons which enable a much higher degree of field modeling. More concretely, in order to have an isochronous revolution frequency, constant for all particle energies, the average magnetic field has to increase in proportion to the relativistic mass increase, $m = m_0\,\gamma = m_0\,/\,[1 - (v\,/\,c)^2]^{1/2}$. Since the corresponding field shape leads to vertical defocussing, one has to give up the rotational symmetry of the magnetic field and introduce field sectors.

Within this limit, however, the cyclotron is one of the most convenient and reliable machines (see also [1]). It can produce pulsed ion beams of average intensities comparable to that of electrostatic accelerators, with the advantage of avoiding difficult insulation problems. With respect to standard linear accelerators of the same energy it has smaller energy width and is more compact. On the other hand, this compactness of a cyclotron leads to the disadvantage of a constrained geometry for injecting ions close to the center of the cyclotron as compared to linear accelerators.

In order to obtain an idea of the dimensions of a cyclotron we apply (1.9) expressing $m$ and v as function of its maximum kinetic energy $T_{max}$ and obtain for nonrelativistic velocities for which $m \approx m_0$, $T \approx m_0\,v_{max}^2\,/\,2$,

$$R_{max} = \frac{\sqrt{2m_0 T_{max}}}{Q\,B} \tag{1.18}$$

Typical magnetic flux densities $B$ in practice are of the order of $10^4$ Gauss = 1 Tesla =1 kg s$^{-2}$ A$^{-1}$. For 10 MeV protons this yields $R_{max} \approx 0.45$ m.

[1]   Joho, W.: "Modern Trends in Cyclotrons, CERN Report 87-10, p. 261, (1987)

Vice versa, if we are interested in the specific energy $T_s$ for a cyclotron of given radius $R_{max}$ and field $B$, we obtain

$$T_s = \frac{T_{max}}{A} = \frac{u\, v_{max}^2}{2} = \frac{u}{2}\left(\frac{Q\,B\,R_{max}}{m}\right)^2 = \frac{u}{2}\left(\frac{Q\,B\,R_{max}}{A\,u}\right)^2 = \frac{1}{2u}\left(\frac{Q\,B\,R_{max}}{A}\right)^2. \quad (1.19)$$

For a cyclotron radius $R_{max} = 1$ m and a magnetic flux density $B = 1$ Tesla we obtain the following specific energies. For a proton with $Q\,/\,|e| = Z = +1$ we obtain $T_s \approx 50$ MeV/$u$. For $^{40}$Ar in a charge state of $Q\,/\,|e| = +5$, we obtain $T_s \approx 0.8$ MeV/$u$. For $^{238}$U in a charge state of $Q\,/\,|e| = +11$, we obtain $T_s \approx 0.1$ MeV/$u$.

Sometimes cyclotrons are characterized by their $k$ value, where $k = (R_{max}\,B)^2\,/\,(2\,u)$, yielding a specific energy $T_s = k\,(Q\,/\,A)^2$.

### Synchrotrons

In cyclotrons the most expensive part is the magnet. According to (1.18) the radius of the pole piece increases with the square-root of the maximum kinetic energy $T_{max}$, that is $R_{max} = \text{const} \cdot T_{max}^{1/2}$. The price $P$ of a cyclotron, which is closely related to the volume of the magnet, increases with the third power of its radius, that is $P \approx \text{const} \cdot T_{max}^{3/2}$. Above some hundred MeV therefore the reduction of the volume of the magnet becomes mandatory. This resulted in the development of the synchrotron in which the accelerated particles follow a circular orbit of constant radius $R$ rather than a spiral of increasing radius. Thus the magnetic field is required only in the region of the orbit and not in the space enclosed by it. The cost of such a ring-shaped magnet is roughly proportional to the radius, whereas for a magnet with circular poles it is roughly proportional to the cube of the radius.

In the synchrotron, as the particles follow a fixed orbit, the vacuum chamber has the shape of a torus. The high frequency acceleration occurs via resonant cavities localized at certain points of the torus.

Similar to the cyclotron, the total energy gain $T$ of the accelerator corresponds to the number $n$ of gap passages in the synchrotron times the energy gain $Q\,U$ per acceleration gap. The time $\tau_{turn}$ required by a particle to complete a full turn must be equal to or be a multiple $N$ of the time period $\tau$ of the high frequency generator, $\tau_{turn} \equiv 2\pi\,R\,/\,v = N\,\tau = N\,2\pi/\,\omega$, or

$$\omega = N\,\frac{v}{R} = N\,\frac{Q\,B}{m}. \quad (1.20)$$

As the energy increases the high frequency as well as the magnetic field must increase during the acceleration cycle. This is exactly the same requirement as for the handling of an ancient stone sling. However, if the particle approaches the velocity of light, the angular frequency $\omega$ will approach the limiting value $N\,c\,/\,R$. In order to keep the particle on a fixed orbit of radius $R$ while the energy increases, the magnetic field must increase proportional with its momentum. According to the relativistic generalization of (1.9), $R = P\,/\,(Q\,B)$, it must increase linearly with the momentum $P = m\,v$ of the particle:

$$B = \frac{P}{Q\,R} = \frac{m\,v}{Q\,R} \quad \text{or} \quad v = \frac{B\,Q\,R}{m}, \quad (1.21)$$

where $m = \gamma\,m_0 = m_0\,/\,[\,1 - v^2\,/\,c^2\,]^{1/2}$ is the relativistic mass of the particle. We thus obtain for the required angular frequency using (1.20) and (1.33)

$$\omega = \frac{N\,Q\,B}{m} = \frac{N\,Q\,B\,E_0}{m_0\,E} \ . \tag{1.22}$$

Then, as long as $E \approx E_0$, a linear relation between $\omega$ and B results. For relativistic velocities B enters indirectly into E. For revealing this dependency, we have to express E as function of P, according to eq. (1.36), $E^2 = E_0{}^2 + c^2\,P^2$, and then $P = Q\,R\,B$ as function of B, according to (1.21). Using eq. (1.22) and eq. (1.32), we obtain the following relation between $\omega$ and B, ruling the operation of a synchrotron:

$$\omega = \frac{N\,Q\,B}{m} = \frac{N\,Q\,B}{m_0}\frac{E_0}{E} = \frac{N\,Q\,B}{m_0\sqrt{1+\left(\frac{c\,Q\,B\,R}{E_0}\right)^2}} \tag{1.23}$$

with the asymptotic limit $\omega \to N\,c\,/\,R$ for $v \to c$.

The particles are usually injected into the synchrotron after being pre-accelerated to a specific energy of a few MeV/u by a linear accelerator. For practical reasons the synchrotron is divided into sectors, each of which serves a different purpose. The curved sections correspond to the location of the bending magnets. The straight sections are used for injection, for high frequency cavities, beam diagnostics, and extraction (Figure 1-13).

To mention but a few trends in present-day accelerator technology, sources for highly-charged ions, radiofrequency quadrupole injectors, superconductive field coils and high frequency resonators, stochastic and electron cooling — improving the beam properties on an individual and a collective basis— are being investigated in order to improve the efficiency and compactness of future accelerators.

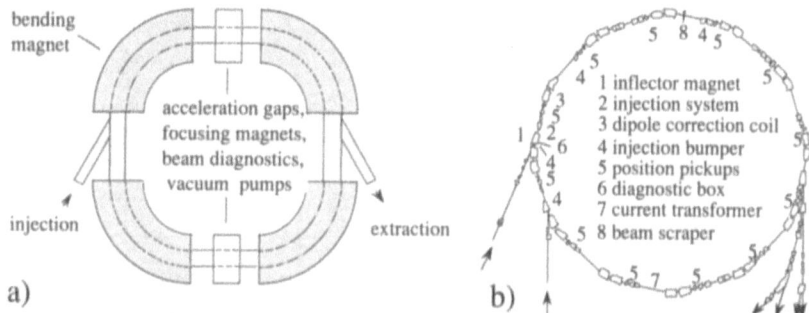

**Figure 1-13** (a) *Principle of synchrotron.* In order to keep the bending radius constant with increasing ion velocity, the magnetic field as well as the gap frequency has to be increased. (b) *Layout of the SIS accelerator*, the heavy ion synchrotron of GSI, D-6100 Darmstadt, with a diameter of roughly 50 m, designed to accelerate $^{238}$U to kinetic energies of roughly 1 GeV/u.

## 1.2.5 Behavior of ions at relativistic energies

For estimating the importance of relativistic effects in ion acceleration and during their interaction with solids, an excursion into the field of relativity may be useful. At low ion energies relativistic effects can often be neglected. However, with increasing ion energy relativistic effects are starting to influence the design of the accelerator and the relations between the ion energy and its velocity and momentum become relativistic. The use of relativistic mechanics is necessary if the particles reach velocities comparable to the velocity of light ($c$ = 2.99792$\cdot$10$^{10}$ cm / s).

An ion of a specific energy of 1 MeV/$u$ has reached only about 5 % of the speed of light and relativistic effects are still very small. At 10 MeV/$u$ it has reached about 15 % of the speed of light and relativistic effects start to be observed. At 100 MeV/$u$ it has reached 43 % of the speed of light and relativistic effects are becoming dominant.

The following relations exist between mass, momentum, energy and velocity of the particle and have to be considered at high energies (Table 1-5):

**Table 1-5**  Fundamental relations for particles at relativistic velocities.

| | |
|---|---|
| mass | $m = \gamma\, m_0$ |
| momentum | $P = \gamma\, m_0\, v$ |
| total energy | $E = m\, c^2$ |
| kinetic energy | $T = E - m_0\, c^2$ |
| rest energy | $E_0 = m_0\, c^2$ |

$$\text{where}\quad \gamma = \frac{1}{\sqrt{1 - \beta^2}}, \qquad \beta = \frac{v}{c}. \tag{1.24}$$

In these definitions $\gamma$ is the relativistic mass ratio $m / m_0$, $\beta = v / c$ is the "relative velocity" measured in units of the light velocity $c$, $m_0$ the mass of a particle at rest, $m$ its mass at velocity v.

Newtons law of classical mechanics, $F = m_0\, dv / dt$, where $F$ is the force acting on the particle and v is the particle velocity, transforms in relativistic mechanics into:

$$F = \frac{dP}{dt} \tag{1.25}$$

which is the fundamental equation of motion in relativistic dynamics, where $P$ is the momentum of the particle, given by

$$P = m\, v = \gamma\, m_0\, v \tag{1.26}$$

In this relation $m_0$ is the "rest mass" which is identical with the mass in classical mechanics. By this definition the momentum still has the formal expression of the product of mass and velocity, as in classical mechanics. However, the mass depends on the velocity. For velocities v small compared with the speed of light $c$, (1.25) and (1.26) reduce to the law of classical mechanics $F = m_0\, dv / dt$.

Eq. (1.25) can be decomposed into a component tangential (subscript "$t$"), and into a component normal (subscript "$n$") to the path of the particle, respectively

$$F_t = \frac{dP}{dt}, \quad \text{where}\ P = |P|,$$

$$F_n = m\, b_r = m\, v\, \omega = m\, \frac{v^2}{R} = P\, \frac{v}{R} \tag{1.27}$$

where $b_r = v\,\omega = v^2/R$ is the radial acceleration corresponding to the velocity v and the angular frequency $\omega$, $P$ is the scalar momentum, and $R$ is the radius of curvature of the particle trajectory.

The kinetic energy $T$ of a particle is related to its mass-increase with respect to the rest mass, according to the relation

$$T = (m - m_0)\, c^2 = (\gamma - 1)\, m_0\, c^2 \ . \tag{1.28}$$

The relative mass excess measured in units of $m_0$ is

$$\frac{m - m_0}{m_0} \equiv \gamma - 1 = \frac{T}{m_0\, c^2} \tag{1.29}$$

which is drawn as function of the specific energy $T_s$ in Figure 1-14, using eqs. (1.28) and (1.24) for obtaining $T = T(\beta)$, and eq. (1.38) for obtaining $\beta = \beta(T_s)$.

By expanding the expression for $\gamma$ from (1.24) into a Taylor series it can be verified that for $\beta^2 \ll 1$ eq. (1.28) reduces to the classical expression $T = m_0\, v^2/2$ .

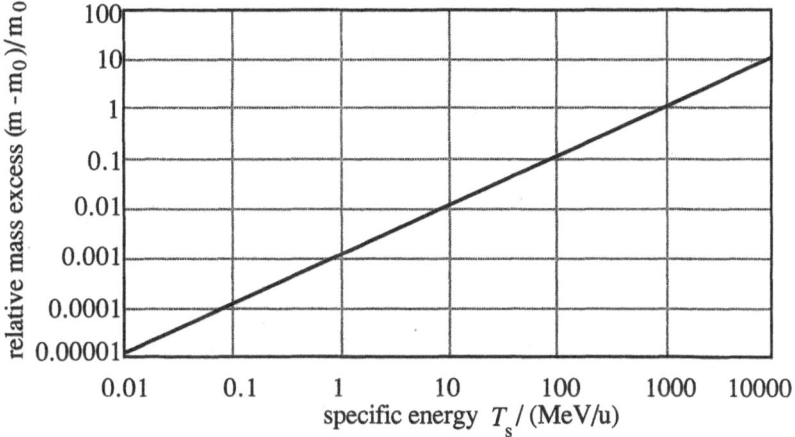

**Figure 1-14** *Relative mass excess as function of specific energy.* Above 100 MeV/u relativistic effects become important.

According to (1.28) a particle with the rest mass $m_0$ has a "rest energy"

$$E_0 = m_0\, c^2 \ , \tag{1.30}$$

corresponding to its energy at zero velocity. One can also define a "total energy" of the particle, according to

$$E = E_0 + T = m\, c^2 = \gamma m_0\, c^2 = \gamma E_0 = \frac{E_0}{\sqrt{1 - \beta^2}} \tag{1.31}$$

From eq. (1.31) we obtain $\gamma = 1 + T/E_0$ and can thus express the mass as function of the kinetic energy $T$:

$$m = m_0 \left(1 + \frac{T}{E_0}\right) \tag{1.32}$$

From (1.32), using $T = E - E_0$ from (1.31) we can express the mass as function of the total energy $E$:

$$m = m_0 \frac{E}{E_0} . \tag{1.33}$$

This leads to $P \equiv m\,v = m_0\,v\,E / E_0 = m_0\,v\,E / (m_0\,c^2)$, or

$$c\,P = E\frac{v}{c} = \beta\,E \tag{1.34}$$

From $E = \gamma E_0$ (1.31) we obtain

$$E^2(1 - \beta^2) = E_0^2 \quad \text{or} \quad E^2 = E_0^2 + \beta^2 E^2 . \tag{1.35}$$

Inserting $\beta^2 E^2 = c^2 P^2 =$ from (1.34) we obtain

$$E^2 = E_0^2 + c^2 P^2 , \tag{1.36}$$

which expresses the total energy $E$ as function of the momentum $P$.

From (1.31) we can express the relative velocity $\beta$ also as function of the total particle energy $E$ by rearrangement of $E = E_0 / (1 - \beta^2)^{1/2}$,

$$\beta = \sqrt{1 - \frac{E_0^2}{E^2}} . \tag{1.37}$$

If we are interested to express $\beta$ rather as function of the specific energy $T_s$ of the accelerator, we have to express $E$ as function of the specific energy $T_s$. Remembering from (1.31) that $E = E_0 + T$, from (1.14) that $m_0 = A\,u$, from (1.30), that $E_0 = m_0\,c^2$, and from (1.15) that $T = A\,T_s$, we obtain:

$$\beta = \sqrt{1 - \frac{1}{\left(1 + \frac{T_s}{u c^2}\right)^2}} , \tag{1.38}$$

represented in Figure 1-15 or its inverse function $T_s = T_s(\beta)$:

$$T_s = u c^2 \left[\frac{1}{\sqrt{1 - \beta^2}} - 1\right] . \tag{1.39}$$

Again, $\beta = v / c$ does not depend on the ion mass but is merely given by the capability of the accelerator to generate ions of the specific energy $T_s$.

The condition for the classical approximation, $\beta^2 \ll 1$, can be written as $E \approx E_0$, which, using $E = E_0 + T$ (1.31), transforms into

$$T \ll E_0 \ , \tag{1.40}$$

which is useful for finding out if accelerated particles still behave nonrelativistically. For example, the rest mass of the electron of $9.1095 \cdot 10^{-28}$ g corresponds to an energy of 0.5110 MeV, the rest mass of the proton to 938.2 MeV. Therefore electrons of 0.5 MeV behave already relativistically, whereas protons of 0.5 MeV still behave quite much classically.

This again stresses the usefulness of the specific-energy scale for which eq. (1.38) reads $T = A \ T_s \ll E_0 = m_0 \ c^2 = A \ u \ c^2$ , or

$$T_s \ll u \ c^2 \approx 10^3 \ (\mathrm{MeV}/u) \ , \quad \text{where } u = 1.6606 \cdot 10^{-24} \mathrm{g} \ . \tag{1.41}$$

Any particle of the same specific energy $T_s$ will have the same velocity. If the specific energy $T_s$ is of the order of $u \ c^2 \approx 10^3$ MeV/$u$ the particle behaves relativistically.

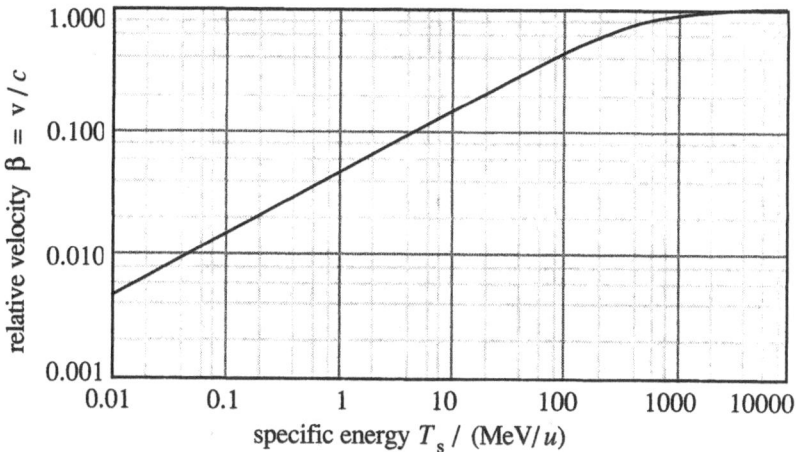

**Figure 1-15** *Relative ion velocity as function of its specific energy.* Relativistic effects are negligible for $T_s \ll 10^3$ MeV/$u$.

# 1.3 Irradiation targets and equipment

Inside and outside the vacuum system of an accelerator different techniques have to be used to irradiate solid targets. Within the accelerator, the particles move under vacuum conditions in order to reduce the probability of collisions with the remaining gas molecules leading to energy-loss and charge-exchange phenomena. For linear accelerators and classical cyclotrons a vacuum of the order of $10^{-6}$ torr (roughly $10^{-4}$ Pascal) is sufficient. For synchrotrons ultrahigh vacuum ($10^{-8}$ to $10^{-10}$ torr) is required, due to the very long paths of the charged particles in the accelerator. Accordingly, the irradiation equipment for internal targets has to be adapted to the vacuum requirements of the accelerator. Special techniques have been developed for controlling the beam distribution over the sample surface and for terminating the irradiation at a given preset areal dose [1].

[1]    Spohr, R.: "Nuclear Track Irradiations at GSI." Nucl. Tracks **4**, 101, (1980).

### Internal targets

Up to specific energies of the order of 1 MeV/$u$ or ion ranges up to 10 μm, internal targets — targets inside the vacuum system of the accelerator — are mandatory, due to the short ion range. For internal targets the accelerated ion beam collides directly with the target without any further interaction with additional matter. Thereby the "contamination" of the ion beam by lower energy ions and by scattered particles from accelerator apertures can be kept small. The target has to be inserted into the vacuum system, positioned and oriented with respect to the direction of the ion beam. After irradiation it has to be removed again from the vacuum system (Figure 1-16). Internal targets may require a provision for target cooling if the energy-deposition rates are high.

### External targets

Beyond specific ion energies of the order of 10 MeV/$u$ or ion ranges beyond 100 μm it is often advantageous to employ a "beam window" in the form of a thin metallic foil (for example a 5-10 μm thick nickel foil). This enables conventional engineering for all parts outside the vacuum, a reasonable approach if large-area samples or large sample numbers have to be irradiated. In external irradiations, the interaction between the ion beam and the window and atmosphere leads to energy loss phenomena in the form of scattered ions of decreased energy and window-imported ions, increasing with the window thickness. At the same time the angular spread and energy width of the ion beam is increased. In many cases this "beam contamination" can be tolerated.

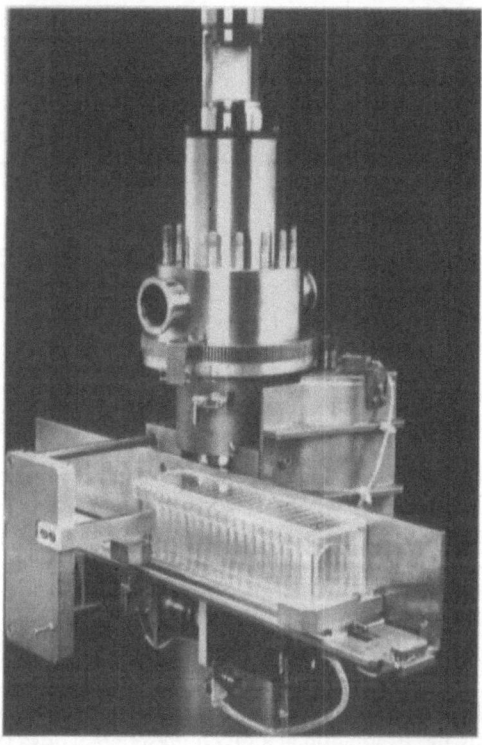

**Figure 1-16** *Sample inlet system* used at GSI Darmstadt for irradiating frame-mounted solids. Maximum sample size 50 mm x 50 mm x 5 mm.

### 1.3.1 Wide-beam irradiation devices

Figure 1-17 shows the principle of an irradiation device for internal irradiations. Before entering the irradiation device, the ion beam is prepared to meet certain approximate conditions (mainly beam intensity, width, and angular spread). The beam diameter is defined by a beam aperture after which it impinges onto the target. During the irradiation the accumulated particle dose is monitored using a nondestructive beam monitor. In contrast to fundamental-research experiments — which are mostly electronic observations of nuclear and atomic processes — ion-track irradiations require a precise control of the dimensions of the ion beam, its lateral distribution, and the accumulated areal dose. The lateral intensity distribution, the beam angle to the surface-normal of the sample, and the accumulated particle dose must be carefully monitored throughout the irradiation. After reaching the desired particle dose, the beam has to be switched-off precisely and rapidly, using a beam shutter. The destructive beam monitor is used for the absolute calibration of the nondestructive beam monitor. In rare cases — if the beam is sufficiently stable in time and space — it may be sufficient to use the destructive beam monitor intermittently, for example before and after the irradiation.

For external irradiations a beam window of approximate thickness of 5-10 µm nickel or stainless steel can be inserted immediately in front of the irradiated sample. In this way, beam windows up to about 5 cm height at arbitrary width can be realized without a supporting grid structure.

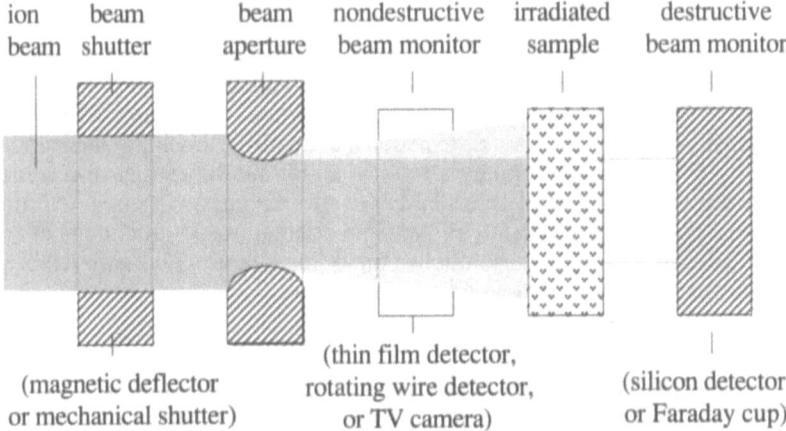

**Figure 1-17** *Principle of wide-beam irradiation facility.* The incident ion beam should meet certain injection conditions, such as intensity, beam cross-section and beam parallelism. The beam-shutter speed defines the precision of the achieved areal doses. The beam aperture defines the irradiated sample area and should produce as few scattered particles as possible. The nondestructive beam monitor should interfere as little as possible with the ion beam. At the same time it should enable to monitor the areal distribution of the ion beam continuously. The irradiation angle is defined by the angle of the sample with respect to the incident ion beam. In crystal irradiations, two angles have to be set precisely, by turning the sample around its normal axis and by turning the sample around an axis in the sample plane. During beam calibrations, the destructive beam monitor is inserted instead of the sample. In thin-sample irradiations the ions penetrate the sample and can be conveniently monitored by the destructive beam monitor alone, avoiding the nondestructive monitor.

## 1.3.2    Scanning ion microbeams

Figure 1-18 gives an idea of a scanning ion microbeam at an accelerator [1].

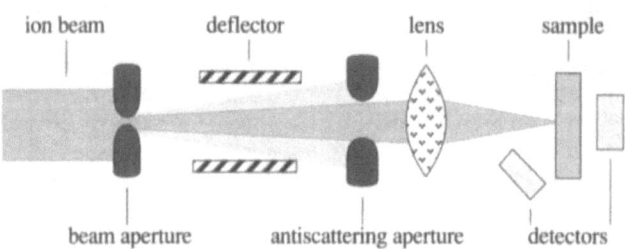

**Figure 1-18** *Principle of ion microbeam device.* In order to reduce the thermal load of the beam aperture the arriving beam is reduced in size before impinging onto it. The beam aperture consists of pairs of highly polished metal rods for reducing the number of scattered ions. Ions transmitted through the beam aperture are focussed onto the sample surface. Scattered particles from the beam aperture are suppressed by several orders of magnitude using an antiscattering aperture. The ion microbeam is scanned over the surface of the sample by means of a beam deflector. The transmitted ions and the emitted particles from the sample are be used for detecting the arrival of individual ions and for various diagnostic purposes. The ion beam can be interrupted using a rapid beam shutter.

While wide-beam irradiations give the possibility to generate ion tracks that are randomly distributed over the exposed sample area, ion microbeams make a locally pre-defined generation of ion tracks possible. For this purpose a focussing lens has to be inserted between the beam aperture and the irradiated sample. Further requirements are the addition of antiscattering beam-apertures and of a beam scanner. Most important is the quality of the ion beam arriving at the beam aperture. Its brightness $I / (\alpha \cdot r)^2$ should be as high as possible. The beam distribution and intensity should be stable in time.

The present resolution limit of an ion microbeam used at a medium-energy heavy-ion accelerator with about 1 MeV/$u$ specific energy is about 0.5 µm . In the near future, it may become possible to generate oriented track structures with a precision of about 0.1 µm. Still higher resolutions require ion beams with improved beam properties, such as available from field-emission ion sources.

# 1.4    Radiation safety

During the generation of ion tracks beyond energies corresponding to the so-called Coulomb threshold of nuclear excitation, radioactive nuclei can be generated. At high ion track densities the generated radioactive isotopes have to decay to a permissible radiation level before the irradiated samples can be processed further. Even though beyond the Coulomb threshold there exists very little radiation hazard from samples irradiated at low track densities, radiation monitoring is a necessary precaution in all irradiations using ion energies exceeding the Coulomb threshold.

Another possibility for radioactive contamination are the alpha and fission sources which by themselves are radioactive and therefore require radiation monitoring. In addition, fission fragments are proton deficient radioactive nuclides and a source of pos-

[1]    Fischer, B.E.: Gesellschaft für Schwerionenforschung (GSI), Planckstr. 1, D-6100 Darmstadt 11.

sible contamination after their implantation into the sample. The following section is mainly based on reference [1]. For actual radiation safety standards — which vary from nation to nation and are updated from time to time, see for example [2].

## 1.4.1    Handling of radioactive sources

### Permissible levels of radiation exposure

The permissible levels of radiation exposure are still under dispute. Direct evidence relating to humans is available from the use of nuclear weapons, radioactive accidents, and long-term exposures to x rays or to the radiations from naturally occurring radioisotopes. In all cases the actual doses received by particular individuals are generally known only very imprecisely. Another complication is that many radiation effects develop very slowly and may not appear for tens of years. Some of the effects may manifest themselves only in the offspring of the irradiated individuals. Because of this limited human experience it is necessary to draw on results from radiation biology. Nevertheless, the fixing of permissible levels is still not free of arbitrary decision. A reasonable — mostly accepted — standpoint is to set the safety limit as closely as possible to the radiation levels in the prenuclear age.

### Units of radiation dose

The following units are used for measuring the number of decays (Becquerel), the absorbed energy (Gray, rad), the number of created ion pairs in air (Coulomb per kg, Roentgen), and the effect of the irradiation on living tissue (Sievert, Rem) (Table 1-6):

**Table 1-6**  Units of radiation dose [3].

> **Becquerel** (Bq), unit of radioactivity, 1 Bq = 1 decay event / s. Official unit <u>before</u> 1986: 1 Curie (Ci) = $3.7 \cdot 10^{10}$ decay events / s.
>
> **Gray** (Gy), unit of absorbed energy dose, 1 Gy = 1 J / kg. Official unit <u>before</u> 1986: 1 rad = $10^{-2}$ Gy = $10^{-2}$ J / kg = 100 erg / g.
>
> **Coulomb per kilogramm (C / kg)**, "unit" of absorbed radiation dose, defined as total positive — or negative — charge generated during an irradiation. Official unit <u>before</u> 1986: 1 Roentgen = $2.58 \cdot 10^{-4}$ C / kg. One Roentgen produces $3.7 \cdot 10^{10}$ ion pairs in air at standard pressure and temperature.
>
> **Sievert (Sv)**, unit of absorbed dose-equivalent, defined as energy dose in Gray, multiplied by a dimensionless quality factor — the so-called relative biological effectiveness (*RBE*) of the particular type of radiation under consideration. 1 Sv = 1 J / kg). Official unit <u>before</u> 1986: 1 Rem (Roentgen equivalent man) = $10^{-2}$ Gy = $10^{-2}$ J / kg.

[1]    "**The Radiochemical Manual.**" Amersham Radiochemical Centre, White Lion Road, Buckinghamshire, HP79LL, UK, (1966).

[2]    Schmatz, H., M. Nöthlichs, H.P. Weber: "Strahlenschutz. Radioaktive Stoffe, Röntgengeräte, Beschleuniger. Handbuch des Strahlenschutzrechts und Erläuterungen." Erich Schmidt Verlag, 2nd edition, (1977).

[3]    Cohen, E., P. Giacomo: "**Symbols, Units, Nomenclature and Fundamental Constants in Physics.**" International Union of Pure and Applied Physics (IUPAP), Revision 1987, 67 pp. (1987).

The "unit" Coulomb per kilogramm is analogous to the old unit Roentgen used as exposure dose for x or γ radiation. One Roentgen is the amount of x or gamma radiation which can generate one electrostatic charge unit of positive — or negative — ions per 0.001239 grams of air. One electrostatic unit of charge corresponds to $1/4.8032 \cdot 10^{-10} = 2.0819 \cdot 10^9$ electrons $= 3.3356 \cdot 10^{-10}$ Coulomb.

The unit Sievert (formerly Rem or Roentgen equivalent man) had to be introduced since the biological effects for equal absorbed doses of different kinds of radiation are not the same, mainly due to the different local distribution of the deposited energy. For x rays, γ rays, β particles and positrons the relative biological effectiveness (RBE) has the value 1. For α particles and heavier ions the RBE can be different from 1, depending on the ion charge and its velocity.

### Control of radiation exposure

Control of radiation exposure is based on recommendations of the International Commission on Radiological Protection (ICRP). This commission recommends the maximum doses which can be considered as being unlikely to cause significant injurious effects. The recommendations which will be of most concern to workers with radioactive material are those relating to occupationally exposed personnel (Figure 1-19). Lower limits are set for the population at large, who are not subjected either to checks of the actual doses they receive, or to regular medical examinations to detect any indications of radiation effects.

| maximum annual doses for occupationally exposed personnel | | |
|---|---|---|
| region of body | category A | category B |
| 1. whole body, bone marrow, gonads, uterus | 50 mJ/kg (5 rem) | 15 mJ/kg (1.5 rem) |
| 2. hands, forearms, feet, legs, ankles including their skin | 600 mJ/kg (60 rem) | 200 mJ/kg (20 rem) |
| 3. skin, if exclusively skin is exposed, except skin of hands, forearms, feet, legs and ankles | 300 mJ/kg (30 rem) | 100 mJ/kg (10 rem) |
| 4. bones, thyroid gland | 300 mJ/kg (30 rem) | 100 mJ/kg 10 rem |
| 5. other organs | 150 mJ/kg (15 rem) | 50 mJ/kg (5 rem) |

**Figure 1-19** *Maximum permissible doses* for occupationally exposed personnel per calendar year, where category A concerns a high and category B a low exposure-level group [1].

### Radiation monitors

For monitoring the radiation level in laboratories where radioactive materials are used, the most useful instrument is a portable battery-operated ionization chamber which is commercially available in a number of forms with sensitivities ranging from fractions of a mRoentgen per hour to the order of hundreds of roentgens per hour.

[1]     Schmatz, H., M. Nöthlichs, H.P. Weber: "Strahlenschutz. Radioaktive Stoffe, Röntgengeräte, Beschleuniger. Handbuch des Strahlenschutzrechts und Erläuterungen." Erich Schmidt Verlag, 2nd edition, (1977).

At high-energy ion accelerators, operating above the Coulomb barrier of nuclear activation, most of the radiation burden comes from neutrons, requiring special monitoring instruments, consisting essentially of an ionization chamber embedded within a proton-rich moderator, as for example polyethylene.

For recording the actual doses received by individual workers the most convenient and widely used device is the film badge. This is essentially a small piece of photographic film, placed in a special holder which can be attached to the worker's clothing. These film badges are collected and processed at intervals of days, weeks, or months, depending on the likely level of exposure. The amount of blackening of the film is compared with standards to determine the total exposure received.

## 1.4.2 Basics of sample activation by accelerated ions

It is a very fortunate circumstance for heavy ion-track generation that already one ion is sufficient for creating a useful microstructure. In contrast to ion implantation, the deposited energy and not the deposited mass leads to the observed effect. The track-development essentially amplifies the size of the damaged zone by several orders of magnitude. The ion-doses required in ion track technology are therefore usually less than $10^{12}$ ions per cm$^2$. This is much smaller than the ion doses required in ion implantation, which are usually larger than $10^{14}$ ions per cm$^2$.

### Coulomb-threshold for nuclear excitation

Track irradiations at specific energies above a few MeV/$u$ can lead to long-lived nuclear excitations. Above a reaction-specific threshold energy, the resulting radioactive nuclei can pose a radiation hazard at high ion-track densities. The onset of nuclear activation corresponds roughly to a central collision in which the projectile and target nuclei touch each other and the attractive nuclear forces become dominant. For this to occur, the projectile has to have sufficient kinetic energy to surmount the barrier of electrostatic repulsion between the projectile nucleus and the target nucleus. Only above this so-called "Coulomb-threshold" the projectile and target nuclei can become internally excited and the excitation energy may lead to the emission of nuclear radiations, such as $\alpha$, $\beta$, $\gamma$, radiation,. induced fission, or fragmentation. Below the Coulomb-threshold, practically no energy is stored in the nuclei.

For obtaining a rough estimate of the Coulomb-threshold we ask for the kinetic energy of the projectile required for a closest approach between the projectile and target nucleus in which the nuclei just contact each other at their periphery. We use the result for the distance of closest approach $b$ in a central collision in which the kinetic energy $T = m\,v^2/2$ corresponds to the potential energy $V = q_1\,q_2/b$, according to eq. (2.40):

$$b = \frac{q_1\,q_2}{\frac{1}{2}\,m_0\,v^2} \quad , \quad \text{where} \quad m_0 = \frac{m_1\,m_2}{m_1 + m_2} = \frac{A_1\,A_2}{A_1 + A_2}\,u \tag{1.42}$$

is the reduced mass, $m_1 = A_1\,u$ the mass of the projectile nucleus, and $m_2 = A_2\,u$ is the mass of the target nucleus, $v$ is the velocity of the projectile-ion, $q_1 = Z_1\,e$ is the charge of the projectile nucleus, and $q_2 = Z_2\,e$ is the charge of the target nucleus. For contact, the distance of closest approach $b$ has to be equal to the sum of the nuclear radii, $r_1 + r_2$, (Figure 1-20).

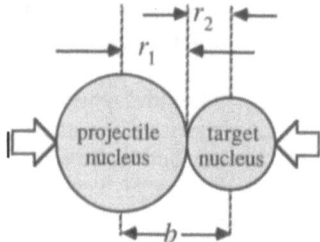

**Figure 1-20** *Basic assumption for rough estimate of Coulomb-threshold.* For inducing a nuclear reaction the distance of closest approach $b$ in a head-on collision (central collision) has to be equal to the sum of the nuclear radii, $r_1 + r_2$. Below the Coulomb threshold usually no long-lived radioactive nuclei are created.

We estimate the radius $r_A$ for a nucleus of atomic number $A$ according to the droplet model on the basis that the nucleus is spherical and composed of $A$ nucleons of radius $r_p$, where $r_p$ is the classical proton radius:

$$V_A \equiv \frac{4\pi}{3} r_A^3 = A \frac{4\pi}{3} r_p^3 \Rightarrow r_A = \sqrt[3]{A}\, r_p \, , \text{ with } r_p \approx 1.2 \cdot 10^{-13} \text{cm} \, . \qquad (1.43)$$

The condition $b = r_1 + r_2$ leads to

$$\frac{q_1\, q_2}{\frac{1}{2} m_0\, v^2} = r_p \left( \sqrt[3]{A_1} + \sqrt[3]{A_2} \right) \Rightarrow v^2 = \frac{2\, q_1\, q_2}{m_0\, r_p \left( \sqrt[3]{A_1} + \sqrt[3]{A_2} \right)} \, , \qquad (1.44)$$

which after multiplication by $m_1$ and division by 2 yields the projectile kinetic energy $T$ required for the touching condition:

$$T = \frac{1}{2} m_1 v^2 = \frac{m_1}{m_0} \frac{q_1\, q_2}{r_p \left( \sqrt[3]{A_1} + \sqrt[3]{A_2} \right)} = \frac{q_1\, q_2}{r_p} \frac{A_1 + A_2}{A_2 \left( \sqrt[3]{A_1} + \sqrt[3]{A_2} \right)} \, . \qquad (1.45)$$

Its main result is, that the Coulomb-threshold energy increases with increasing nuclear charge (atomic number) of the projectile, $q_1$, and of the target, $q_2$. Slightly counteracting is the second factor in which the simultaneous increase of the nuclear radii is reflected, decreasing the required distance of closest approach. If we give $T$ in specific energy units, we have to dived by $A_1$ and obtain a formula which is symmetric with respect to the exchange of the projectile and the nucleus (Figure 1-21):

$$T_s = \frac{q_1\, q_2}{r_p} \frac{A_1 + A_2}{A_1 A_2 \left( \sqrt[3]{A_1} + \sqrt[3]{A_2} \right)} \qquad (1.46)$$

which indicates the advantage of heavier ions in comparison to lighter ions in shifting the Coulomb-threshold to higher specific energies and thus ion ranges.

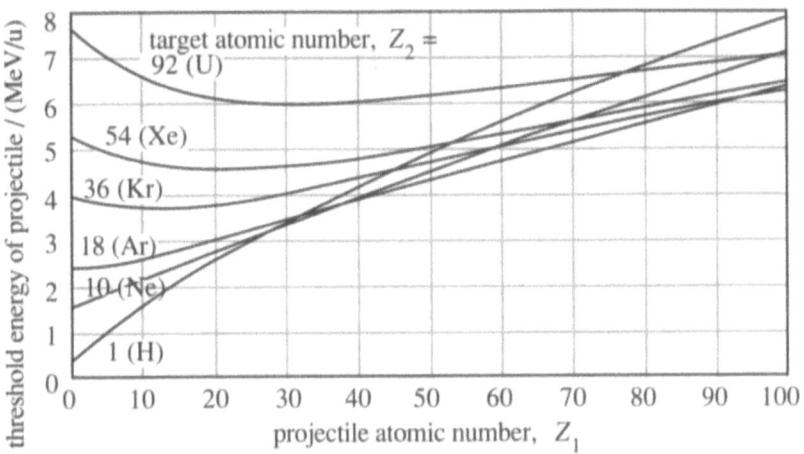

**Figure 1-21** *Rough estimate of Coulomb-threshold.* While light projectile ions on light targets have a low threshold energy, heavy ions have a high threshold energy, almost independent of the used target material. This is one reason for prefering track generation by heavy ions. Other reasons are their higher energy density along the track and their smaller angular straggling, especially in light target materials. The curves should be considered only as a guideline, since short-lived nuclear excitations may exist even below the threshold.

### Cross-section for nuclear activation

A rough estimate on the probability for nuclear activation in track irradiations beyond the Coulomb-threshold can be obtained from the nuclear cross sections. Thereby one assumes that the projectile ion possesses sufficient energy to neglect the repulsive forces between the projectile and target nuclei completely (Figure 1-22). This yields an approximate limit of the fragmentation cross-section approached for relativistic energies.

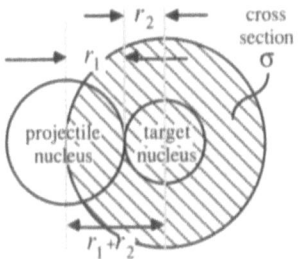

**Figure 1-22** *Rough definition of nuclear reaction cross-section.*

If $N$ is the number-density of the target nuclei, the probability $P_a$ for nuclear activation per unit path length is the sum of all nuclear cross sections per unit volume:

$$P_a \le N \pi (r_1 + r_2)^2 = N \pi r_p^2 \left( \sqrt[3]{A_1} + \sqrt[3]{A_2} \right)^2, \qquad (1.47)$$

where $r_p \approx 1.2 \cdot 10^{-13}$ cm is the proton radius, $N$ is the number of target nuclei per unit volume, $A_1$ and $A_2$ are the atomic mass numbers of projectile and target nucleus, respectively. The inverse of the activation probability $P_a$ corresponds to the mean-free-path without interaction and is represented in Figure 1-23. Thereby we assume $N = 1/V_{atom} = N_A / V_m = \rho / (A_2 u)$, where $V_{atom} = V_m / N_A$ is the average volume of one target atom, $V_m$ is the molar volume, $N_A$ is the Avogadro number, and $\rho$ is the mass density of the target material.

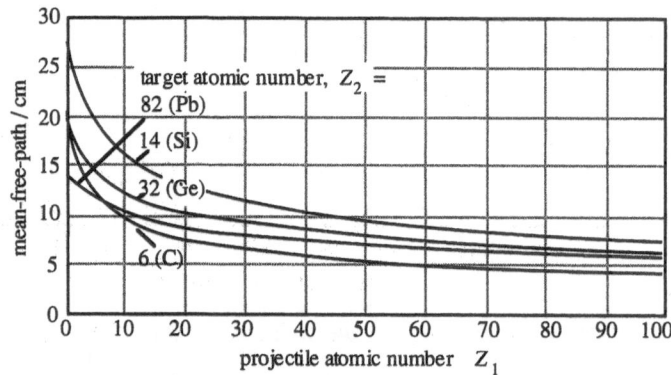

**Figure 1-23** *Rough estimate of mean-free-path for inducing nuclear reactions.* This rough estimate determined from classical nuclear radii represents an approximate limit of the fragmentation cross-section ultimately reached for relativistic projectile energies.

# 2    Energy-loss phenomena

### Basic steps in track creation

On the way from the ion irradiation to the observation of the developed track three distinct fields exist, associated with different degrees of complexity (Figure 2-1):

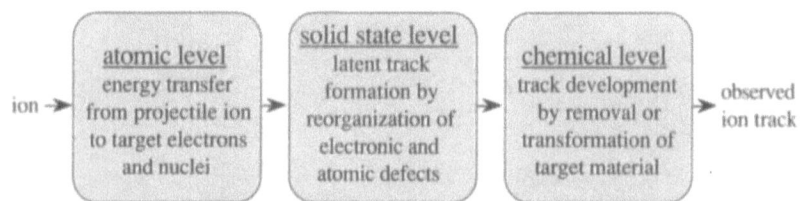

**Figure 2-1** *Three-step approach for track creation.* Each step, represented by a "black box" translates the input variables into a resulting effect. The first step is treated in this chapter.

The starting point of the track formation process is the transfer of energy from the projectile-ion to the target electrons and nuclei in binary encounters involving only the projectile and one other particle at a time. This first step is understood much better than any of the successive steps, involving many particles, in which the energy input of the projectile-ion is converted in a nonlinear way into a developable radiation damage of the latent track and ultimately into the observed track.

The main goals of track formation theory are to understand the threshold of track formation, the magnitude and spatial distribution of the observed effect, and the saturation of the observed effect with increasing energy input.

The chapter "Energy-loss of ions in matter" gives a classical-physics introduction to this field and is based mainly on the references [1], [2], [3], [4], and [5]. Interpolated experimental energy-losses (stopping powers) and ranges for pure elements are tabulated in references [6], [7], and [8]. Their precision is between 10% and 30%, depending on the ion/target combination and the experimentally accessible energy range.

[1]    Bonderup, E.: "**Penetration of Charged Particles Through Matter.**" Lecture Notes, Institute of Physics, University of Aarhus, Denmark, 229 pp. (1974).
[2]    Flügge, S.: "**Lehrbuch der theoretischen Physik.**" Springer, Berlin, **3**, 42, (1961).
[3]    Granzer, F.: "Strahlenschädigung in Kristallen, Teil I: Primärprozesse bei der Wechselwirkung geladener Teilchen mit Materie". Inst. Angew. Phys., Univ. Frankfurt/M, ca. 108 pp., (1971).
[4]    Sigmund, P.: 1975: "Energy Loss of Charged Particle in Solids", in "**Radiation Damage Processes in Materials**", C.H.S. Dupuy (editor), Nato Advanced Study Institutes Series E: Applied Sciences - No. 8, pp. 1 - 117 (1975).
[5]    Ziegler, J.F., J.P. Biersack, U. Littmark: "The Stopping and Range of Ions in Solids." Pergamon Press, New York, 321 pp., (1985).
[6]    Northcliffe, L.C., R.F. Schilling: "**Range and Stopping-Power Tables for Heavy Ions.**" Nuclear Data Tables A7, 233 pp., (1970).
[7]    Ziegler, J.F.: "**The Stopping and Ranges of Ions in Matter.**" **1-5**, Pergamon Press, London, (1977).
[8]    Hubert, F., A. Fleury, R. Bimbot, D. Gardes: "**Range and Stopping-Power Tables for 2.5 - 100 MeV/Nucleon Heavy Ions in Solids.**" Ann. Phys. (Paris) **5**, 1-214, (1980).

**Observed phenomena**

The observed effects of the ion passage through the solid depend, first, on its mass $M_1$, its atomic number — or nuclear charge number — $Z_1$, and its kinetic energy $E_1$ and, second, on the atomic mass $M_2$ and the nuclear charge number $Z_2$ of the target material. Within the target, the ion assumes the effective charge number $Z_{eff} \leq Z_1$, which — or its associated screening length [1]— is a crucial fitting parameter. The passage of the ion through a solid causes three groups of phenomena to occur (Figure 2-2) [2].

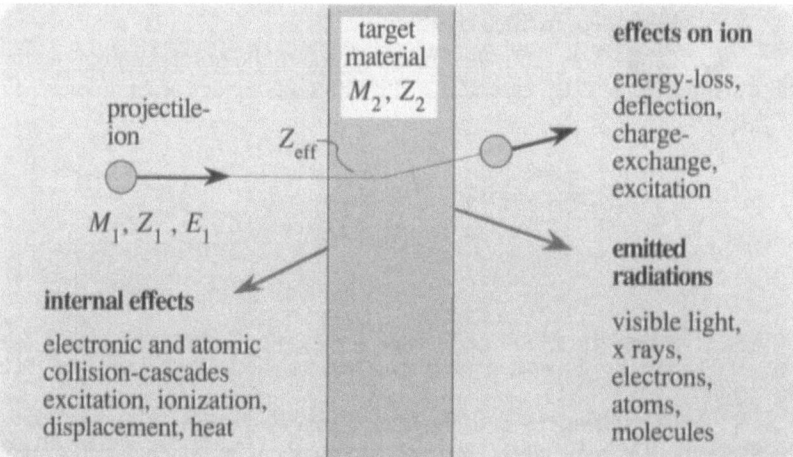

**Figure 2-2** *Observed effects in ion-solid interactions* for thin targets. The projectile (index "1") has the mass $M_1$, the nuclear charge $Z_1$, and the kinetic energy $E_1$. It behaves like a point charge of charge number $Z_{eff} \leq Z_1$ within the target. The target (index "2") has mass $M_2$, and nuclear charge $Z_2$.

1. *Effects on the ion.* Most readily observed are the energy-loss effects of the ion, due to the easy detection of charged particles of high energies. The ion undergoes changes in its kinetic energy. It may become electronically excited. It may loose or pick up further electrons and thereby change its charge-state. And finally, the transfer of linear momentum in the collision process leads to the deflection of the ion from its original direction.

2. *Emitted radiations.* Somewhat more difficult to observe are secondary radiations of much lower energies, created by the ion passage in the solid. The observation of x rays, visible light, electrons, atoms, and molecules can give valuable information about internal target processes, inaccessible to a direct observation. Thereby the depth from which information is recovered decreases with increasing interaction between the emitted particles and the solid, decreasing from x rays to visible light, electrons, atoms, and molecules.

3. *Internal effects.* The third group of effects are internal processes hidden within the target. They are directly not accessible, but the associated secondary radiations can give valuable clues.

[1]    Brandt, W., M. Kitagawa, Phys. Rev. **25B**, 5631, (1982).
[2]    Armbruster, P., J.C. Jousset, J. Remillieux (chairmen): "**Symposium on Swift Heavy Ions in Matter.**" Caen, May 16-19, 1989, Radiation Effects and Defects in Solids, **110**, Nos. 1-2, (1989).

### Shape of the energy-loss function

Figure 2-3 gives the qualitative shape of the energy-loss function for a projectile-ion in a solid containing the basic features of all energy-loss curves. The figure provides a first rough idea for classical energy-loss theory described in this chapter.

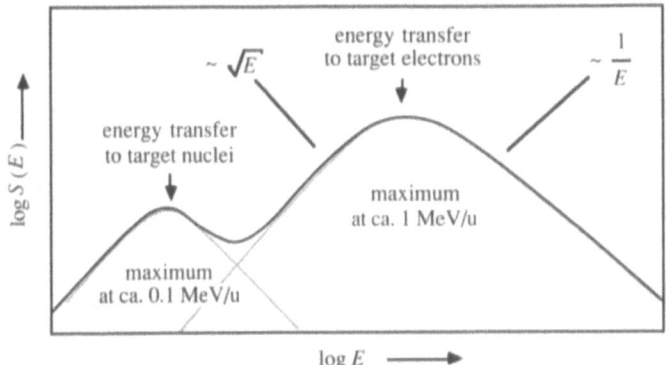

**Figure 2-3** *Qualitative shape of stopping function S(E) of a projectile-ion in solid matter as function of the kinetic energy E of the projectile-ion. At low energies of the order of 0.1 MeV/u and less, the energy-transfer to the atomic nuclei prevails (Nuclear stopping range). Above an energy of the order of 0.1 MeV/u the energy-transfer to the individual electrons of the irradiated solid prevails (Electronic stopping range). This so-called electronic energy-loss region is mainly responsible for the creation of ion tracks.*

On its way through the solid the ion looses energy. With decreasing energy the energy-loss function, or stopping power $S(E) \equiv (dE / dx)$, sweeps over many orders of magnitude. Therefore the function $S(E)$ is often represented on double-logarithmic scales. Its rough shape is characterized by a maximum around 1 MeV/u, the so-called "Bragg peak", corresponding to the maximum of the electronic energy-loss function. To higher energies the function decreases as $1/E$ — the range of Bethe theory [1] — and to lower energies it decreases as $E^{1/2}$ — the range of Lindhard-Scharff-Schiøtt (LSS) theory [2] —. The transition region around the maximum energy-loss is still the domain of semi-empirical theories based on experiments. At low ion energies, shortly before the end of the ion range, nuclear stopping becomes prevalent. It has a maximum roughly around 0.1 MeV/u. Normally the energy-loss function $S(E)$ is split up into two terms, the electronic (index "e") and the nuclear (index "n") stopping term:

$$S(E) \equiv S_e(E) + S_n(E) \iff \left(\frac{dE}{dx}\right) \equiv \left(\frac{dE}{dx}\right)_e + \left(\frac{dE}{dx}\right)_n. \tag{2.1}$$

While $S(E)_e$ corresponds already roughly to the quasi-continuous retardation of one single projectile-ion, $S(E)_n$ reflects only the average retardation of many projectile-ions, since the individual scattering processes correspond to abrupt changes of the momentum and energy of the projectile. Stopping power $S(E)$, energy loss $(dE / dx)$ and linear energy transfer (LET) are used in practice as synonyms. Alternatively, for $S$ the term "stopping cross section" $\equiv (dE / dx) / N$, is used in the literature, where $N$ is the number of target atoms per unit volume.

[1]    Ahlen, S.P., Rev. Mod. Phys. Vol. **52**, 121, (1980).
[2]    Lindhard, J., M. Scharff, H.E. Schiøtt, Mat. Fys. Medd. Dan. **33**, No. 14, (1963).

### Electronic stopping

Looking somewhat closer at the first step of track creation, the energy-loss of ions in matter can be further resolved into three distinct phases (Figure 2-4).

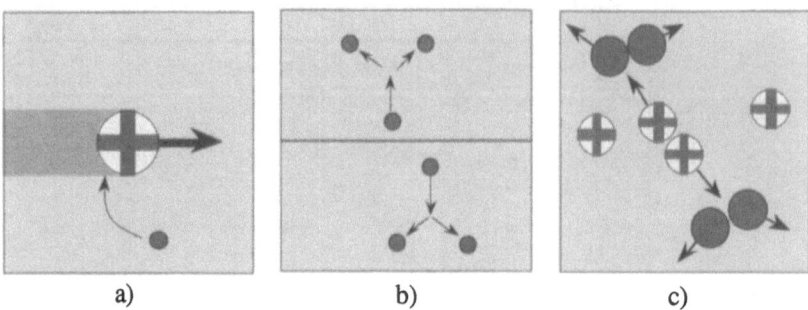

a)                              b)                              c)

**Figure 2-4** *Principal processes in track creation.* (**a**) Energy-transfer from the positive ion to a target electron, (**b**) Electronic collision cascades, triggered by accelerated target electrons. (**c**) In insulating solids the resulting positive charge cloud expands by Coulomb repulsion and leads to an atomic collision cascade.

a)  *Energy transfer.* During the first step, energy is transferred from the ion to individual electrons of the target material. The corresponding energy-loss of the ion and its deflection can be observed directly by determining the energy and angle of the ion before and after a thin target. Energy-loss phenomena therefore are treated quite satisfactory, theoretically and experimentally.

b)  *Electronic collision cascades.* The second step is the formation of electronic collision-cascades (delta electrons) which carry away energy and negative charge — corresponding to an electric current — from the atoms along the projectile trajectory up to a distance of the order of about 1 μm [1].

c)  *Atomic collision cascades.* The third step is the formation of atomic collision-cascades, resulting from the Coulomb repulsion of the remaining ions along the projectile-ion path. The atomic cascade has a much shorter range than the electronic cascade. Its diameter is of the order of 0.01 μm and can be treated by transport theory or by Monte-Carlo computer codes [2] [3], [4], ultimately leading to excited or displaced atoms — so-called point defects.

### Nuclear stopping

As the projectile energy decreases, the electron shells surrounding the target nuclei become more and more impenetrable for the intruding projectile. In the limiting case the impact process resembles finally the collision of two billiard balls of comparable masses. With increasing energy the electronic shells of both atoms are penetrating each other increasingly and above about 0.1 MeV/u one observes practically pure Rutherford scattering. The low energy processes are known as "nuclear stopping" prevailing at low ion energies.

[1]    Schou, J.: "Transport Theory for Kinetic Emission of Secondary Electrons from Solids." Phys. Rev. B, **22**, No. 5, pp. 2141-2174, (1980).
[2]    Biersack, J.P., L.G. Haggmark, Nucl. Instr. Meth. **170**, 257, (1980).
[3]    Biersack, J.P., Fusion Technology, **6**, 475, (1984).
[4]    Biersack, J.P., W.G. Eckstein, Appl. Phys. **A34**, 73, (1984).

Nuclear stopping accounts for the possibility of a direct transfer of momentum and energy from the projectile-ion to the target nuclei or atoms. This energy-transfer mechanism is considerably less smooth than the energy-transfer to the target electrons, which are much lighter than the projectile-ions. It takes away much larger lumps of energy from the primary projectile, leading to an irregular path of the ion with frequent kinks. In the nuclear stopping process, the target atoms receive large amounts of the available kinetic energy. The resulting knock-on atoms can be treated as new input particles, starting their own electronic or atomic collision-cascades, in a similar way as the projectile-ions and leading to defect clusters even in metals. In most cases of track creation, however, the projectile-ion energy is so high that nuclear stopping can be considered as a small effect.

### Track-related observation techniques

The energy-loss of the ion is directly related to the thickness of the target. Therefore an immediate application of the energy-loss function is the determination of the thickness of a thin film using alpha particles. Also, the angular straggling of a parallel ion beam by transmission through a thin target, is directly related with the target thickness. The knowledge of the angular straggling function can be used to widen the ion beam in ion track irradiations. From the knowledge of the energy-loss function, $(dE / dx)$, it is principally possible to calculate the range of the ions in a thick target in which the ions are stopped.

Table 2-1 gives a survey of the observed effects and some practical applications related to the field of ion tracks.

**Table 2-1** Observed interaction phenomena and their track-related applications [1].

| phenomena | track-related applications |
| --- | --- |
| **effects caused on ion** | |
| energy-loss of projectile-ion | energy reduction |
| | ion-range determination, |
| | range straggling, |
| | window and target thickness, |
| angular deflection of ion | angular straggling, |
| | ion-beam widening, |
| charge exchange | stripping process |
| **emitted radiations** | |
| light emission | ion-beam profile monitoring, |
| electron emission | electronic collision-cascade, |
| | ion-beam intensity monitoring, |
| x ray emission | target-composition analysis, |
| emission of atoms and molecules | atomic collision-cascade, |
| | track mechanism, |
| | surface analysis, |

[1]     Groeneveld, K.O.E.: "Nuclear Track Formation Related Electron Production and Transport From Ion Penetration Through Solids." 14th International Conference on Solid State Nuclear Track Detectors, Lahore, Pakistan, 2-6 April, 1988.

# 2.1    Energy-transfer to target electrons

There exist four steps in calculating the energy-loss of ions in matter. First, the energy-loss of ions in matter is reduced to binary-encounter collisions between the projectile-ion and the target electrons. Second, the transferred kinetic energy $T$ is calculated as function of the scattering angle $\psi$ of the target electron in the laboratory reference-frame. Third, the scattering angle $\psi$ is expressed as function of the collision parameter $p$. Fourth, the energy-transfer is integrated over all accessible impact parameters $p$.

## 2.1.1    Binary-encounter model

According to the binary-encounter model — valid for sufficiently high velocity and sufficiently heavy projectile-ions — it is possible to reduce the energy-transfer of a projectile-ion in solid matter to a sequence of independent collision events in which the projectile-ion interacts only with one single particle at a time. According to the binary-encounter model, simultaneous collisions between three and more particles at a time are completely neglected. The binary-encounter model is also applicable to secondary processes in which the knocked-on particles are treated again as projectiles in their own right. In this way the formation of extended collision-cascades — in which the number of knocked-on particles multiplies rapidly — can be treated successfully using computer codes.

### Symmetry and conservation laws

The binary-encounter model is based on the conservation theorems for energy, linear momentum and angular momentum, observables that are invariable (or invariant) with respect to certain changes in the reference frame (or symmetries), such as time, translation, and rotation. Implicitly assumed are furthermore the conservation of mass and charge. The following section is mainly based on reference [1].

**Translation symmetry and conservation of momentum.** The equation of motion of a particle of Mass $M$, moving in one dimension in a potential $V(x)$ is

$$M \ddot{x} \equiv M \frac{d^2 x}{dt^2} = - \frac{dV}{dx} \tag{2.2}$$

If $V(x)$ = const (independent of $x$) or expressed in terms of symmetry, if $V(x)$ is invariant with respect to a translation of the coordinate system in which $x$ is replaced by $x + a$, or $x \rightarrow x\ a$, than we have $M (d^2 x / dt^2) = -(dV / dx) = 0$. Integration over the time $t$ yields

$$P \equiv M \dot{x} = M \frac{dx}{dt} = \text{const} , \tag{2.3}$$

which means the observable $P$, the linear momentum, is conserved, or in other words is an invariant quantity. This result can be directly transferred from the one-dimensional problem into three dimensions, $P = M \mathbf{v} = \text{const}$.

[1]    Elliot, J.P. , P.G. Dawber: "**Symmetry in Physics**", vol. 1: "Principles and Simple Applications", Oxford University Press, New York, (ca. 280 pp.), (1979).

**Rotational symmetry and conservation of angular momentum.** The equation of motion of a particle of mass $M$ in two dimensions in a potential $V(x, y)$ in two dimensions is

$$M \ddot{x} = -\frac{\partial V}{\partial x} \, , \tag{2.4}$$

$$M \ddot{y} = -\frac{\partial V}{\partial y} \, . \tag{2.5}$$

If $V(x, y)$ is invariant with respect to a turning (or a rotation) of the coordinate system around the origin, that means if $V(x, y)$ is a centrally symmetric (or central) potential, then

$$\frac{\partial V (r, \vartheta)}{\partial \vartheta} = 0 \, . \tag{2.6}$$

This is just a more precise definition of a central potential in two dimensions, where the potential $V$ is represented in polar coordinates $(r, \vartheta)$. If we transfer this statement into the $(x, y)$-coordinate system, we obtain

$$\frac{\partial V}{\partial \vartheta} = \frac{\partial V}{\partial x} \frac{\partial x}{\partial \vartheta} + \frac{\partial V}{\partial y} \frac{\partial y}{\partial \vartheta} = -y \frac{\partial V}{\partial x} + x \frac{\partial V}{\partial y} \, . \tag{2.7}$$

Thereby we used the geometrical relations between the $(x, y)$ and the $(r, \vartheta)$ coordinate system (Figure 2-5), leading to

$$dx = -r \, d\vartheta \sin \vartheta \Rightarrow \frac{\partial x}{\partial \vartheta} = -r \sin \vartheta = -y \, , \tag{2.8}$$

$$dy = r \, d\vartheta \cos \vartheta \Rightarrow \frac{\partial y}{\partial \vartheta} = r \cos \vartheta = x \, . \tag{2.9}$$

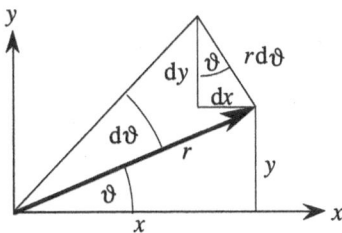

**Figure 2-5**   Coordinate transformation from Cartesian coordinates, $(x, y)$, to polar coordinates, $(r, \vartheta)$.

Inserting eqs. (2.4) and (2.5) in (2.7) yields

$$\frac{\partial V}{\partial \vartheta} = -y \, (-M \ddot{x}) + x \, (-M \ddot{y}) \, , \quad \text{or} \tag{2.10}$$

$$\frac{\partial V}{\partial \vartheta} = M (y \ddot{x} - x \ddot{y}) \ . \tag{2.11}$$

This equation can be transformed in the following way:

$$\frac{d}{dt} (y \dot{x} - x \dot{y}) = (y \ddot{x} - x \ddot{y}) \ . \tag{2.12}$$

Connecting eqs. (2.6), (2.11), and (2.12) in series thus yields

$$0 = \frac{\partial V}{\partial \vartheta} = M (y \ddot{x} - x \ddot{y}) = \frac{d}{dt} M (y \dot{x} - x \dot{y}) \ , \ \text{or} \tag{2.13}$$

$$\frac{\partial I}{\partial t} = 0 \ , \ \text{with} \ I \equiv M (x \dot{y} - y \dot{x}) = \text{const} \ . \tag{2.14}$$

The quantity $I$, defined in (2.14), is a conserved quantity, called the angular momentum. The eqs. (2.13) and (2.14) can also be read in the following way: the angular independence of the potential $V$ is equivalent with the conservation of the angular momentum $I$. In three dimensions the angular momentum is a vector, since any angular momentum can be decomposed into three independent components. It is the vector product of the vector $r$ and the linear momentum $P = M \mathbf{v}$ (Figure 2-6):

$$I \equiv r \times P = r \times M \dot{r} \ \Rightarrow \ I = |I| = M r \mathbf{v} \sin(r, \mathbf{v}) \ . \tag{2.15}$$

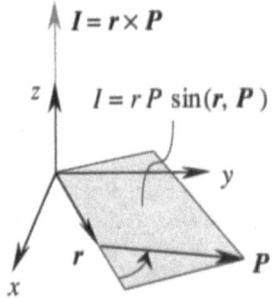

**Figure 2-6**  Definition of angular momentum.

If we rotate the reference frame about (0,0,0) such that $I$ points along the $z$-axis, than

$$|I| \equiv I = |M (x, y, 0) \times (\dot{x}, \dot{y}, 0)| = |M (0, 0, x\dot{y} - y\dot{x})| = M (x\dot{y} - y\dot{x}) \ , \tag{2.16}$$

as in the two-dimensional case. The magnitude $I \equiv |I|$ in polar coordinates with the $z$-axis pointing in the direction of the angular momentum vector is obtained by subtraction of the terms (Figure 2-5)

$$x \dot{y} = r \cos \vartheta \frac{d}{dt} (r \sin \vartheta) = r \cos \vartheta (\dot{r} \sin \vartheta + r \cos(\vartheta) \dot{\vartheta}) \ \text{and} \tag{2.17}$$

$$y\,\dot{x} = r\,\sin\vartheta\,\frac{d}{dt}\,(r\,\cos\vartheta) = r\,\sin\vartheta\,(\dot{r}\,\cos\vartheta - r\,\sin(\vartheta)\,\dot{\vartheta}),\ \text{yielding}\quad (2.18)$$

$$x\,\dot{y} - y\,\dot{x} = r\,\dot{r}\,(\cos\vartheta\sin\vartheta - \sin\vartheta\cos\vartheta) + r^2(\cos^2(\vartheta)\,\dot{\vartheta} + \sin^2(\vartheta)\,\dot{\vartheta}),\quad (2.19)$$

$$x\,\dot{y} - y\,\dot{x} = r^2\,\dot{\vartheta}\,(\cos^2(\vartheta) + \sin^2(\vartheta))\ .\qquad\qquad (2.20)$$

Applying the identity $\cos 2\vartheta + \sin 2\vartheta \equiv 1$ and multiplying the result with the mass $M$, we obtain:

$$I = M\,(x\,\dot{y} - y\,\dot{x}) = M\,r^2\,\dot{\vartheta} = M\,r^2\,\omega,\ \text{where}\ \omega \equiv \dot{\vartheta}\qquad (2.21)$$

is the angular frequency. The quantity $I = M\,r^2\,\omega$ can be concretely interpreted as the product of the particle mass $M$ and twice the area $dF$ covered by the radial vector $r$ per unit time (Figure 2-7):

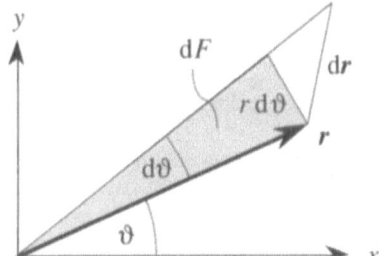

**Figure 2-7** Kepler's second law of the conservation of angular momentum: The radius vector from the origin to the moving particle sweeps out equal areas in equal intervals of time.

$$I = M\,r^2\,\dot{\vartheta} = M\,r^2\,\frac{d\vartheta}{dt} = 2\,M\,\frac{dF}{dt} = \text{const},\ \text{where}\ dF = \frac{1}{2}\,r^2\,d\vartheta\ .\quad (2.22)$$

This result is Kepler's second law that in a central potential or central field-of-force the radius vector from the origin to the moving particle sweeps out equal areas in equal intervals of time.

### Cross-section

Another important basis of the binary encounter model is the idea of the cross-section.which gives the probability for the occurrence of a specific event. The event $A$ corresponds to the detection of the particle in a certain detector. For example, projectiles passing within an impact-parameter range $\Delta p$ are registered within an angular interval $\Delta\varphi$. There exist several ways to define the cross-section [1].

[1]     Sigmund, P.: 1975: "Energy Loss of Charged Particle in Solids", in "**Radiation Damage Processes in Materials**", C.H.S. Dupuy (editor), Nato Advanced Study Institutes Series E: Applied Sciences - No. 8, pp. 1 - 117 (1975).

**Classical definition.** From this point of view, the infinitesimal area $\sigma_A$ is an effective area per target atom. The probability for the occurrence of the event $A$ corresponds to the area of a circular ring of width $\Delta p$, representing the "differential cross-section" for events in the angular range $\Delta\varphi$ (Figure 2-8). The event $A$ occurs if and only if the projectile trajectory aims at a point within this area. In most cases the problem has cylindrical symmetry and $\sigma_A$ is a circular ring of width $\Delta p$ at the distance $p$ from the target nucleus. Therefore

$$\sigma_A = 2\pi p \,\Delta p \; . \tag{2.23}$$

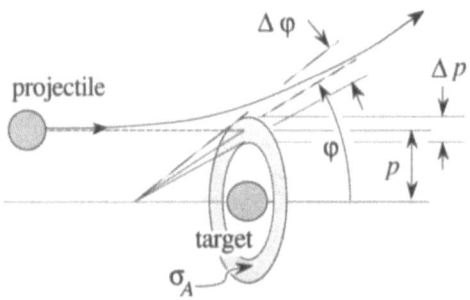

**Figure 2-8** *Definition of classical cross-section.* Projectiles within the impact-parameter range $\Delta p$ are scattered into the angular interval $\Delta\varphi$. The probability of this process to occur corresponds to the differential cross-section $\sigma_A = 2\pi \, p \, \Delta p$.

**Probability definition.** Assuming a thin target of surface $S$ and thickness $\Delta x$, containing $N'$ target particles per volume element, the probability $P_A$ for the occurrence of the event $A$ in a collision experiment is given by

$$P_A = \frac{\text{"opaque" area}}{\text{total area}} = \frac{n\,\sigma_A}{S} = N'\,\Delta x\;\sigma_A \;\Rightarrow\; \sigma_A = \frac{P_A}{N'\,\Delta x}\;, \tag{2.24}$$

where $n = N'\,S\,\Delta x$ is the total number of target particles and $N'\,\Delta x$ the number of target particles per unit surface area.

### Energy-loss and cross-section

The calculation of the energy-loss function $(dE/dx)$ is based, first, on the probability of a certain energy-loss $T$ per target particle and, second, on the number $n$ of target particles in a target of unit area and thickness $\Delta x$ (Figure 2-9).

**Energy-loss per target particle.** Each impact parameter $p$ corresponds to a certain angular deflection $\varphi$ of the projectile and thus to a certain energy-transfer $T$. The probability of such an energy-transfer $T$ corresponds to a cross-sectional area $\sigma$. By summing — or integrating — over all accessible energy-transfers (Figure 2-9 a) or impact parameters or deflection angles one obtains the average energy-loss of the projectile-ion per target particle,

$$<T> = \frac{\sigma_1 T_1 + \sigma_2 T_2 + \dots + \sigma_m T_m}{\sigma_1 + \sigma_2 + \dots + \sigma_m}\;, \quad \text{or} \tag{2.25}$$

$$<T> = \frac{\int_0^{p_{max}} 2\pi\, p\ dp\ T\,(p\,)}{\int_0^{p_{max}} 2\pi\, p\ dp} \quad . \tag{2.26}$$

The summation or integration has to consider all cross-sectional areas $\sigma$ up to a maximum impact parameter $p_{max}$, corresponding to a minimum energy-transfer $T_{min}$. Beyond the maximum impact parameter $p_{max}$ (below the minimum energy-transfer $T_{min}$) it is impossible to impart energy to the target particle. This low-energy "cut-off" prevents the summation or integration to diverge to infinity in Coulomb interactions without screening. It is the result of the binding energy of the target particles within the target which leads to minimum energy quanta required for excitation.

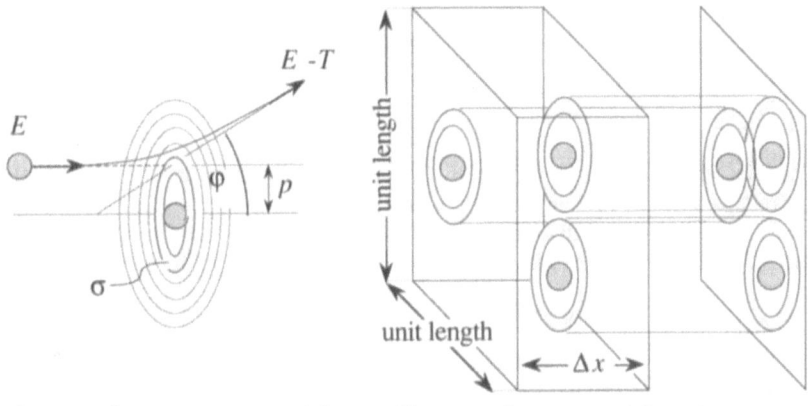

a) energy-loss per target particle        b) energy-loss per path length $\Delta x$

**Figure 2-9** *Principal steps for calculating the energy-loss function* $(dE/dx)$. **(a)** Each impact parameter $p$ corresponds to a certain angular deflection $\varphi$ of the projectile and thus to a certain energy-transfer $T$. By integrating over all accessible energy-transfers $T$ (or impact parameters or deflection angles) one obtains the average energy-loss $<T>$ of the projectile-ion per target particle. **(b)** By multiplying the average energy-loss $<T>$ per target particle with the number $n = N\,\Delta x$ of target particles present in a target of unit area and thickness $\Delta x$ one obtains the total energy-loss $\Delta E$ in the target. Division by $\Delta x$ yields the energy-loss function $(dE/dx)=N <T>$. This step is equivalent to projecting all available cross-sections onto a plane.

**Energy-loss per unit path length.** In order to determine the energy-loss due to an ensemble of $n$ target particles contained within a target of unit area and thickness $\Delta x$, it is assumed that the target thickness is sufficiently small, so that the projectile undergoes not more than one scattering process within the thickness $\Delta x$. In this case no overlap occurs between the cross-sections of different atoms and all cross-sections can be thought to be projected for example onto the exit plane of the thin target (Figure 2-9 b). We obtain for the energy-loss

$$\Delta E = n <T> = N\,\Delta x <T> \ , \ \text{with } n = N\,\Delta x \ . \tag{2.27}$$

In this equation $N$ is the number of target particles per unit volume. By dividing this equation through $\Delta x$, we obtain for $\Delta x \to 0$ the stopping power $S(E)$ or energy-loss function $(dE/dx)$:

$$S(E) \equiv \frac{dE}{dx} = \lim_{\Delta x \to 0} \frac{\Delta E}{\Delta x} = N <T> \ . \tag{2.28}$$

In order to obtain the energy-loss of the projectile-ion per unit path length, $(dE/dx)$, one has therefore to multiply the average energy-loss $<T>$ per target particle by the number $N$ of target particles per volume element.

### Elastic versus inelastic collisions

In an elastic scattering process the sum of the kinetic energies of the particles before and after the collision is constant. Two particles with no internal degrees of freedom — such as two electrons — can only scatter elastically on each other, as long as no other particles are created. Also, if the particles — such as an ion colliding with an atom — possess internal degrees of freedom, but the kinetic energy of the ion is too small for excitation of internal degrees of freedom we obtain an elastic scattering process. In this case the kinetic energy of the projectile $E_1$ before the collision corresponds exactly to the sum of the kinetic energies $E_1' + E_2'$ of the two particles after the collision:

$$E_1 = E_1' + E_2' \ . \tag{2.29}$$

The elastic process dominates at low kinetic energies. In every-day life the elastic collision corresponds to the collision of two billiard balls.

If internal degrees of freedom exist — such as for a projectile-ion and a target atom — and if they are exited during the collision, one obtains instead the relation

$$E_1 > E_1' + E_2' \ . \tag{2.30}$$

The inelastic process dominates at high energies, in which internal degrees of freedom can be excited. It corresponds in every-day life to a collision between two bodies which can be plastically deformed, such as two cars in a traffic accident.

In both cases, the conservation of momentum gives an additional set of three equations which can be used to simplify the physical description of the problem:

$$p_1 = p_1' + p_2' \ . \tag{2.31}$$

## 2.1.2    Impact parameter and scattering angle

### Central collision

There exist very restricted conditions for a central collision. In this case the energy-transfer from the projectile to the target, can be directly calculated from the conservation of energy and momentum before and after the collision. Also the minimum distance-of-approach between two charged particles can be determined much easier in a central collision than in any off-axis condition. The results can be applied directly to the off-axis collision problem yielding an upper limit for the energy and momentum transfer. The central collision geometry is given in Figure 2-10. The projectile is characterized by charge $q_1$, mass $m_1$, velocity $v_1$ before and velocity $v_1'$ after the collision. The corresponding target parameters are $q_2$, $m_2$, and $v_2'$, since the target velocity before the collision is zero.

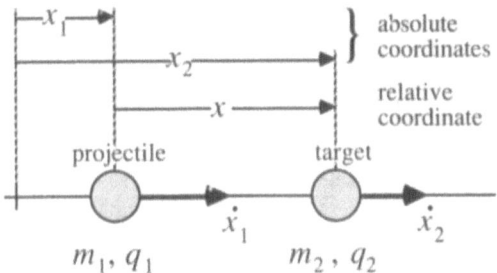

**Figure 2-10** Central collision between two particles.

**Transferred energy.** In order to determine the energy-transfer from the projectile to the target, it is sufficient to apply the conservation of linear momentum and energy sufficiently long before and sufficiently long after the collision, when the interaction between the particles is negligible. This leads to the equations

$$m_1 v_1 = m_1 v_1' + m_2 v_2' \quad \text{and} \quad m_1 v_1^2 = m_1 v_1'^2 + m_2 v_2'^2 \;, \tag{2.32}$$

respectively. These two equations contain two unknown variables $v_1'$ and $v_2'$ and therefore can be solved unambiguously. Since we are interested in the energy-transfer from the projectile to the target, we solve the equations for $v_2'$. Eliminating $v_1'$ from the equations yields

$$v_2' = 2 \frac{m_1}{m_1 + m_2} v_1 \;, \quad \text{or} \quad T = \frac{m_2 v_2'^2}{2} = 4 \frac{m_1 m_2}{(m_1 + m_2)^2} E \;. \tag{2.33}$$

where $T$ is the energy-transfer from the projectile to the target and $E = m_1 v_1^2 / 2$ the kinetic energy of the projectile before the collision. Obviously, the transferred energy does not depend on the interaction potential between the particles, represented by the Coulomb potential between the charges $q_1$ and $q_2$.

**Distance of closest approach.** While the conservation of energy and linear momentum gave sufficient constraint to solve the problem of the energy-transfer between the particles, one has to take into account the specific shape of the interaction potential $V(x_2 - x_1)$ if one wants to determine the distance of closest approach between the colliding particles. The fundamental equations-of-motion for the two particles are:

$$m_1 \ddot{x}_1 = -\frac{\partial V_1}{\partial x_1} = -\frac{q_1 q_2}{(x_2 - x_1)^2} \;, \quad \text{with } V_1 = \frac{q_1 q_2}{|x_2 - x_1|} \;, \tag{2.34}$$

$$m_2 \ddot{x}_2 = -\frac{\partial V_2}{\partial x_2} = -\frac{q_1 q_2}{(x_2 - x_1)^2} \;, \quad \text{with } V_2 = \frac{q_1 q_2}{|x_2 - x_1|} = V_1 \;. \tag{2.35}$$

The eqs. (2.34) and (2.35) are coupled via the interaction potential $V(x_2 - x_1)$, which is a function of the difference of the particle coordinates and can be reduced to just one differential equation by introducing the relative coordinate $x \equiv x_2 - x_1$. For this purpose we divide the equations by the masses $m_1$ and $m_2$, respectively, subtract the first equation from the second one, and obtain:

$$\ddot{x} = \left(\frac{1}{m_1} + \frac{1}{m_2}\right)\frac{q_1 q_2}{x^2} = \frac{m_1 + m_2}{m_1 m_2}\frac{q_1 q_2}{x^2} \quad, \text{ where } x \equiv x_2 - x_1 \,. \qquad (2.36)$$

It can be written in the standard form of an equation of motion for a particle of the "reduced mass" $m_0 \equiv m_1 m_2 / (m_1 + m_2)$ if we multiply eq. (2.36) by $m_0$:

$$m_0 \ddot{x} = \frac{q_1 q_2}{x^2} \quad, \quad \text{where } m_0 \equiv \frac{m_1 m_2}{(m_1 + m_2)} \,. \qquad (2.37)$$

By multiplying the equation with the relative velocity $(dx/dt)$ we obtain a total differential

$$m_0 \ddot{x}\dot{x} = \frac{q_1 q_2}{x^2}\dot{x} \quad \Leftrightarrow \quad \frac{d}{dt}\left(\frac{1}{2}m_0 \dot{x}^2\right) = \frac{d}{dt}\left(-\frac{q_1 q_2}{x}\right) \,, \qquad (2.38)$$

which can be directly integrated and yields

$$\frac{1}{2}m_0 \dot{x}^2 + \frac{q_1 q_2}{x} = \frac{1}{2}m_0 v_1^2 = \text{const} \quad \Leftrightarrow \quad T_0 + V_0 = E_0 \,, \qquad (2.39)$$

whereby the integration-constant is the reduced kinetic energy $T_0 \equiv m_0 v_1^2 / 2$ before the collision, determined from eq. (2.39) for $t \rightarrow -\infty$, $V_0 = V$ is the usual potential energy, and $E_0$ is the reduced total energy of the system. In terms of the reduced units the equation just represents the familiar equality between the total energy and the sum of the kinetic and potential energy for the reduced system.

The distance of closest approach $b$ follows directly, if we realize that at the point-of-return the relative velocity $(dx/dt)$ vanishes.

$$b = \frac{q_1 q_2}{\frac{1}{2}m_0 v^2} \quad, \text{ where } m_0 = \frac{m_1 m_2}{m_1 + m_2} \,, \qquad (2.40)$$

in which the index "1" of the original projectile velocity $v_1$ has been omitted. The equation reads: At the point of closest-approach $b$ the reduced kinetic energy is equal to the reduced potential energy.

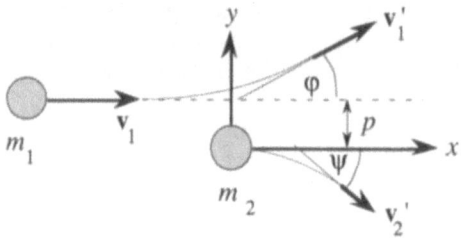

**Figure 2-11** *Collision parameters in laboratory coordinate system.* The collision takes place between a projectile of mass $m_1$ and velocity $v_1$ and a target particle of mass $m_2$ at rest, located at the origin of the laboratory coordinate system. There exist two scattering angles in the laboratory system, the angle $\varphi$ for the projectile, and the angle $\psi$ for the target.

### Laboratory system

The laboratory coordinate system is fixed with respect to the position of the observer in the laboratory and it's origin coincides with the target nucleus before the collision takes place (Figure 2-11). The ultimate results of the scattering process are expressed in terms of the laboratory system, although — for convenience — one has to use the center-of-mass system as an intermediate step in the calculations.

### Calculation of transferred energy

The kinetic energy $T$ transferred from the projectile to the target particle can be determined on the basis of the conservation of energy and linear momentum. It can be expresses as function of any one of the three parameters, $p$, $\varphi$, or $\psi$. First, the impact parameter $p$, second, the scattering angle $\varphi$ of the projectile in the laboratory system and, third, the scattering angle $\psi$ of the target particle. The first step of energy-loss theory usually is to calculate the transferred energy $T(\psi)$ as function of the scattering angle $\psi$ of the target particle. The next step is the determination of the function $\psi(p)$, representing the scattering angle $\psi$ of the target particle as function of the impact parameter $p$ .

From the conservation of linear momentum we obtain two equations, one equation for the $x$-component of the linear momentum, and one equation for the $y$-component of the linear momentum:

$$m_1 v_1 = m_1 v_1' \cos\varphi + m_2 v_2' \cos\psi, \tag{2.41}$$

$$0 = m_1 v_1' \sin\varphi - m_2 v_2' \sin\psi . \tag{2.42}$$

From the conservation of energy we obtain one additional equation,

$$m_1 v_1^2 = m_1 v_1'^2 + m_2 v_2'^2 \Rightarrow m_1 v_1'^2 = m_1 v_1^2 - m_2 v_2'^2 . \tag{2.43}$$

Since eqs. (2.41) to (2.43) contain four parameters in a system of three independent equations, the system is not completely defined, but it is possible to express all parameters as function of only one parameter. This leaves freedom to choose the impact parameter $p$ (alternatively $\varphi$ or $\psi$) as the independent variable. Since the goal is to calculate the energy-transfer $T$ as function of the impact parameter $p$, we will express the scattering angle $\psi$ of the target particle as function of the impact parameter. The first step of the calculation is the elimination of the projectile scattering angle $\varphi$, using the identity $\cos^2\varphi + \sin^2\varphi \equiv 1$. For this purpose we arrange the terms containing $\varphi$ on the left side of the eqs. (2.41) and (2.42) and obtain after quadrating and summing of the equations

$$m_1^2 v_1^2 = m_1^2 v_1'^2 - 2 m_1 v_1' m_2 v_2' \cos\psi + m_2^2 v_2'^2 , \tag{2.44}$$

whereby we applied the identity $\cos^2\varphi + \sin^2\varphi \equiv 1$ also to the last term on the right side. The next step is the elimination of the parameter $v_1'$. For this purpose we use the third defining equation, eq. (2.43), multiply it by the factor $m_1$ and subtract eq. (2.44):

$$0 = - 2 m_1 v_1' m_2 v_2' \cos\psi + m_2^2 v_2'^2 + m_1 m_2 v_2'^2 , \tag{2.45}$$

which after division by the factor $m_2 v_2'$ and rearrangement yields a relation between $v_2'$ and $\psi$

$$v_2' = 2 \frac{m_1}{m_1 + m_2} v_1 \cos\psi , \tag{2.46}$$

from which the transferred energy $T$ can be determined.

### Resulting energy-transfer

On the basis of the velocity $v_2'(\psi)$ of the target particle, eq. (2.46), the kinetic energy $T$, transferred from the projectile to the target particle, can be calculated:

$$T=\frac{m_2 v_2'^2}{2}=\frac{m_2}{2}\,4\frac{m_1^2}{(m_1+m_2)^2}\,v_1^2\cos^2\psi=4\frac{m_1 m_2}{(m_1+m_2)^2}\,\frac{m_1 v_1^2}{2}\cos^2\psi. \quad (2.47)$$

The essence of this equation becomes clearer visible if we realize that $m_1 v_1^2/2 = E$ is the kinetic energy of the projectile before the collision

$$T=T_{max}\cos^2\psi,\ \text{with}\ T_{max}=4\frac{m_1 m_2}{(m_1+m_2)^2}\,E\begin{cases}\rightarrow 2m_2 v^2\ \text{for}\ m_2\ll m_1\\[2mm]\rightarrow\dfrac{m}{2}v^2\ \text{for}\ m_1=m_2=m\end{cases} \quad (2.48)$$

The maximum energy-transfer $T_{max}$ corresponds to a central collision and is invariant with respect to an exchange of the masses. For example, its value is unchanged if we compare the case $m_1 = 3$ u, $m_2 = 5$ u with the case $m_1 = 5$ u, $m_2 = 3$ u, where $u = 1.6606\cdot10^{-24}$ g is the atomic mass unit. We can simplify eq. (2.48) by introducing the fraction $\varepsilon = T/E$ of the kinetic energy-transfer from the projectile to the target, as well as introducing the mass ratio $\mu = m_1/m_2$. We obtain (Figure 2-12.):

$$\varepsilon=\frac{T}{E}=\varepsilon_{max}\cos^2\psi,\ \text{with}\ \varepsilon_{max}=4\frac{\mu}{(1+\mu)^2}\ \text{and}\ \mu=\frac{m_1}{m_2}. \quad (2.49)$$

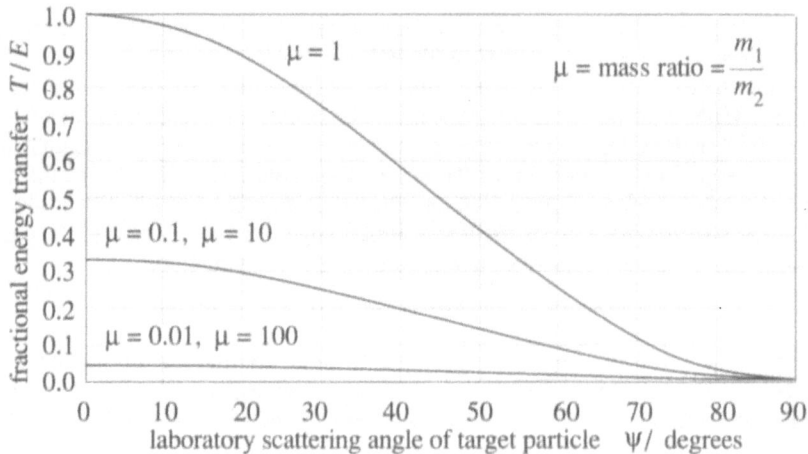

**Figure 2-12** *Energy transfer in elastic collisions.* Fraction $T/E$ of energy transferred from the projectile to the target in an elastic collision as function of the laboratory scattering angle $\psi$ of the target. The fraction of transferred energy is symmetric with respect to a mass exchange $m_1 \rightarrow m_2, m_2 \rightarrow m_1$, it has a maximum for equal masses, $m_1 = m_2$, and it decreases with increasing scattering angle $\psi$.

### Center-of-mass system

**Basic relations.** The description of the collision between the projectile and the target can be simplified using the center-of-mass system, short the "c.m.-system", characterized here by the subscript "c". The center-of-mass of a system of an arbitrary number of bodies is defined as that point for which the first moment of the masses vanishes,

$$\sum_i m_i r_{i,\,c} = 0 \; . \tag{2.50}$$

This is equivalent with the definition of the center-of-mass $r_c$:

$$r_c = \frac{\sum_i m_i r_i}{\sum_i m_i} \; , \tag{2.51}$$

which can be shown by multiplying with $\sum m_i$, and rearrangement:

$$\sum_i m_i r_c = \sum_i m_i r_i \;\; \Rightarrow \;\; \sum_i m_i \, (r_i - r_c) = 0, \;\; \text{with } r_i - r_c \equiv r_{i,\,c} \; . \tag{2.52}$$

Reducing eq. (2.50) now to the collision of two bodies we have

$$m_1 r_{1,\,c} + m_2 r_{2,\,c} = 0 \; , \;\; \text{and} \tag{2.53},$$

$$r_c = \frac{m_1 r_1 + m_2 r_2}{m_1 + m_2} \; , \tag{2.54}$$

which is geometrically represented in Figure 2-13.

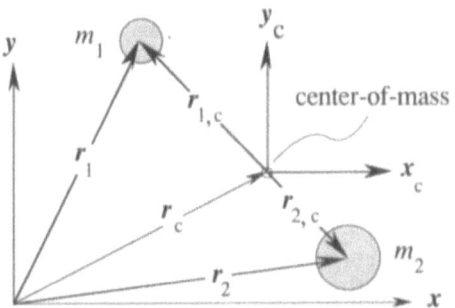

**Figure 2-13** *Definition of center-of-mass system.* The position vectors $r_{1,c}$ and $r_{2,c}$ of the colliding bodies in the center-of-mass system are antiparallel. This holds a-fortiori for the velocities $v_{1,c}$ and $v_{2,c}$ and the linear momenta $P_{1,c}$ and $P_{2,c}$.

From Figure 2-13 we read the defining relations between the position vectors in the laboratory and in the c.m. system:

$$r_{1,c} = r_1 - r_c \;\; , \;\;\; r_{2,c} = r_2 - r_c \; . \tag{2.55}$$

If we differentiate eq. (2.55) with respect to time, we obtain:

$$v_{1,c} = v_1 - v_c \;\; , \;\;\; v_{2,c} = v_2 - v_c \; . \tag{2.56}$$

Due to the initial conditions of the binary collision, $v_1 = v$, $v_2 = 0$ for $t = -\infty$, we obtain from eq. (2.56) the following relations for the c.m velocities:

$$v_{1,c} = v - v_c \quad , \quad v_{2,c} = -v_c \quad , \quad \text{for } t = -\infty . \tag{2.57}$$

If we differentiate eq. (2.54) with respect to time, we obtain $v_c$:

$$v_c = \frac{m_1 v_1 + m_2 v_2}{m_1 + m_2} \quad , \tag{2.58}$$

which is constant, due to the conservation of the total linear momentum $m_1 v_1 + m_2 v_2$. In order to determine the value of $v_c$ we remember that for $t = -\infty$ we have $v_1 = v$, $v_2 = 0$ and obtain

$$v_c = \frac{m_1 v}{m_1 + m_2} \tag{2.59}$$

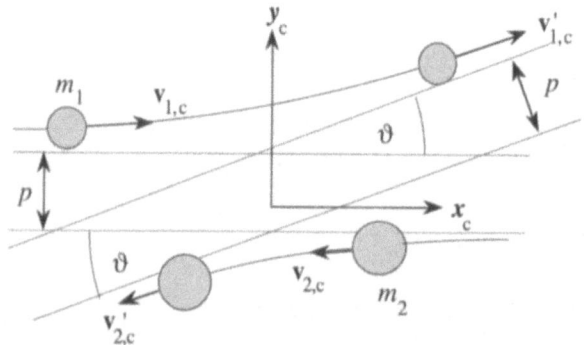

**Figure 2-14** *Definition of c.m. scattering angle $\vartheta$. Collision as observed in the c.m. system. Due to the symmetry of the collision in the the c.m. system, there exists only one impact parameter $p$ and only one scattering angle in the c.m. system, the angle $\vartheta$, the angle between the original and the final direction of the antiparallel velocities.*

If we differentiate eq. (2.53) with respect to time, we obtain the statement that in the c.m. system the velocities and the linear momenta are antiparallel:

$$m_1 \dot{r}_{1,c} + m_2 \dot{r}_{2,c} = 0 \;\Leftrightarrow\; m_1 v_{1,c} + m_2 v_{2,c} = 0 \;\Leftrightarrow\; P_{1,c} + P_{2,c} = 0 . \tag{2.60}$$

As we see also, the absolute values of the linear momenta are equal and we can write

$$|P_{1,c}| = |P_{2,c}| \equiv P_c . \tag{2.61}$$

Since the collision is assumed to be elastic, the sum of the kinetic energies $T$ of the two particles in the c.m. system before and after the collision is unchanged:

$$T = T' \quad\Leftrightarrow\quad \frac{P_c^2}{2m_1} + \frac{P_c^2}{2m_2} = \frac{P_c'^2}{2m_1} + \frac{P_c'^2}{2m_2} \quad \text{or} \tag{2.62}$$

$$\frac{P_c^2}{2}\left(\frac{1}{m_1} + \frac{1}{m_2}\right) = \frac{P_c'^2}{2}\left(\frac{1}{m_1} + \frac{1}{m_2}\right) \;\Rightarrow\; P_c = P_c' \quad , \quad \text{or} \tag{2.63}$$

$$|P_{1,\,c}| = |P_{2,\,c}| = |P'_{1,\,c}| = |P'_{2,\,c}| \,. \qquad (2.64)$$

Therefore the c.m. velocities have the same magnitudes before and after the collision:

$$|v_{1,\,c}| = |v'_{1,\,c}| = |v - v_c| \quad \text{and} \quad |v_{2,\,c}| = |v'_{2,\,c}| = |v_c| \,. \qquad (2.65)$$

Since the scattering angle $\vartheta$ in the c.m. system is the angle between the original and the final direction of the particle velocities, the description of the particle collision becomes more symmetric and therefore much simpler in the c.m. system. While we had the two scattering angles $\varphi$ and $\psi$ in the laboratory system, we have only one scattering angle $\vartheta$ in the c.m. system, due to the antiparallelity of the particle velocities throughout the collision (Figure 2-14).

**Relations between laboratory and c.m. angles.** We can now construct a geometric diagram of the collision from which we can directly read the relations between the velocities and angles in the laboratory and in the c.m.-system. For this purpose we use the geometric interpretation of the eqs. (2.57), (2.60), and (2.65) (Figure 2-15).

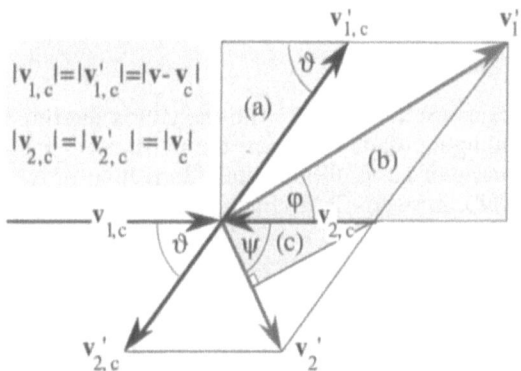

**Figure 2-15** *Scattering diagram*, describing the geometric relations between the laboratory scattering angles $\varphi$ and $\psi$ and the only one c.m. scattering angle $\vartheta$.

Triangle (b) contains the angle $\varphi$. Its tangent can be expressed as the ratio between the left side of triangle (a), $v_{1,c}' \sin \vartheta = (v - v_c) \sin \vartheta$, and the sum of the uppermost side of the velocity parallelogram, $v_c$, and the upper side of triangle (a), $(v - v_c) \cos \vartheta$. We obtain:

$$\tan \varphi = \frac{(v - v_c) \sin \vartheta}{v_c + (v - v_c) \cos \vartheta} \,. \qquad (2.66)$$

If we insert $v_c = m_1 v / (m_1 + m_2)$ from eq. (2.59), we obtain the relation between the c.m. scattering angle $\vartheta$ and the first laboratory angle $\varphi$:

$$\tan \varphi = \frac{m_2 \sin \vartheta}{m_1 + m_2 \cos \vartheta} \,. \qquad (2.67)$$

The triangle (c) contains the angles $\psi$, 90°, and $\vartheta/2$, from which we obtain the relation the c.m. scattering angle $\vartheta$ and the laboratory angle $\psi$:

$$\psi = \frac{\pi - \vartheta}{2} .$$

(2.68)

From the triangle (c) we read the relation between $v_2'$, $\vartheta$, and $v_c = - v_{2,c}$ which aims at the calculation of the kinetic energy transferred from the projectile to the target particle:

$$v_2' = 2 v_c \sin\frac{\vartheta}{2} .$$

(2.69)

**Transferred energy.** In order to determine the energy-transfer from the projectile to the target we use eqs. (2.69) and (2.59) and obtain:

$$T = \frac{1}{2} m_2 v_2'^2 = \frac{1}{2} m_2 \left(2 v_c \sin\frac{\vartheta}{2}\right)^2 = 4\frac{m_1 m_2}{(m_1 + m_2)^2} \frac{1}{2} m_1 v^2 \sin^2\frac{\vartheta}{2} , \text{ or}$$

(2.70)

$$T = T_{max} \sin^2\frac{\vartheta}{2} , \text{ with } T_{max} = 4\frac{m_1 m_2}{(m_1 + m_2)^2} E , \quad E = \frac{1}{2} m_1 v^2 .$$

(2.71)

This equation connects the transferred energy $T$ with the c.m. scattering angle $\vartheta$, in a similar way as the corresponding eq. (2.48) connects the transferred energy $T$ with the laboratory angle $\psi$. On the basis of eq. (2.68) a simple derivation of eq. (2.71) is obtained by inserting eq. (2.68) into eq. (2.48), using the transformation $\cos \psi = \cos[(\pi - \vartheta) / 2] = \sin(\vartheta / 2)$.

### Coulomb collision

It is obvious that the c.m. scattering angle $\vartheta$ will depend on the specific type of the interaction force — for example on its strength and on its radial dependence — and on the impact parameter. We will determine the scattering angle $\vartheta$ as function of the impact parameter $p$ for a Coulomb potential, an interaction potential which decreases as $1/r$ with the mutual distance between the colliding particles. The calculation is performed in the relative coordinate system, the origin of which is fixed in the center of one of the particles (Figure 2-16).

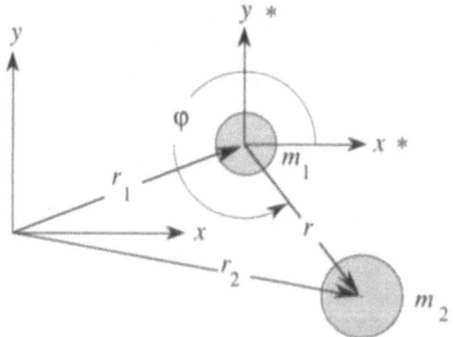

**Figure 2-16** *Definition of relative coordinate system,* fixed in particle 1 and used for determining the c.m. scattering angle $\vartheta$ in a central (here: Coulomb) field.

The fundamental equations of motion in the laboratory system for the projectile and target are:

$$m_1 \ddot{r}_1 = -\frac{q_1 q_2}{(r_2 - r_1)^2} \frac{r}{r} \quad ,$$

(2.72)

$$m_2 \ddot{r}_2 = \frac{q_1 q_2}{(r_2 - r_1)^2} \frac{r}{r} \quad , \text{ where } r = r_2 - r_1 .$$

(2.73)

Similar to the calculation of the distance of closest approach in the central collision, the system of two differential equations can be reduced to one joint differential equation — the reduced equation — by introducing the relative coordinate system. By dividing eqs. (2.72) and (2.73) through the respective masses and subtraction we obtain:

$$\ddot{r}_2 - \ddot{r}_1 = \left(\frac{1}{m_2} + \frac{1}{m_1}\right) \frac{q_1 q_2}{(r_2 - r_1)^2} \frac{r}{r} \quad , \text{ or}$$

(2.74)

$$m_0 \ddot{r} = \frac{q_1 q_2}{r^2} \frac{r}{r} \quad , \text{ where } m_0 = \frac{m_1 m_2}{m_1 + m_2} .$$

(2.75)

This equation corresponds to the equivalent problem of a mass $m_0$ scattered at the origin ' according to Figure 2-17.

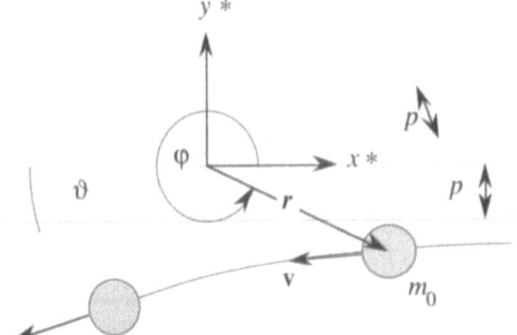

**Figure 2-17** *Equivalent scattering problem* for the reduced mass $m_0$, scattered at the origin.

For integrating this differential equation we multiply by $(dr/dt)$ and obtain:

$$m_0 \ddot{r} \dot{r} = \frac{q_1 q_2}{r^2} \frac{r \dot{r}}{r} \quad \Leftrightarrow \quad \frac{d}{dt}\left(\frac{1}{2} m_0 \dot{r}^2\right) = \frac{d}{dt}\left(-\frac{q_1 q_2}{r}\right) .$$

(2.76)

This represents the familiar fact that the increase in the kinetic energy corresponds to the decrease in potential energy:

$$\frac{dT}{dt} = -\frac{dV}{dt} \quad , \text{ where } T = \frac{1}{2} m_0 \dot{r}^2 \text{ and } V = \frac{q_1 q_2}{r} .$$

(2.77)

Integration yields:

$$T + V = E \quad , \quad \text{where } E = \frac{1}{2} m_0 \, v^2 \tag{2.78}$$

is determined from the boundary condition, that for $t = -\infty$ we have $(dr/dt) = -(dr_1/dt) = -v$. This represents an intermediate result, just confirming the well-known conservation of energy in an elastic collision. In order to find the solution of the fundamental equation of motion in relative coordinates, satisfying the boundary conditions, and determine the c.m. scattering angle $\vartheta$, we write down the x-component of eq. (2.75):

$$m_0 \ddot{x} \equiv m_0 \frac{d}{dt} \dot{x} = \frac{q_1 q_2}{r^3} x \equiv \frac{q_1 q_2}{r^3} r \cos\varphi \ . \tag{2.79}$$

Multiplication by $dt$ and integration from $t = -\infty$ to $t = +\infty$ yields

$$m_0 \dot{x}(t) \Big|_{-\infty}^{+\infty} \equiv m_0 \, v \, (\cos\vartheta - 1) = q_1 q_2 \int_{-\infty}^{+\infty} \frac{\cos\varphi}{r^2} \, dt \ . \tag{2.80}$$

On the right side of this equation we can — due to the conservation of angular momentum $I$ in a central field according to eq. (2.22) — replace the $1/r^2$ by the angular frequency $(d\varphi / dt)$, a trick which makes the right side integrable:

$$I = m_0 \, r \dot{r} \, \sin(r, \dot{r}) = m_0 \, r^2 \dot{\varphi} = \text{const} = -m_0 \, p \, v, \quad \text{for} -\infty \le t \le +\infty, \tag{2.81}$$

whereby the constant $I = -m_0 \, p \, v$ is determined according to Figure 2-17 for the boundary condition $t = -\infty$. From eq.(2.81) we obtain:

$$\frac{1}{r^2} = -\frac{1}{p \, v} \dot{\varphi} = -\frac{1}{p \, v} \left(\frac{d\varphi}{dt}\right) \tag{2.82}$$

which after insertion into the integral of eq. (2.80) enables its integration and yields:

$$m_0 \, v \, (\cos\vartheta - 1) = +\frac{q_1 q_2}{p \, v} \int_0^{\pi+\vartheta} \cos\varphi \ d\varphi = -\frac{q_1 q_2}{p \, v} \sin\vartheta \ , \tag{2.83}$$

whereby the negative direction of integration $-d\varphi$ and the limits are taken according to Figure 2-17. The eq. (2.83) can be simplified by collecting the terms containing the scattering angle $\vartheta$ on the left side of the equation and by applying the addition theorems of the angular functions:

$$\frac{\cos\vartheta - 1}{\sin\vartheta} \equiv \frac{\cos\frac{\vartheta}{2}\cos\frac{\vartheta}{2} - \sin\frac{\vartheta}{2}\sin\frac{\vartheta}{2} - \left(\cos^2\frac{\vartheta}{2} + \sin^2\frac{\vartheta}{2}\right)}{2\sin\frac{\vartheta}{2}\cos\frac{\vartheta}{2}} \equiv -\tan\frac{\vartheta}{2} \ . \tag{2.84}$$

We obtain as final result the connection between the impact parameter $p$ and the c.m. scattering angle $\vartheta$ in a collision with Coulomb interaction (Figure 2-18):

$$\tan\left(\frac{\vartheta}{2}\right) = \frac{b}{2p} \ , \quad \text{where } b = \frac{q_1 q_2}{\frac{1}{2} m_0 \, v^2} \quad \text{and } m_0 = \frac{m_1 m_2}{m_1 + m_2} \ . \tag{2.85}$$

The factor $b$ is the distance of closest approach in the central collision, introduced according to eq. (2.40), and $m_0$ is the reduced massz. As we see from the equation, the scattering angle $\vartheta$ decreases with increasing impact parameter $p$. From the factor $b$ we recognize, that the scattering angle increases with increasing charge of the colliding particles and decreases with the increasing kinetic energy $E = m_1\, v^2 / 2$ of the projectile.

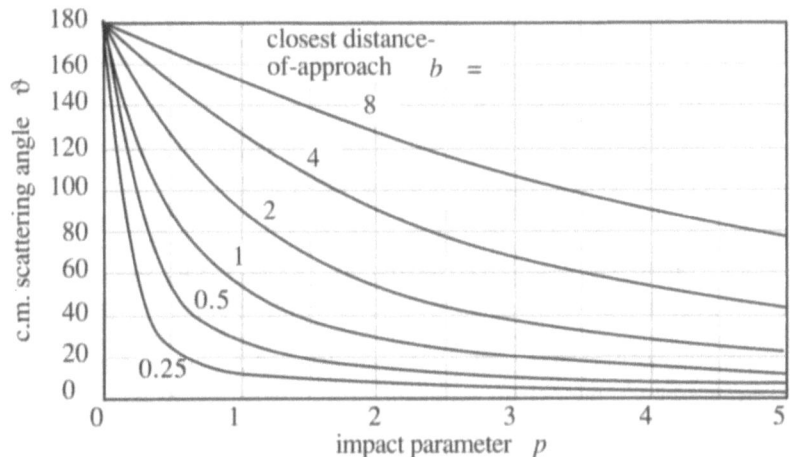

**Figure 2-18** *Scattering angle and impact parameter in elastic collisions.* The parameter $b$ reflects the ratio between the strength of the Coulomb interaction $q_1\, q_2$ and the kinetic energy $E = m_1\, v^2 / 2$ of the projectile. The impact parameter $p$ and the distance of closest approach are measured in the same units of length, for example in nm.

## 2.1.3 Transferred kinetic energy

For obtaining the transferred energy $T$ as function of the impact parameter $p$, we connect first $p$ with $\vartheta$ and then $\vartheta$ with $T$. If we insert eq. (2.85), connecting the impact parameter $p$ and the scattering angle $\vartheta$, into eq. (2.71),

$$T = \frac{1}{2}\, m_2\, v_2'^{\,2} = 4\frac{m_1\, m_2}{(m_1 + m_2)^2}\, \frac{1}{2}\, m_1\, v^2 \sin^2\frac{\vartheta}{2}\,. \tag{2.86}$$

connecting the scattering angle $\vartheta$ and the transferred energy $T$, we obtain the transferred energy $T$ as function of the impact parameter $p$. For this purpose we have to use the identity $\sin^2\alpha = \tan^2\alpha / (1 + \tan^2\alpha)$ and obtain

$$T = \frac{4}{m_2}\, \frac{m_1^2\, m_2^2}{(m_1 + m_2)^2}\, \frac{1}{2}\, v^2 \sin^2\!\left(\frac{\vartheta}{2}\right) = \frac{4}{m_2}\, m_0\, \frac{1}{2}\, m_0\, v^2 \sin^2\!\left(\frac{\vartheta}{2}\right), \tag{2.87}$$

$$T = \frac{4}{m_2}\, m_0\, \frac{q_1\, q_2}{b}\, \sin^2\!\left(\frac{\vartheta}{2}\right) = \frac{4}{m_2}\, m_0\, \frac{q_1\, q_2}{b}\, \frac{\left(\dfrac{b}{2p}\right)^2}{1 + \left(\dfrac{b}{2p}\right)^2}\,, \quad \text{or} \tag{2.88}$$

$$T = 2\frac{q_1^2 q_2^2}{m_2 v^2} \frac{1}{p^2 + \left(\frac{b}{2}\right)^2} , \text{ where } b = \frac{q_1 q_2}{\frac{1}{2} m_0 v^2} \text{ and } m_0 = \frac{m_1 m_2}{m_1 + m_2} . \tag{2.89}$$

This important equation connects the impact parameter $p$ with the transferred energy $T$. Its limiting cases are $T \to 0$ for $p \to \infty$ and $T \to T_{max} = 2 m_0^2 v^2 / m_2$ for $p \to 0$, in other words, for a central collision. For a heavy ion colliding with a much lighter target electron we have $m_2 = m_e \ll m_1$ and obtain $T_{max} \approx 2 m_2 v^2 = 2 m_e v^2$ (eq. (2.48)).

## 2.1.4 Energy-loss per unit length-of-path

After having defined the cross-section already above, we will now proceed to calculate the energy-loss cross-section classically.

If we know the transferred kinetic energy $T$ as function of the impact parameter $p$ we can write down the probability of an energy-transfer from the projectile to a target particle in the interval $(T , T + dT)$, corresponding to the impact-parameter interval $(p, p + dp)$ (Figure 2-8). The differential energy-transfer $dT$ corresponds to the energy-transfer $T(p)$ multiplied by the probability for its occurrence, the cross-section $2\pi p \, dp$:

$$dT = T(p) \, 2\pi p \, dp . \tag{2.90}$$

By summing (integrating) over all target particles contained within a thin target of thickness $dx$ and unit surface area (cf. Figure 2-9) we obtain the energy-loss of the projectile over the distance $dx$, whereby we specialize now on the energy-transfer from a heavy projectile-ion to much lighter target electrons. Thereby we follow the convention to count the energy-loss negative if the energy of the projectile is decreased and characterize this energy-loss to electrons, the "electronic energy-loss", by the index "$e$".

$$(dE)_e = - N_e \, dx \int_0^{p_{max}} dT = - N_e \, dx \int_0^{p_{max}} T(p) \, 2\pi p \, dp , \tag{2.91}$$

where $N_e = Z_2 N = Z_2 \rho / (A_t u)$ is the number of target electrons per unit volume, $Z_2$ its atomic number, $N$ the number-density of the target material, $\rho$ the density of the target material, $A$ the atomic mass number of the target material, and $u$ the atomic mass unit. The integration thereby has to be terminated — at this stage of the argumentation rather artificially — at a maximum impact parameter, corresponding to a minimum energy-transfer. Otherwise the integral becomes infinite due to the term $p^2$ in the denominator of $T$. This condition — the cut-off energy — corresponds quantum mechanically to the first accessible excitation state of a bound system (a bound electron or a plasmon).

In order to calculate the electronic energy-loss $(dE / dx)_e$ of the projectile, we divide eq. (2.91) by $dx$, remember that $T$ is a function of $p^2$, and write $2 p \, dp$ as $d(p^2)$:

$$\left(-\frac{dE}{dx}\right)_e = N_e \int_0^{p_{max}} T(p^2) \, 2\pi p \, dp = \pi N_e \int_0^{p_{max}} T(p^2) \, d(p^2) . \tag{2.92}$$

We substitute in $T$ the term $[ p^2 + (b/2)^2 ]$ by $x$. This leads to $dx = d(p^2)$, $x_{min} = (b/2)^2$, and $x_{max} = p_{max}^2 + (b/2)^2$. For sufficiently large $p_{max}$ (sufficiently small cut-off energy) we have $x_{max} \approx p_{max}^2$. We obtain:

$$\left(-\frac{dE}{dx}\right)_e = \pi N Z_2 \frac{2\,(Z_1\,e)^2\,e^2}{m_e\,v^2} \int\limits_{\left(\frac{b}{2}\right)^2}^{p_{max}^2} \frac{dx}{x} \quad , \text{ or} \tag{2.93}$$

$$\left(-\frac{dE}{dx}\right)_e = \pi N Z_2 \frac{2\,(Z_1\,e)^2\,e^2}{m_e\,v^2} \left[\ln\left(p_{max}^2\right) - \ln\left(\frac{b}{2}\right)^2\right], \text{ or} \tag{2.94}$$

$$\left(-\frac{dE}{dx}\right)_e = 4\pi \frac{Z_1^2\,e^4}{m_e\,v^2} N Z_2 \ln\left(\frac{2\,p_{max}}{b}\right), \text{ where } b \approx \frac{2\,Z_1\,e^2}{m_e\,v^2}, \tag{2.95}$$

This formula contains already the main features of the electronic energy-loss function (stopping power) according to Bohr [1] and can be used to approximate experimental data for nonrelativistic ion energies above the nuclear stopping regime. Thereby $Z_1$ has to be replaced by the effective charge number of the projectile-ion $Z_{eff} < Z_1$, which is a function of the projectile velocity, and $p_{max}$ corresponds roughly to the inverse of the average electron binding energy $I$ (ionization potential) of the target material.

### 2.1.5 Cut-off energy

At first sight the introduction of a low-energy cut-off appears as an artificial trick, introduced to avoid the divergence of the logarithm function in the electronic stopping power. What is the physical significance of the cut-off energy? As it turns out, an elastically bound electron cannot always be excited by the pulling force of a rapidly moving ion passing by at a large impact parameter. Even in the case of free electrons, moving on a background of stationary ions, a minimum energy is required to excite collective oscillations, so-called plasmons.

#### Momentum approximation

The momentum approximation is applied if the projectile-ion passes at a large distance (impact parameter $p$) from the target electron, which itself is elastically bound to the (much heavier) target nucleus (Figure 2-19).

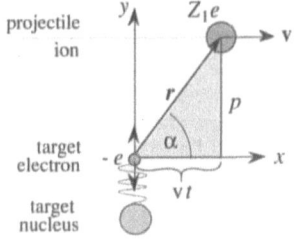

**Figure 2-19** *Concept of an elastically bound electron according to momentum approximation.* The coordinate system is fixed with respect to the (heavy) target nucleus. Its origin is located at the zero-position of the oscillating electron.

[1]    Bohr, N., K. Dan. Vidnesk. Selsk. Mat. Fys. Medd. **18**, No. 8, (1948).

According to the fundamental equation of motion, the force in $y$ direction, $F_y$, is equal to the momentum change per unit time in $y$-direction. Integration leads to the total momentum change in $y$ direction:

$$F_y = \frac{dP_y}{dt} \quad \Rightarrow \quad \Delta P_y = \int_{-\infty}^{+\infty} F_y \, dt \, , \quad \text{where } F_y = \frac{Z_1 e \, e}{r^2} \cos \alpha \cdot \quad (2.96)$$

If we insert $r = [\, p^2 + (v \, t)^2 \,]^{1/2}$ and $\cos \alpha = p \, / \, r$ we obtain

$$\Delta P_y = \int_{-\infty}^{+\infty} \frac{Z_1 \, e^2 \, p}{\left[ p^2 + (v \, t)^2 \right]^{3/2}} \, dt = Z_1 e^2 p \int_{-\infty}^{+\infty} \frac{dt}{p^3 \left[ 1 + \left( \frac{v}{p} t \right)^2 \right]^{3/2}} \cdot \quad (2.97)$$

For standardization of the integral we substitute $\xi = v \, t \, / \, p$, $d\xi = v \, dt \, / \, p$, and obtain

$$\Delta P_y = Z_1 e^2 \frac{1}{p \, v} \int_{-\infty}^{+\infty} \frac{d\xi}{\left[ 1 + \xi^2 \right]^{3/2}} = 2 \, Z_1 e^2 \frac{1}{p \, v} \int_{0}^{+\infty} \frac{d\xi}{\left[ 1 + \xi^2 \right]^{3/2}} \cdot \quad (2.98)$$

$$\Delta P_y = 2 \frac{Z_1 \, e^2}{p \, v} \left[ \frac{x}{\sqrt[2]{1 + x^2}} \right]_0^{\infty} = \frac{2 \, Z_1 \, e^2}{p \, v} \cdot \quad (2.99)$$

This result can be read in the following way: The momentum transfer $\Delta P_y$ corresponds to the thrust of a constant force $Z_1 e^2 / p^2$, acting during the time $\Delta t \equiv \tau = 2 \, p \, / \, v$, required for traversing the distance $2 \, p$. In this case we have $\Delta P_y = F_y \, \Delta t = 2 \, Z_1 \, e^2 / (p \, v)$. This simplified interpretation is quite practical, can be easily memorized, and is shown symbolically in Figure 2-20.

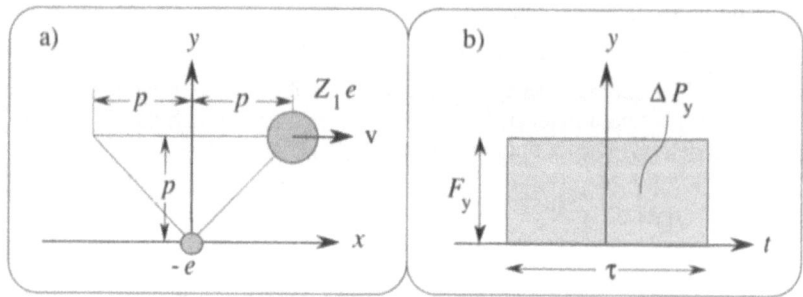

**Figure 2-20** *Graphical equivalent of momentum approximation.* (a) Symbolic geometry. (b) Corresponding time-force integral. The thrust of the projectile-ion passing at the distance $p$ from the target electron corresponds to a constant force $F_y(p) = Z_1 \, e^2 / p^2$, acting during the time $\tau = 2 \, p \, / \, v$.

### Elastic cut-off energy

The physical relevance of a cut-off energy can be classically interpreted using the model of an elastically bound electron. As it turns out, the energy-transfer between the projectile-ion and the elastically bound electron vanishes rapidly with increasing impact parameter, due to the incommensurability of the (long) duration of the impact process with the (short) duration of the time period of the harmonic oscillator. Quantum-mechanically this phenomenon corresponds to a minimum quantum of excitation-energy required to transfer energy to a bound system. We assume the simplified geometry of Figure 2-20 and apply a constant interaction force $F_y$, acting for a time $\tau$, according to the momentum approximation. For simplicity, we start the interaction at time $t = 0$. This corresponds to a shift of the time scale by $-\tau/2$.

For all times $t \leq 0$, the electron is resting at the origin $y = 0$:

$$y(0) = 0, \quad \dot{y}(0) = 0 . \tag{2.100}$$

Within the time interval $0 \leq t \leq \tau$ we have the fundamental equation of motion:

$$m\ddot{y} = -m\omega^2 y + F_y, \quad \text{for } 0 \leq t \leq \tau, \tag{2.101}$$

where $m$ is the mass of the target electron, $F_y = Z_1 e^2/p$ is the constant interaction force in $y$-direction, and $\omega = 2\pi\nu$ is the angular frequency of the harmonic oscillator with the frequency $\nu$. The differential equation can be simplified by introducing the shifted variable $y_1(t) \equiv y(t) - F_y/(m\omega^2)$. With this new variable we obtain:

$$m\ddot{y}_1 = -m\omega^2 y_1, \quad \text{with } y_1(0) = -\frac{F_y}{m\omega^2} \quad \text{and} \quad \dot{y}_1(0) = 0 . \tag{2.102}$$

It has the solution (Figure 2-21)

$$y_1(t) = -\frac{F_y}{m\omega^2}\cos\omega t \Rightarrow y(t) = \frac{F_y}{m\omega^2}\left(1 - \cos\omega t\right) \text{ for } 0 \leq t \leq \tau . \tag{2.103}$$

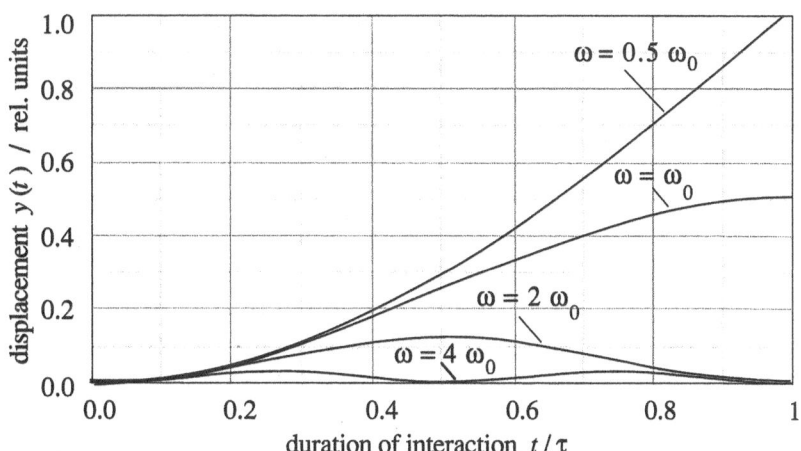

**Figure 2-21** *Energy transfer to a harmonically bound electron.* Excitation of a harmonic oscillator, according to momentum approximation. The force acts during the time interval $0 \leq t \leq \tau$. With increasing inherent angular frequency $\omega$ of the harmonic oscillator the maximum displacement amplitude decreases rapidly. $\omega_0 = 2\pi/(2\tau)$ corresponds to the duration $\tau$ of the applied force. The displacement is given in units of $(F_y/m)/40$.

We can determine the transferred energy using eq. (2.103):

$$T = T_{kin} + T_{pot} = \frac{1}{2} m \left[ \dot{y}^2 + \omega^2 y^2 \right]_\tau = \frac{F_y^2}{m \, \omega^2} \left( 1 - \cos \omega \tau \right) . \qquad (2.104)$$

For $\omega \tau \ll 1$, corresponding to a quasi-free electron or a fast projectile, we can give the transferred energy exactly. For $\omega \tau \gg 1$, corresponding to a strongly bound electron or a slow projectile, we can at least give an upper limit for the transferred energy.

$$T = \frac{(F_y \tau)^2}{2m} \text{ for } \omega \tau \ll 1 \ , \ T \le \frac{2 F_y^2}{m \, \omega^2} = \frac{(F_y \tau)^2}{2m} \left( \frac{2}{\omega \tau} \right)^2 \text{ for } \omega \tau \gg 1 . \quad (2.105)$$

If we compare the two limiting cases we see that for high elastic binding and long interaction times ($\omega \tau \gg 1$) the electron mass increases apparently by the factor $(\omega \tau / 2)^2$. In other words, for $\omega \tau \gg 1$, the effective mass of a harmonically bound electron appears to be much larger than its real mass. This can be qualitatively understood since not only the electron, but also the nucleus to which it is bound starts oscillating in the case of a strong harmonic binding between the electron and its nucleus.

The maximum impact parameter for which energy can be transferred corresponds roughly to the condition $\omega \tau \approx 1$, for which the transferred energy starts to depend on the angular frequency of the harmonically bound electron and starts to decrease rapidly as $1 / \omega^2$ for $\omega \tau \gg 1$. We set the cut-off condition arbitrarily to $\omega \tau \approx 2$ or $\tau = 2 / \omega$. Expressed in other words, the maximum impact parameter $p_{max}$ corresponds to a collision time $\tau$ which is comparable to the characteristic time $2 / \omega$ of the harmonically bound electron. Since according to the momentum approximation (cf. Figure 2-20) $\tau = 2 p / v$, we obtain for the corresponding maximum impact parameter $p_{max}$

$$p_{max} = \frac{v \, \tau}{2} \approx \frac{v}{2} \frac{2}{\omega} = \frac{v}{\omega} . \qquad (2.106)$$

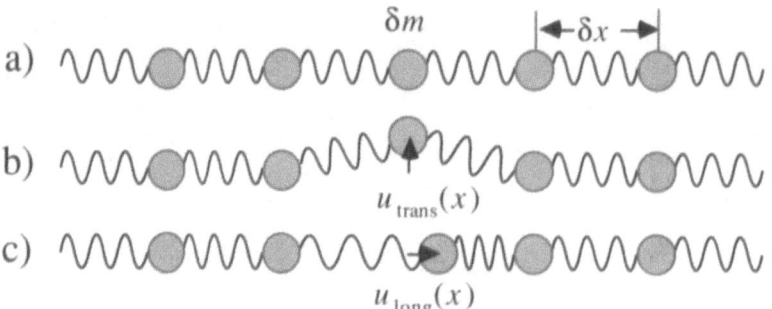

**Figure 2-22** *Derivation of elastic-wave equation.* (**a**) Quiescent state. (**b**) Transversal excitation. (**c**) Longitudinal excitation.

**Plasmon cut-off energy**

In metals, the conduction electrons can be considered as a gas of quasi-free particles, their electric charge being neutralized by the fixed positive ions forming the metal lattice. Similar to a gas, energy can be pumped into the system of quasi-free electrons in the form of compression and dilatation waves. The associated energy is known as plasmon energy. The minimum plasmon energy thereby corresponds to a cut-off energy and represents another natural limit for the otherwise diverging integral of eq. (2.93).

**Wave equation.** A system of elastically coupled particles is usually capable of two modes of excitation, transversal and longitudinal waves (Figure 2-22).

For exciting a transversal or a longitudinal wave the mass $\delta m$ must be shifted transversally by a distance $u_{trans}(x)$ or longitudinally by a distance $u_{long}(x)$, respectively. Considering in the following only the longitudinal waves, the mass $\delta m$ will be accelerated by the difference between the force $F_r$ pulling it to the right and the force $F_l$ pulling it to the left. Each of these forces $F_r$ and $F_l$ depend in turn on the difference between the positions of two particles of mass $\vartheta m$. We obtain as fundamental equation of motion:

$$F \equiv F_r - F_l = S \frac{u(x+\delta x) - u(x) - \delta x}{\delta x} - S \frac{u(x) - u(x-\delta x) - \delta x}{\delta x} = \delta m \frac{\partial^2 u}{\partial t^2} \quad , \quad (2.107)$$

where $S$ is the elastic coupling constant, according to Hooke's law ($F = S\,\Delta x\,/\,x$, where $F$ = force, $x$ = length of the elastic spring). If we make the transition from discrete particles to an elastic continuum we have to make the transition $\delta x \to 0$ with the constraint that the average mass density is simultaneously maintained, $\delta m\,/\,\delta x \equiv \rho_m$ = const. We obtain the wave equation

$$\rho_m \frac{\partial^2 u}{\partial t^2} = S \frac{\partial^2 u}{\partial x^2} \quad \text{or} \quad \frac{\partial^2 u}{\partial t^2} = c^2 \frac{\partial^2 u}{\partial x^2} \quad \text{with} \quad c = \sqrt{\frac{S}{\rho_m}} \quad . \quad (2.108)$$

whereby under closer consideration $c$ turns out as the phase velocity of the excited wave. This wave equation is solved by displacement waves of the form

$$u(x,t) = f(x)\,g(t) = C \sin(k\,x + \alpha)\,\sin(\omega t + \beta) \quad , \quad \text{or} \quad (2.109)$$

$$u(x,t) = C \sin\left(\frac{2\pi}{\lambda} x + \alpha\right) \sin\left(\frac{2\pi}{T} t + \beta\right) \quad . \quad (2.110)$$

where $k = 2\pi\,/\,\lambda$ is the x-component of the wave vector (in three dimensions), $\omega = 2\pi\,/\,T$ is the angular frequency, $\lambda$ is the wavelength, $T$ is the oscillation period, $\alpha$ and $\beta$ are arbitrary phase factors.

**Excitation of plasma oscillations.** A collective of free electrons moving on a background of fixed ions is capable to take up energy from an external electric field. For deriving the plasmon frequency we observe Figure 2-23.

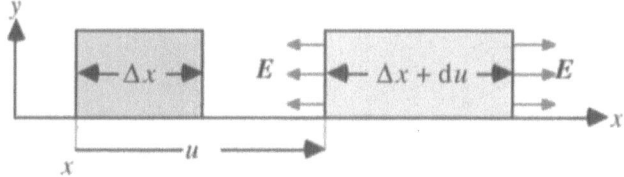

**Figure 2-23** *Derivation of electronic-plasma wave-equation.*

In the quiescent state the one-dimensional volume element $\Delta x$ is located as the position $x$. For exciting a plasma oscillation the volume element $\Delta x$ is shifted over a distance $u$ and is thereby simultaneously dilated by $du \ll \Delta x$. This leads to an uncompensated (positive) electric space charge $\rho_e$, corresponding to the difference of the electric charge density after and before dilatation:

$$\rho_e = \rho_{e,after} - \rho_{e,before} = \frac{\Delta x}{\Delta x + du}\, n_0(-e) - n_0\,(-e) \rightarrow e n_0\, \frac{\partial u}{\partial x} \qquad (2.111)$$

for $\Delta x \rightarrow 0$, where $n_0$ is the number of electrons per unit volume in the quiescent state and $e$ is the magnitude of the elementary charge. Due to the uncompensated positive charge in the dilated volume element we have, according to the Poisson equation a field strength $E$, associated with the space charge $\rho_e$:

$$\mathrm{div}\left(\mathrm{grad}\,\varphi\right) \equiv \nabla\left(\nabla\varphi\right) = \nabla\left(-E\right) = \frac{\partial}{\partial x}\left(-E\right) = -4\pi\,\rho_e = -4\pi\,e\,n_0\,\frac{\partial u}{\partial x}\ . \qquad (2.112)$$

Integration yields $E(x) = 4\pi\,e\,n_0\,u(x)$. We can now write down the fundamental equation of motion, according to the scheme $m\,(\partial^2 u / \partial t^2) = F$, where $F$ is the force causing the acceleration, and obtain:

$$n\,\delta V\,m_e\,\frac{\partial^2 u}{\partial t^2} = -n\,\delta V\,e\,E = -n\,\delta V\,e\,4\pi\,e\,n_0\,u\ ,\ \text{or} \qquad (2.113)$$

$$\frac{\partial^2 u}{\partial t^2} = -\omega_0^2\,u\ ,\ \text{where}\quad \omega_0 = \sqrt{\frac{4\pi\,e^2 n_0}{m_e}} \qquad (2.114)$$

is the plasma frequency.

**Maximum impact-parameter for excitation of plasmons.** Around the ion trajectory the quasi-free target electrons are contracted only up to a certain maximum distance. As it turns out this maximum distance corresponds to the minimum energy required for the excitation of plasma oscillations [1]. For deriving the corresponding relation we observe Figure 2-24.

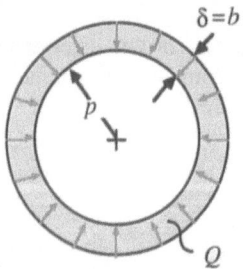

**Figure 2-24** *Derivation of maximum impact parameter for excitation of plasma oscillations.* The ion trajectory passes through the axis of symmetry.

[1]     Bonderup, E.: "Penetration of Charged Particles Through Matter." Lecture Notes, Institute of Physics, University of Aarhus, Denmark, 229 pp. (1974).

Due to the attraction between the positive charge of the projectile-ion and the negative charge of the quasi-free electrons the electrons gain momentum towards the axis of symmetry and move by a distance $\delta$, counteracted by the build-up of an electric polarization field which pulls them back. We use the momentum approximation for calculating the distance $\delta$ of this shift. During the time $\tau = 2\,p\,/\,v$ the electrons are accelerated by the constant force $Z_1\,e^2\,/\,p^2$ and move by a distance $\vartheta = (a\,/\,2)\,\tau^2$, where $a = (d^2\vartheta\,/\,dt^2)$ is the acceleration:

$$\delta = \frac{1}{2}\frac{\partial^2 p}{\partial t^2}\,\tau^2 = \frac{1}{2}\frac{Z_1\,e^2}{p^2}\frac{1}{m}\left(\frac{2p}{v}\right)^2 = \frac{(Z_1\,e)\,e}{\frac{1}{2}\,m\,v^2} = b\ , \tag{2.115}$$

recognizing $\vartheta$ alternatively as $b$, the familiar distance of closest approach for $m_2 \equiv m \ll m_1$. The collective movement of electrons into the cylinder with radius $p$ causes a build-up of a counteracting polarization field $E$, which can be determined from the influx of negative charge into the cylinder. The negative charge influx is given by

$$Q = -2\pi p\,e\,n_0\,\delta \ \Rightarrow\ \rho_e = \frac{Q}{\pi p^2} = \frac{-2\,e\,n_0\,\delta}{p}\ . \tag{2.116}$$

The polarization field $E_{pol}$ corresponds to this charge influx, obtained by integration of the Poisson equation, where $d\sigma$ is the vector corresponding to the surface element:

$$\text{div}\,E_{pol} = \lim_{\text{Vol.}\to 0}\frac{\int E_{pol}\,d\sigma}{\text{Vol.}} = \frac{-E_{pol}\,2\pi\,p}{\pi\,p^2} = -2\frac{E_{pol}}{p} = -4\pi\,\rho_e\ , \tag{2.117}$$

$$E_{pol} = +2\pi\,p\,\rho_e = 2\pi\,p\,\frac{-2\,e\,n_0\,\delta}{p} = -4\pi\,e\,n_0\,\delta = -4\pi\,e\,n_0\frac{Z_1\,e^2}{\frac{1}{2}mv^2}\ . \tag{2.118}$$

We obtain equilibrium between the attractive force $Z_1\,e^2\,/\,p^2$ of the projectile-ion and the repulsive force of the electric polarization field $E_{pol}$ if we set

$$E_{pol} = -4\pi\,e\,n_0\frac{Z_1\,e^2}{\frac{1}{2}\,m\,v^2} \overset{!}{=} E_{projectile} = \frac{F_y}{-e} = \frac{-Z_1\,e}{p^2}\ . \tag{2.119}$$

This corresponds to a condition for the maximum impact parameter $p_{max}$ for which the projectile-ion is still capable to pull electrons through the cylinder wall toward the passing ion:

$$p_{max}^2 = \frac{1}{2}\frac{m\,v^2}{4\pi\,n_0\,e^2} = \frac{1}{2}\frac{v^2}{\omega_0^2} \ \Rightarrow p_{max} = \frac{1}{\sqrt{2}}\frac{v}{\omega_0}\ ,\ \text{where } \omega_0 = \sqrt{\frac{4\pi\,n_0\,e^2}{m}} \tag{2.120}$$

is the plasma frequency. This result indicates that even quasi-free electrons on a background of fixed ions posses a low-energy cut-off which is associated with the plasma frequency $\omega_0$.

## 2.1.6    Bohr's energy-loss relation

In the classical energy-loss formula (2.95) we have assumed that the projectile-ion interacts with its full nuclear charge number $Z_1$ with the target electrons. This holds only for fully stripped ions moving with relativistic energies ($T_s \gg 10$ MeV/u) through the solid target material. Otherwise the nuclear charge $Z$ is at least partially shielded by electrons bound to the projectile nucleus. In order to obtain a more realistic picture of the energy-loss process at lower velocities ($T_s \leq 10$ MeV/u) we have to consider the charge exchange between the projectile-ion and the solid, which represents a dynamic equilibrium between the binding force of the projectile-ion and the stripping force due to the collisions. The electrons surrounding the projectile nucleus thereby are considered as an electron cloud surrounding a heavy attractive center. The more loosely bound outer electrons have a higher tendency to get lost due to scattering processes with the quasi-stationary electrons of the target material. The result is an effective charge number $Z_{\text{eff}}$ of the projectile-ion which is smaller than the nuclear charge number $Z_1$ and is a function of the projectile-ion energy.

## 2.1.7    Charge-state of projectile-ion

### Properties of the electron gas

**Electron binding-energy in the hydrogen atom.** We can obtain a feeling for the principal behavior of the electron gas in the case of the hydrogen atom, if we minimize the total energy of this two-body system,

$$E = T + V = \frac{p^2}{2m} + \frac{(-e)\,e}{r} \tag{2.121}$$

under the additional condition that, according to the Heisenberg uncertainty relation $\Delta p\,\Delta r \approx p\,r \approx h / (2\pi)$. This yields:

$$\frac{dE}{dr} = 0 \ \text{ for } \ r = a_0 = \frac{h^2}{4\pi^2 m\,e^2} \ \Rightarrow \ E_0 = -\frac{1}{2}\frac{4\pi^2 e^4\,m}{h^2}, \tag{2.122}$$

which accidentally gives the exact values of the Bohr radius, $a_0 = 0.529177 \cdot 10^{-8}$ cm, and the hydrogen electron binding energy $E_0 = -13.6$ eV (Figure 2-25).

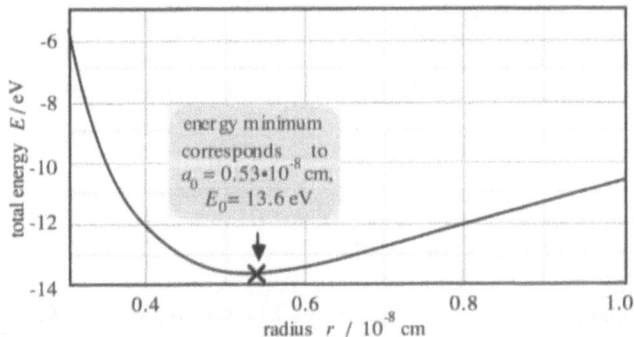

**Figure 2-25** Electron binding energy of hydrogen as function of orbit radius. With increasing radius the kinetic energy decreases and the potential energy increases. The total-energy minimum corresponds to the Heisenberg uncertainty relation, $\Delta p\,\Delta r \approx p\,r \approx h / (2\pi)$.

**Electron density in a box potential.** The wave equation of a quasi-free electron in a box potential in one dimension is given by:

$$\frac{\partial^2 u}{\partial x^2} + k^2 u = 0, \text{ where } k = \frac{2\pi}{\lambda} = \frac{2\pi p}{h} \qquad (2.123)$$

and has the standing-wave or stationary solutions and energies depending on one quantum number $n$:

$$u_n = \left(\frac{2}{L}\right)^{1/2} \sin\frac{n\pi x}{L} \quad \text{and} \quad E_n = \frac{p^2}{2m} = \frac{h^2 k^2}{8\pi^2 m} = \frac{h^2}{8mL^2} n^2, \qquad (2.124)$$

respectively (Figure 2-26).

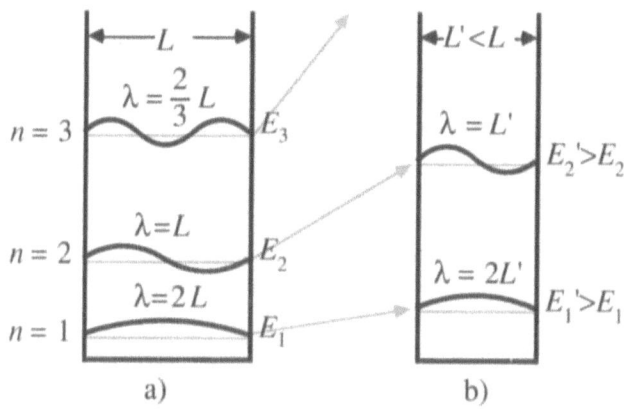

**Figure 2-26.** *Basic behavior of quantum states in a box potential.* (**a**) *Wide box — many narrow-spaced energy levels.* (**b**) *Narrow box — few wide-speced energy levels.*

Accordingly in a three-dimensional cubic box:

$$\nabla^2 u \equiv \frac{\partial^2 u}{\partial x^2} + \frac{\partial^2 u}{\partial y^2} + \frac{\partial^2 u}{\partial z^2} + k^2 u = 0, \qquad (2.125)$$

with solutions that depend on a set of three quantum numbers, $(n_1, n_2, n_3)$:

$$u_{n_1 n_2 n_3} = \left(\frac{2}{L}\right)^{3/2} \sin\frac{n_1\pi x}{L} \sin\frac{n_2\pi y}{L} \sin\frac{n_3\pi z}{L}. \qquad (2.126)$$

The energy of such a state depends also on the set $(n_1, n_2, n_3)$ of quantum numbers:

$$E_{n_1 n_2 n_3} = \frac{p^2}{2m} = \frac{h^2 k^2}{8\pi^2 m} = \frac{h^2}{8m} \frac{1}{L^2} \left(n_1^2 + n_2^2 + n_3^2\right) \equiv \frac{h^2}{8mL^2} n^2. \qquad (2.127)$$

For a large energy, that is for a large quantum number $n$, the number $z$ of allowed energy states of the electron up to the quantum number $n \equiv (n_1^2, n_2^2, n_3^2)^{1/2}$, corresponding to state energies up to $E_n$, is equal to the volume of one octant of a sphere with radius $n$ in the $(n_1, n_2, n_3)$-space (Figure 2-27).

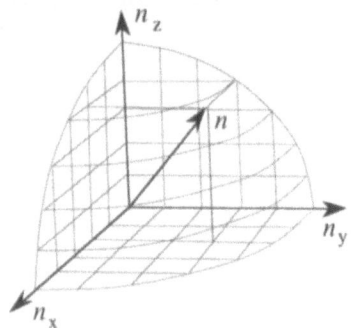

**Figure 2-27** *Determination of the state-density* of a cubic box in $(n_1, n_2, n_3)$-space.

The number of states in the quantum number interval $(n, dn)$ corresponds to the volume of the spherical shell between $n$ and $dn$. The allowed quantum states between $n$ and $n + dn$ correspond to energies between $E$, and $E + dE$. In order to relate the number interval $(n, dn)$ with the energy interval $(E, dE)$ we solve eq. (2.127) for $n$:

$$n = \frac{2L}{h} \sqrt{2mE} \quad \Rightarrow \quad dn = \frac{L}{h} \sqrt{\frac{2m}{E}} \; dE \; . \tag{2.128}$$

In addition we have to take into account that, according to the Pauli principle, there exist two allowed states of opposite spin per quantum state. We obtain therefore

$$dz = \frac{1}{8} \, 4\pi n^2 \, dn \; 2 = \pi n^2 \, dn = 8\pi \frac{L^3}{h^3} \, m \, \sqrt{2mE} \; dE \; . \tag{2.129}$$

If now count all allowed energy states of interaction-free electrons moving in the box potential up to a maximum energy $E_F$, corresponding to the Fermi energy of quasi-free electrons moving in a metal, we have to integrate $dz$ over all energies up to $E_F$ and obtain:

$$z = \int_0^{z_{max}} dz = \frac{8\pi}{3} \frac{L^3}{h^3} (2mE_F)^{3/2} \equiv n_e \, L^3 \; , \tag{2.130}$$

which corresponds to the number of allowed electron states per volume element, $n_e$, times the total volume of the box, $L^3$. Since the Fermi energy is the highest allowed energy and therefore corresponds to the electrons of the highest linear momentum, according to $E_F = p_{max}^2 / (2m)$, we obtain the following relation between the maximum linear momentum $p_{max}$ in a three-dimensional box potential and the number $n_e$ of electrons per volume element by solving eq. (2.130) for the electron density $n_e$:

$$n_e = 2 \left( \frac{4\pi}{3} p_{max}^3 \right) \frac{1}{h^3} \; , \quad \text{or} \quad N_e = 2 \frac{\left( \frac{4\pi}{3} p_{max}^3 \right) L^3}{h^3} \; . \tag{2.131}$$

This important result can be read in the following way: The total number $N_e$ of electrons in the box potential is twice the total volume of the available phase space,

$\{(4\pi\, p_{max}^3)\,/\,3\}\, L^3$, divided by the cell size $h^3$, which is the volume required by one electron. In other words there exist exactly two allowed electron states per phase-space volume element of size $h^3$, in accordance with the the Pauli principle. The result can be generalized for a slowly varying potential well of arbitrary shape and to many-electron systems like atoms in which the individual electrons move in the slowly varying average potential of the nucleus and the other electrons. This is the basis of the Thomas-Fermi atom model.

### Thomas-Fermi atom model

It is assumed that the electrons move in the smoothly varying, average potential of the other electrons surrounding the atomic nucleus, similar as a gaseous atmosphere surrounds a stellar body. It is assumed that, according to the Pauli principle, there exist exactly two allowed electron states per phase space cell of volume $h^3$. For a bound electron we have the condition:

$$E = T + V \leq 0 \Rightarrow \frac{p^2}{2m} - e\,\varphi(r) \leq 0 \Rightarrow \frac{p_{max}^2}{2m} - e\,\varphi(r) = 0 , \qquad (2.132)$$

the limiting condition for the maximum linear momentum $p_{max}$ of an electron that is still bound to the nucleus. If we insert the result $p_{max} = (3\, n_e\, \pi^2)^{1/3}\,(h\,/\,2\pi)$ from the rearranged eq. (2.131) into (2.132) we obtain a relation between the electrostatic potential $\varphi(r)$ and the number density $n_e$ of the electrons:

$$n_e = \left(\frac{8\pi^2 m\, e\, \varphi(r)}{h^2}\right)^{3/2} \frac{1}{3\,\pi^2} , \quad \text{in other words,} \; n_e = n_e(\varphi(r)) . \qquad (2.133)$$

If we remember, that — according to the Poisson equation — the potential $\varphi(r)$ itself depends on the charge density $\rho_e = -\,e\, n_e$, we obtain a closed circular relation. The Poisson equation reads in this case:

$$\text{div}\left(\text{grad} \bullet \varphi(r)\right) \equiv \Delta\varphi(r) = -\,4\pi\,\rho(r) = 4\pi\, e\, n_e(r), \text{ or } \varphi(r) = \varphi(n_e(r)). \quad (2.134)$$

recognizing the mutual interdependence between $n_e$ and $\varphi$, we obtain a raw form of the Thomas-Fermi differential equation by inserting $n_e$ from (2.133) into (2.134):

$$\Delta\varphi(r) = \frac{4\,e}{3\,\pi} \left(\frac{8\pi^2 m\, e\, \varphi(r)}{h^2}\right)^{3/2} , \qquad (2.135)$$

which can be simplified by the substitutions $\varphi(r) \equiv (Z\, e\,/\,r)\, u(r)$, $r \equiv a\, x$. The factor $(Z\, e\,/\,r)$ reflects the potential in close vicinity of a nucleus of charge number Z. The factor $a = a_0\,[\,9\pi^2\,/\,(128\, Z)\,]^{1/3}$, where $a_0 = h^2\,/\,(4\pi^2\, m\, e^2) = 0.529177 \bullet 10^{-8}$ cm is the Bohr radius, represents a scaling factor which transforms all atoms with different nuclear charge numbers $Z$ into the one standard Thomas-Fermi atom. The reduced Thomas-Fermi differential equation is:

$$\frac{d^2 u}{dx^2} = \frac{u^{3/2}}{\sqrt{x}} . \qquad (2.136)$$

The Thomas-Fermi differential equation relates the curvature $(d^2 u\,/\,dx^2)$ of $u$ with given values of $u$ and $x$. Its solution therefore can be obtained by numeric integration with different starting tangents, starting from the boundary condition $u(0) = 1$. For neutral atoms, the correct solution is that for which the boundary conditions $u(x) = 0$ as well as $u'(x) = 0$ are fulfilled for $x \to \infty$ (Figure 2-28).

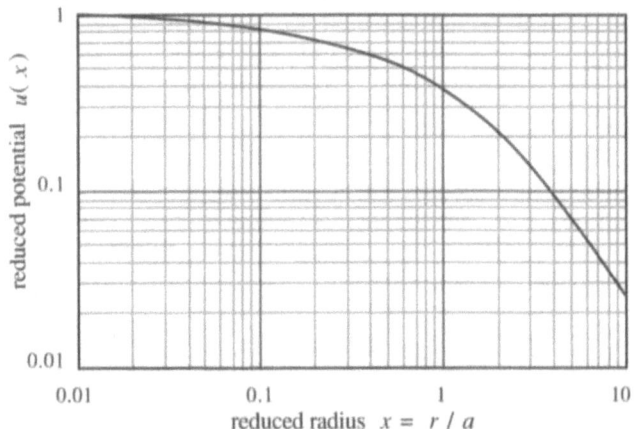

**Figure 2-28.** *Solution of Thomas-Fermi differential equation.*

Only the solution with the asymptotic behavior $u \to 144 / x^3$ for $x \to \infty$ meets these conditions simultaneously [1]. This is a somewhat too slow decay of a probability-density function with increasing radius as compared to quantum mechanical density functions which decay exponentially with the radius for $r \to \infty$.

The radially symmetric Thomas-Fermi model of the atom is universal in the sense that it does not reflect specific properties of the element, such as its nuclear charge number Z. The model is quite adequate for the description of the stripping process. The reduced potentials corresponding to the solution $u(x)$ of the Thomas-Fermi differential equation are shown in Figure 2-29.

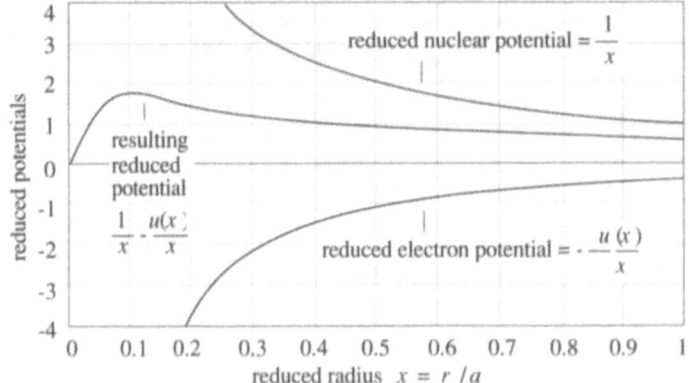

**Figure 2-29** *Reduced potentials.* Nuclear, electron, and resulting potential according to the Thomas-Fermi atom model. The resulting potential, $[1 - u(x)] / x$, represents a potential-energy well for the negatively charged electrons.

---

[1]     Flügge, S.: "**Lehrbuch der theoretischen Physik**." Springer, Berlin, **3**, 42, (1961).

### Bohr's stripping criterion

According Bohr's stripping criterion, all electrons with orbital velocities $v_e$ smaller than the velocity of the projectile-ion, $v$, will be stripped-off in the target. According to the virial theorem, they correspond to the electrons outside a critical radius $R$ (Figure 2-30). (A refined theory is given in references [1], [2]).

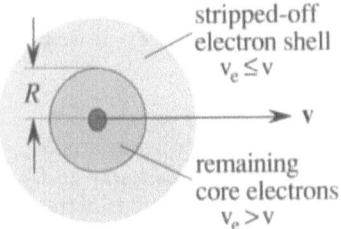

**Figure 2-30** *Bohr's stripping criterion* for calculating the charge state of projectile-ions in a solid. All outer electrons with velocities smaller than the projectile velocity are removed.

### Virial theorem

According to Bohr's classical atomic model we assume now that the electrons move on periodic orbits. For such orbits the electrons return periodically to the same values of the radius $r$ and the velocity $v$. If the duration of this time period is $\tau$, we have:

$$r(0) = r(\tau) \, , \ \ v(0) = v(\tau) \ \Rightarrow \ r(0) \, v(0) = r(\tau) \, v(\tau) \, , \ \text{or} \tag{2.137}$$

$$m \, r(\tau) \, v(\tau) - m \, r(0) \, v(0) = 0 \ \Leftrightarrow \ (rp)_0^{\tau} = \int_0^{\tau} d(rp) = 0 \, . \tag{2.138}$$

Under the condition that the electron is accelerated by a conservative force, derived from a potential, $F = -\,\text{grad}\,V \equiv -\nabla V$, the integral can be reformulated as:

$$\int_0^{\tau} d(rp) = \int_0^{\tau} \frac{d(rp)}{dt} \, dt = \int_0^{\tau} \left\{ (\dot{r}\,p) + (r\dot{p}) \right\} dt = \int_0^{\tau} \left\{ 2T - (r\nabla V) \right\} dt = 0, \tag{2.139}$$

which says, that the time average of twice the kinetic energy $T = p^2 / (2\,m) = (dr/dt)\,p/2$ is equal to the time average of the so-called "virial", $(r\nabla V)$:

$$2 <T> \, = \, <(r\nabla V)> \, . \tag{2.140}$$

For an electron of charge $-\,e$ moving in a Coulomb field of potential $\varphi = Z_{\text{eff}}\,e\,/\,r$ with the potential energy $V = -\,Z_{\text{eff}}\,e^2\,/\,r$ we obtain the result:

$$2 <T> \, \equiv m <v_e^2> \, = \, \left< \left( r\frac{Z_{\text{eff}}\,e^2\,r}{r^3} \right) \right> \, = \, \left< \frac{Z_{\text{eff}}\,e^2}{r} \right> \, = \, <\text{-}V> \, . \tag{2.141}$$

This says, that for a Coulomb field the sum of twice the time average $<T>$ of the kinetic energy and the time average $<V>$ of the potential energy is zero.

[1]    Brandt, W., M. Kitagawa, Phys. Rev. **25B**, 5631, (1982).
[2]    Ziegler, J.F., J.P. Biersack, U. Littmark: "The Stopping and Range of Ions in Solids." Pergamon Press, New York, 321 pp., (1985).

Restricting the model still further, we assume now that the electron orbits are not only periodic in time but that the electrons move on circles, in other words $r = $ const and $v_e = $ const, eq. (2.141) is written:

$$m\, v_e^2 = \frac{Z_{eff}\, e^2}{r} \Leftrightarrow v_e = \sqrt{\frac{e^2}{m\, r}\, Z_{eff}} \Leftrightarrow Z_{eff} = m\, v_e^2\, \frac{r}{e^2}\,, \qquad (2.142)$$

which will enable us to determine the effective charge number $Z_{eff}$ if we know the electron velocity $v_e$ for a given radius $r$.

### Resulting charge state

Starting from a neutral projectile-atom, the effective charge $Z_{eff}$ of the projectile corresponds to the stripped-off charge cloud outside $R$, removed by collisions between the electrons of the projectile-ion and the quasi-free electrons of the target material:

$$- Z_{eff}\, e = \int_R^\infty \rho(r)\, 4\pi\, r^2\, dr\,, \quad \text{where } \rho(r) = - \frac{\Delta\varphi}{4\pi} \text{ and } \varphi = \frac{Z\, e}{r}\, u(x). \qquad (2.143)$$

Here $u(x) \equiv u(r/a)$ is the numeric solution of the Thomas-Fermi differential equation. The effective charge number $Z_{eff}$ thus is a function of the lower limit of integration, in other words of the core radius $R$. Expressing the delta operator in spherical coordinates $\Delta = (d^2/dr^2) + (2/r)\,(d/dr)$, the integral can be reformulated as:

$$Z_{eff} = Z \int_X^\infty x\, \frac{d^2 u}{dx^2}\, dx = Z \left[ X\, \frac{du(X)}{dx} - u(X) \right], \text{ since } \lim_{X \to \infty} u(X) = 0,$$

$$\text{where } X = \frac{R}{a}\,, \ a = a_0 \left( \frac{9\,\pi^2}{128\, Z} \right)^{1/3},\ a_0 = \frac{h^2}{4\pi^2 m\, e^2}\,. \qquad (2.144)$$

Inserting eq. (2.144) into eq. (2.142) yields:

$$v_e = \sqrt{\frac{e^2}{m\, R}\, Z_{eff}} = \sqrt{\frac{e^2}{m\, a X}\, Z \int_X^\infty x\, \frac{d^2 u}{dx^2}\, dx} \ \underset{\substack{\text{Bohr} \\ \text{criterion}}}{\overset{!}{=}}\ v\,, \qquad (2.145)$$

where the reduced core radius $X = R/a$ appears in two different functional dependencies, first, through the virial condition, according to eq. (2.142) and, second, according to eq. (2.144), through the reduced potential $u(x)$ of the Thomas-Fermi atom model.

We have therefore the somewhat perplexing situation that for calculating $Z_{eff}$ as function of v we have first to determine $Z_{eff}$ as function of $X$, according to eq. (2.144), and then insert the obtained value of $X$ together with the obtained value of $Z_{eff}$ into eq. (2.145), according to the scheme:

$$X \longrightarrow \boxed{Z_{eff} = Z \int_X^\infty x\, \frac{d^2 u}{dx^2}\, dx} \xrightarrow{Z_{eff}} \boxed{v = \sqrt{\frac{e^2}{m\, a X}\, Z_{eff}}} \longrightarrow v\,,$$

$$(2.146)$$

from which a table of values ($X$, $Z_{eff}$, v) can be established for determining $Z_{eff}$ as function of v by interpolation. According to the scheme of eq. (2.146), the cut-off value $X$ is used as the input variable and in a first step the effective charge $Z_{eff}$ is determined from it. In the second step, the corresponding velocity v is determined, using $X$ and the just determined $Z_{eff}$ as the input variables. Since in most cases not the ion velocity but its specific energy — quantifying its velocity — is given, we represent in Figure 2-31 the corresponding relation between the specific energy $T_s$ and the effective charge $Z_{eff}$.

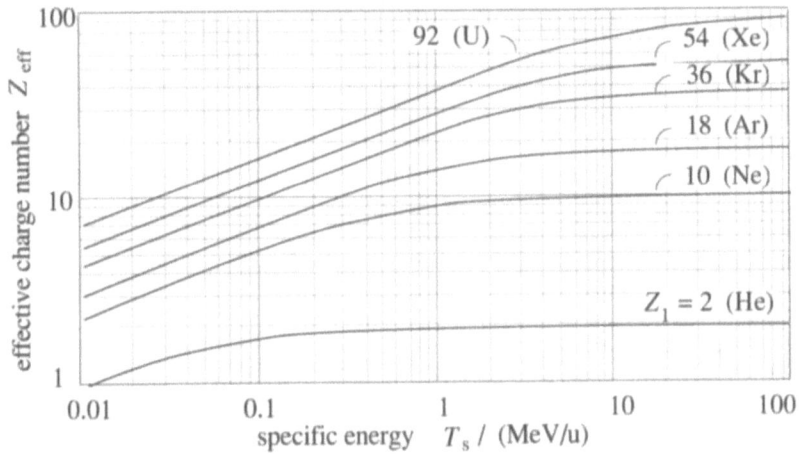

**Figure 2-31** *Effective charge number $Z_{eff}$ as function of specific energy $T_s$ for projectile-ions with different nuclear charge number $Z_1$.*

One of its applications is to give mean charge states of ion beams after passing through a stripper foil in an accelerator. Its main use, however, is in connection with the classical Bohr's formula (2.95) to improve the energy-loss model by replacing $Z_1$ by $Z_{eff}$.

## 2.1.8 Charge-corrected energy-loss relation

An analytical approximation of the relation between the projectile velocity v and its effective charge number $Z_{eff}$ in a solid is given by the semi-empirical formula [1], [2], [3], replacing the nuclear charge number $Z_1$ in eq. (2.95) :

$$Z_{eff} = Z \left(1 - e^{-\frac{v}{v_0} Z^{-2/3}}\right). \tag{2.147}$$

In this equation $v_0 = 2\pi\, e^2\,/\,h = 2.1847 \cdot 10^8$ cm / s is the Bohr velocity. The first-order term of eq. (2.147) reads $Z_{eff} \approx Z_1^{1/2}\, v\,/\,v_0$ and is also due to Bohr [4]. Inserting $Z_{eff}$ according to eq. (2.147) into eq. (2.95) leads to a refined relation describing the electronic energy-loss function $(dE\,/\,dx)_e$ or electronic stopping power $S(E)_e \equiv (dE\,/\,dx)_e$, where the index "e" points to "electronic" in order to distinguish these terms from the cor-

[1] Heckmann, H.H., B.L. Perkins, W.G. Simon, F.M. Smith, W. Barkas: "Ranges and Energy Loss Processes of Heavy Ions in Emulsion." Phys. Rev. **117**, pp. 544-556, (1960).
[2] Betz, H.D., Rev. Mod. Phys. **44**, 465, (1972).
[3] Schmidt-Böcking, H.: "Penetration of Heavy Ions Through Matter." in H. Bethge (editor), **Experimental Meth. in Heavy Ion Physics**, Springer Verlag, Berlin, pp. 81-149, (1978).
[4] Bohr, N. Phys. Rev. **58**, 654, (1940).

responding nuclear energy-loss terms:

$$\left(-\frac{dE}{dx}\right)_e = 4\pi \frac{Z_{eff}^2 e^4}{m_e v^2} N Z_2 \ln\left(\frac{2 p_{max}}{b}\right), \text{ where } b \approx \frac{Z_{eff} e^2}{\frac{1}{2} m_e v^2}. \tag{2.148}$$

Since $v^2 = E / m_1$, where $m_1$ is the mass of the projectile-ion, its most prominent feature is the $(Z_{eff}^2 / E)$-dependency, which for relativistic energies approaches $Z^2 / E$. The logarithmic factor thereby is a relatively slowly varying factor. In contrast to the effective charge $Z_{eff}$ of the projectile-ion, the target nuclear charge $Z_2$ appears only linearly.

Better energy-losses (stopping powers) for pure elements are tabulated in references [1], [2], and [3]. A review on heavy ion stopping is given in reference [4].

The argument of the logarithm contains the ratio between the maximum impact parameter $p_{max}$ and the distance of closest approach $b$ divided by the factor 2. It can be represented in different ways, depending on the physical concept of the cut-off condition. One of the resulting parameters thereby is usually treated as a fitting parameter.

$$\frac{2 p_{max}}{b} = 2 p_{max} \frac{m_e v^2}{2 q_1 q_2} = \frac{2 m_e v^2}{2 \frac{q_1 q_2}{p_{max}}} = \frac{T_{max}}{2 \frac{q_1 q_2}{p_{max}}} \rightarrow \frac{T_{max}}{I} = \frac{2 m_e v^2}{I}, \tag{2.149}$$

whereby $I$ — derived quantum mechanically by Bethe in 1930 [5], [6] — corresponds to the average excitation energy of the target material and is usually treated as a fitting parameter. Another transformation is obtained if we insert as maximum impact parameter the result of eq. (2.120) representing $p_{max}$ as function of the plasma frequency $\omega_0$:

$$\frac{2 p_{max}}{b} \approx \frac{2}{b} \frac{v}{\omega_0} = \frac{2 v}{\omega_0} \frac{m_e v^2}{2 q_1 q_2} = \frac{m_e v^3}{q_1 q_2 \omega_0} = \frac{m_e v^3}{Z_{eff} e^2 \omega_0}, \tag{2.150}$$

in which $\omega_0$ is treated as a fitting parameter. Therefore we have three approximately equivalent representations of the argument of the logarithm function:

$$\frac{2 p_{max}}{b} \Leftrightarrow \frac{m_e v^3}{Z_{eff} e^2 \omega} \Leftrightarrow \frac{2 m_e v^2}{I}, \tag{2.151}$$

where $\omega$ and $I$ are the average plasmon frequency and the average excitation energy of the target material, respectively.

[1]    Northcliffe, L.C., R.F. Schilling: "**Range and Stopping-Power Tables for Heavy Ions.**" Nuclear Data Tables **A7**, 233 pp., (1970).
[2]    Ziegler, J.F.: "**The Stopping and Ranges of Ions in Matter.**" **1-5**, Pergamon Press, London, (1977).
[3]    Hubert, F., A. Fleury, R. Bimbot, D. Gardes: "**Range and Stopping-Power Tables for 2.5 - 100 MeV/Nucleon Heavy Ions in Solids.**" Ann. Phys. (Paris) **5**, 1-214, (1980).
[4]    Geissel, H.:"Heavy-Ion Stopping." in **Semiclassical Descript. of Atomic and Nuclear Collisions**, J. Bang and J. de Boer (editors), Elsevier Science Publishers, Amsterdam., pp. 431-462, (1985).
[5]    Bethe, H.A., Ann. Phys. (Leipzig), **5**, 325 (1930).
[6]    Ahlen, S.P.:"Theoretical and Experimental Aspects of the Energy Loss of Relativistic Heavily Ionizing Particles." Rev. Mod. Phys. **52**, No.1, pp.121-174, (1980).

If we evaluate eq. (2.148) for a thin target consisting of aluminum, using eq. (2.147) for determination of $Z_{eff}$ and the average excitation energy $I = 10$ eV, according to eq. (2.151), in the logarithm, we obtain Figure 2-32, which very roughly corresponds to the observed stopping power.

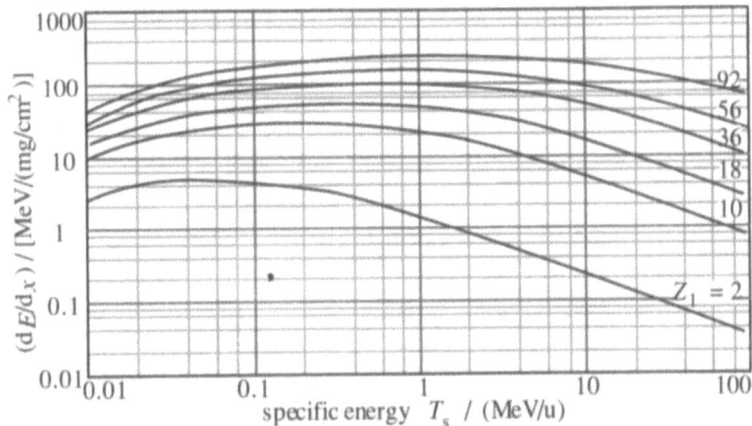

**Figure 2-32** *Approximate stopping power* $S(E) \equiv (dE / dx)$ for projectile-ions of different nuclear charge number $Z$ in aluminum according to eq. (2.148).

# 2.2   Secondary energy-loss effects

## 2.2.1   Energy-loss in multi-elemental targets

For target materials containing more than one element, such as chemical compounds, one can treat the energy-loss, as if it occurred in separate thin targets consisting of the corresponding pure elements and inserted in arbitrary sequence into the accelerator, whereby all targets contain identical number-densities of atoms, according to the scheme:

$$\text{(scheme)} \tag{2.152}$$

Such a scheme is known as "Bragg rule" [1], [2]. For example, a compound target $A_m B_n$ has the same energy-loss as the sum of the energy-losses in the pure elemental targets $A$ and $B$ multiplied by the corresponding numbers $m$ and $n$, respectively:

$$\left(\frac{dE}{dx}\right)_{A_m B_n} = m \left(\frac{dE}{dx}\right)_A + n \left(\frac{dE}{dx}\right)_B . \tag{2.153}$$

[1]   Bragg, W.H., R. Kleeman, Phil. Mag. **10**, 318, (1905).
[2]   Sautter, C.A., E.J. Zimmermann, Phys. Rev. **131**, 1611, (1965).

### 2.2.2 Energy-straggling and angular straggling

It should be emphasized that the energy-loss formula (2.148) describes only the average energy-loss of many ions passing through the target. Since the energy-loss is the result of stochastic processes one will always observe a widened energy distribution $P(E)$ of finite width after the ions have penetrated through matter, which in general is not symmetrical (Figure 2-33).

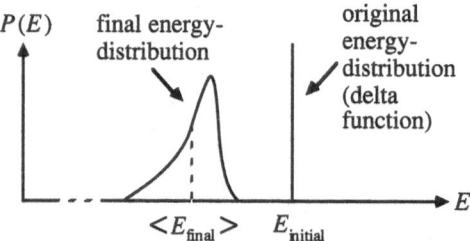

**Figure 2-33** *Energy straggling of ions.*

Another feature of the penetration of ions through matter is the angular deflection of the projectile. A well-collimated incoming beam will always obtain a widened angular distribution after penetration through the target which for symmetry reasons will be axially symmetric around the original ion beam direction. In other words, the average deflection is zero. While a calculation of $(-dE/dx)$ can be performed as soon as the energy-loss cross-sections in eq. (2.25) are known, further contemplation of statistical nature is needed for a characterization of the distribution $P(E)$ and of the angular distribution [1]. In practice, however, frequently angular straggling can be neglected. Figure 2-34 gives an impression of the parallelism of etched ion tracks in mica with about 1 $\mu$m diameter and about 100 $\mu$m length.

### 2.2.3 Energy-transfer to target nuclei

The energy-transfer from the projectile-ion to the target nuclei can be formally treated exactly as the energy-transfer from the projectile-ion to the target electrons. Thereby only the target electron has to be replaced by the much more massive target nucleus. The following basic formula is obtained:

$$\left(-\frac{dE}{dx}\right)_n = 4\pi \frac{Z_1^2 Z_2^2 e^4}{m_2 v^2} N \ln\left(\frac{2 p_{max}}{b}\right) \quad , \text{ where} \tag{2.154}$$

$$b = \frac{q_1 q_2}{\frac{1}{2} m_0 v^2} \quad , \text{ and } \quad m_0 = \frac{m_1 m_2}{m_1 + m_2} \quad . \tag{2.155}$$

[1]    Sigmund, P.: "Energy Loss of Charged Particles in Solids." in **"Radiation Damage Processes in Materials."** C.H.S. Dupuy (Editor), Nato Advanced Study Institute Series, Series E, No. 8, Nordhoff, Leyden, pp. 1-117, (1975).

In eq. (2.154) $p_{max}$ corresponds to a maximum impact parameter for nuclear scattering events. In comparison to the electronic energy-loss according to eq. (2.95) or eq. (2.148), the forefactor before the logarithm function in eq. (2.154) is dramatically different. The target nuclear charge number $Z_2$ appears here in the second power. Since in the denominator the electron mass $m_e$ is replaced by the much larger mass $m_2$ of the target nucleus, the nuclear energy-loss is in general much smaller than the electronic energy-loss. At low energies, the nuclear energy-loss may become dominant (see Figure 2-3). The Coulomb interaction (Rutherford scattering) has to be replaced by the interaction of two (neutral) atoms, corresponding to a screened Coulomb potential like that shown in Figure 2-28, corresponding to eq. (2.135) and eq. (2.136).

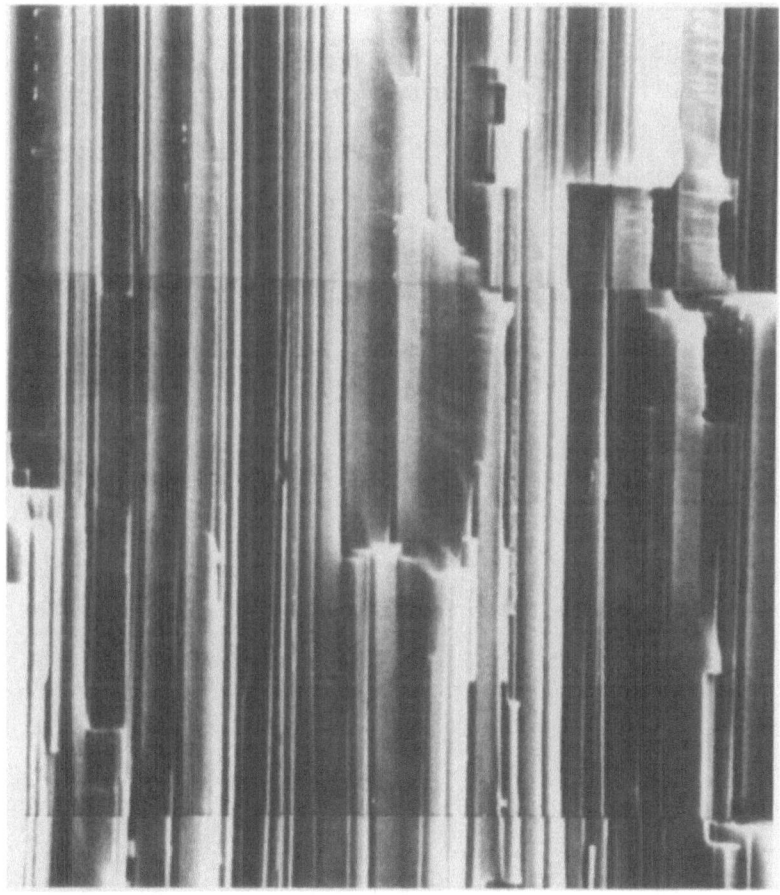

**Figure 2-34** *Parallelism of etched ion tracks in mica*. In the shown case angular straggling is very small. This holds generally for heavy ions in light targets. In practical applications in which only a short fraction of the ion range is used, angular straggling can often be neglegted. Track diameter about 1 μm. Track length about 100 μm.

## 2.2.4　　Calculation of ion range

The most direct application of energy-loss data is the determination of ion ranges in solids. A projectile-ion of initial energy $E_0$, moving through the target material by a distance $\Delta x$, looses the energy $\Delta E$, according to the scheme:

$$E_0 \longrightarrow \boxed{\phantom{x}} \longrightarrow E = E_0 + \Delta E \ , \text{ where } \Delta E = \left(-\frac{dE}{dx}\right)\Delta x \ . \quad (2.156)$$

Vice versa, the target thickness $\Delta x$ can be determined by measuring the energy-loss $\Delta E$, if we know the energy-loss function $(dE / dx)$, according to:

$$\Delta x = \left(-\frac{dE}{dx}\right)^{-1}\Delta E \quad \rightarrow \quad dx = -\left(\frac{dE}{dx}\right)^{-1} dE \ . \quad (2.157)$$

Integration over all values of the energy $E$ between the original energy $E_0$ and zero energy yields the total range $R$ of the projectile-ion in the target material:

$$R = -\int_{E_0}^{0}\left(\frac{dE}{dx}\right)^{-1} dE = \int_{0}^{E_0}\left(\frac{dE}{dx}\right)^{-1} dE \ . \quad (2.158)$$

Experimental ranges for pure elements are tabulated in the references [1], [2], and [3].

[1]　Northcliffe, L.C., R.F. Schilling: "**Range and Stopping-Power Tables for Heavy Ions.**" Nuclear Data Tables **A7**, 233 pp., (1970).

[2]　Ziegler, J.F.: "**The Stopping and Ranges of Ions in Matter.**" **1-5**, Pergamon Press, London, (1977).

[3]　Hubert, F., A. Fleury, R. Bimbot, D. Gardes: "**Range and Stopping-Power Tables for 2.5 - 100 MeV/Nucleon Heavy Ions in Solids.**" Ann. Phys. (Paris) **5**, 1-214, (1980).

# 3 Formation of the latent track

## 3.1 Track core — atomic defects

As result of the electronic and atomic collision-cascades close to the ion path a cloud of interstitial atoms and vacancies is formed. At larger distances, the electronic collision-cascade leads to excited atoms and molecules prone to chemical reactions. Their local distribution can be obtained from computer simulations [1] and defines the starting condition for diffusion processes and long-term secondary reactions. Ultimately the atomic defects reorganize in the form of a track core, a heavily disturbed zone of about 0.01 μm diameter along the ion path. It involves the diffusion of many particles in a highly disturbed solid and can be roughly described by semi-empirical models. The electronic defects lead to chemically activated sites (radicals) up to distances of about 1 μm.

### 3.1.1 Coulomb explosion model

As result of the primary ionization along the projectile-ion trajectory an ion cloud is formed. The corresponding electrons are emitted to large distances. In metals the ion cloud is neutralized immediately after the passage of the projectile ion. However, in insulators the return of the electrons to the ion cloud is inhibited if not prohibited due to electron traps. The ion cloud, containing a large amount of stored electrostatic energy expands explosively and becomes the driving force for an atomic collision-cascade [2].

A high charge density in the vicinity of the projectile-ion trajectory — sufficient to initiate a collective expansion mechanism — has been observed in ion-induced sputtering experiments [3], [4]. More recently, even in metallic alloys the signature of an ion explosion has been found [5], [6], [7]. In order to determine the energy stored within the ion cloud, the linear density as well as the radial distribution of the formed ions along the projectile-ion trajectory has to be known.

[1]   Paretzke, H.G. : "Radiation Track Structure Theory." In **Kinetics of Nonhomogeneous Processes.**" G.R. Freeman (editor), John Wiley, New York, pp.89-170 (1986).

[2]   Groeneveld, K.O., E. Schopper, S. Schumann: "Atomic Displacement Effects from Heavy Ion Induced Coulomb Explosions." **Proceedings of the 10th International Conference on Solid State Nuclear Track Detectors,** Lyon, 2-6 July 1979, Pergamon Press, Oxford, p. 81-86,1980.

[3]   Wien, K., Institut für Kernphysik, Technische Hochschule Darmstadt, D-6100 Darmstadt.

[4]   Wien, K.: "Fast Heavy Ion Induced Desorption." Radiation Effects and Defects in Solids,**109**, pp. 137-167, (1989).

[5]   Klaumüntzer, S., Ming-dong Hou, G. Schumacher: "Coulomb Explosions in a Metallic Glass Due to the Passage of Fast Heavy Ions." Physical Review Letters, **57**, pp. 850-853, (1986).

[6]   Hou, Ming-dong, S. Klaumüntzer, G. Schumacher: "Inelastic Deformations of Metallic Glasses Induced by the Electronic Energy Loss of Fast Ions." Nuclear Instruments and Methods in Physical Research, **B19/20**, pp. 16-20, (1987).

[7]   Klaumüntzer, S., Changlin Li, G. Schumacher: "Plastic Flow of Borosilicate Glass Under Bombardment with Heavy Ions." Appl. Phys. Lett. **51** (2), pp. 97-99, (1987).

### 3.1.2      Atomic collision-cascade

The minimum energy of an atom to be removed from its original site and be displaced to another site is the "displacement energy". The magnitude of the displacement energy — of the order of 10 eV — determines the size of the atomic collision cascade. With increasing displacement energy fewer and fewer atoms can be removed from their original site and ultimately — for very hard crystals — no latent track is formed.

### 3.1.3      Thermal-spike model

The energy-loss mechanism of the projectile-ion leads to electronic and atomic collision-cascades. While the electronic collision-cascade has a long range and its thermal effects can be neglected, the atomic collision-cascade deposits its energy in the close vicinity of the ion trajectory. The thermal spike model [1], [2] (Figure 3-1) replaces the complex process of the atomic collision-cascades by an abrupt temperature rise in an in- finitesimal cylindrical volume around the ion trajectory at the time-of-passage $t = 0$.

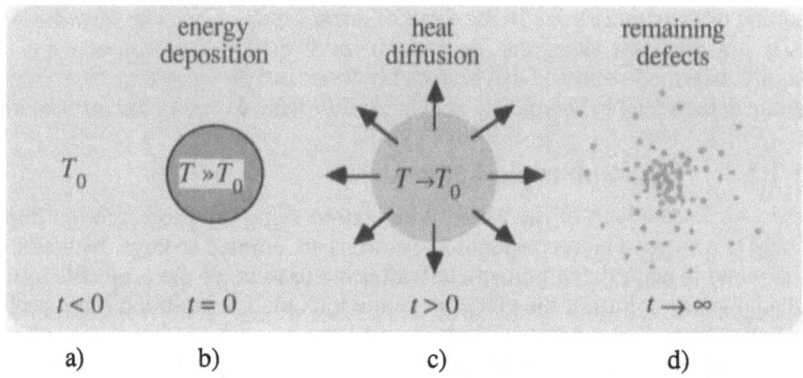

a)                  b)                        c)                        d)

**Figure 3-1**  *Basic steps of thermal-spike model.* The projectile-ion passes vertically into the drawing plane. (**a**) Undisturbed solid at temperature $T_0$. (**b**) At the time-of-passage the tem- perature within a small cylinder rises rapidly to a much higher temperature $T \gg T_0$. (**c**) After the passage of the ion, defects are thermally created while the thermal energy gradually dif- fuses away radially from the ion trajectory. (**d**) Remaining are "frozen" defects.

After the passage of the ion, for $t > 0$, the thermal energy diffuses away from the ion trajectory. Neglecting the influence of the associated shock wave which is promi- nent for tracks close to the surface [3], the thermal spike creates defects via thermal activa- tion which are remaining as "frozen defects" along the ion trajectory due to the rapid quenching of the temperature. The activation of a given volume-element at distance $r$ from the ion trajectory corresponds to its thermal history for times $t > 0$. The starting point of the thermal spike model is the thermal-diffusion differential-equation derived in the fol- lowing.

[1]      Vineyard, G.H.: "Thermal Spikes and Activated Processes", Radiation Effects, **29**, 245-248 (1976).
[2]      Chadderton, L.T.: "On the Anatomy of a Fission Fragment Track." in "**Solid State Nuclear Track Detectors.**" Proceedings of the 14th International Conference, Lahore, Pakistan, 2-6 April 1988, Pergamon Press, 11-29, (1988).
[3]      Ronchi, C.: "The Nature of Surface Fission Tracks in $UO_2$." J. Appl. Phys. **44**, No. 8, pp. 3575- 3585, (1973).

### Stationary thermal diffusion

The heat flow $I$ between two thermal reservoirs at temperatures $T$ and $T'$, connected by a heat conductor of cross section $\sigma$ and length $L$ is proportional to the temperature difference $(T' - T)$, proportional to the cross section $\sigma$, and inversely proportional to the length $L$ of the thermal conductor:

$$I \equiv \frac{Q}{t} = -\kappa \frac{T' - T}{L} \sigma ,$$

(3.1)

where $Q$ is the thermal energy ("heat"), and $\kappa$ is the thermal conductivity — the transferred energy per unit time, per unit temperature difference, and per unit distance. In differential form it can be written:

$$dI \equiv \frac{dQ}{t} = -\kappa \frac{\partial T}{\partial x} d\sigma, \quad \text{or} \quad j \equiv \frac{dQ}{d\sigma} \frac{1}{t} = -\kappa \frac{\partial T}{\partial x} .$$

(3.2)

In a three dimensional isotropic medium the three heat currents in $x$, $y$, and $z$ - direction through the volume element depend on the corresponding gradients $(\partial T / \partial x)$, $(\partial T / \partial y)$, and $(\partial T / \partial z)$ and are independent of each other. Therefore the current density $j$ becomes a vector $j$, proportional but antiparallel to the thermal gradient $\nabla T$ and proportional to the thermal conductivity $\kappa$:

$$j = -\kappa \nabla T .$$

(3.3)

This is the stationary thermal diffusion-equation suited for time-independent problems in which the stored thermal energy per volume element of the thermal conductor does not appear explicitly.

### Dynamic thermal diffusion

The thermal-energy change $dQ$ of a one-dimensional volume element $dx$ is proportional to the specific heat $c_v$ of the volume element times its volume $dx$ times the temperature change $dT$. The specific heat (or thermal capacity) $c_v$ is defined as the increase of the thermal energy $dQ$ per unit temperature change $dT$ per unit volume, $c_v = dQ / dT$. The thermal energy change $dQ$ within the one-dimensional volume-element $dx$ corresponds to the net flow of heat arriving in the volume element $dx$ during the time $dt$. It is the sum of the heat energies $[Q(x) - Q(x+dx)]$ entering the volume element per unit time from left and right, which, according to the stationary thermal diffusion eq. (3.3) are given by the current-densities at $x$ and at $x + dx$, multiplied by the time interval $dt$:

$$dQ = c_v \, dx \, dT$$
$$= Q(x) - Q(x+dx)$$
$$= \left( j(x) - j(x+dx) \right) dt ,$$

(3.4)

which can be written as:

$$c_v \, dT = -\left( \frac{dj}{dx} \right) dt .$$

(3.5)

In a three-dimensional isotropic medium the thermal currents in in $x$, $y$, and $z$ - direction are simultaneously accumulated within the volume element $d\tau \equiv dx\,dy\,dz$. The three thermal currents can therefore be linearly superposed and eq. (3.4) can be written as:

$$c_v\,dT = -\left[\left(\frac{dj_x}{dx}\right) + \left(\frac{dj_y}{dy}\right) + \left(\frac{dj_z}{dz}\right)\right]dt = -(\nabla \cdot j)\,dt \ , \ \text{or} \tag{3.6}$$

$$c_v\frac{\partial T}{\partial t} = \nabla\left(\kappa\,\nabla T\right) \Rightarrow c_v\frac{\partial T}{\partial t} = \kappa\,\Delta T \ \text{ for } \kappa = \text{const} \ . \tag{3.7}$$

The eq. (3.7) is the dynamic differential-equation for the diffusion of heat. For $\kappa = $ const which can be comprehended in the following simple way: The increase of the thermal energy observed in the volume element per time interval $dt$ is equal to the thermal conductivity $\kappa$ multiplied by the divergence of the temperature $T$.

### Solution of dynamic thermal diffusion equation

We assume in the following — somewhat artificially — that the specific heat as well as the thermal conductivity are independent of the temperature and constant throughout the medium. Therefore $c_v = c_{v,0}$ and $\kappa = \kappa_0$ are constants. This assumption yields the partial differential equation:

$$\frac{\partial T}{\partial t} = \frac{\kappa_0}{c_{v,0}}\,\Delta T \quad \text{or} \quad \frac{\partial T}{\partial t} = \text{const }\Delta T \tag{3.8}$$

with the standard solution [1]:

$$T(r,t) = \frac{A}{t}\,e^{-B\frac{r^2}{t}} + T_0 \ , \ \text{where} \ A = \frac{S(E)}{4\pi\,\kappa_0} \ , \ B = \frac{c_{v,0}}{4\,\kappa_0} \ . \tag{3.9}$$

The constant $B$ is determined by inserting $T(r,t)$ from eq. (3.9) into the differential equation (3.8), using the cylindrical-coordinate representation of the divergence operator, $\Delta = (\partial^2/\partial r^2) + (1/r)(\partial/\partial r)$. The constant $A$ is determined by integrating the deposited thermal energy, $\Delta Q = c_{v,0}\,\Delta T$, for a fixed value of the time $t$ (taking for simplicity $t = 1$) from $r = 0$ to $r = \infty$, and equating this total thermal energy to the energy input per unit length-of-path. The total thermal-energy input per unit path-length is given by the stopping power $S(E) = (dE/dx)$ of the projectile in the target material.

For obtaining an approximate picture of a thermal-spike process in glasses or in dielectric crystals, we assume that the initial temperature is negligible in comparison with the temperatures reached during the thermal-spike process. Assuming a specific heat $c_{v,0} = 4 \cdot 10^7$ erg K$^{-1}$ cm$^{-3}$, representative for water, a thermal conductivity $\kappa_0 = 4 \cdot 10^4$ erg s$^{-1}$ cm$^{-1}$ K, representative for dielectric crystals and glasses, and an energy input $S(E) = 10^{11}$ eV / cm $= 1.6 \cdot 10^{-1}$ erg / cm, representative for a medium-mass ion around its maximum energy-loss, we obtain $A = 3 \cdot 10^{-7}$ K s, and $B = 2.5 \cdot 10^2$ s / cm$^2$. The resulting thermal spike is shown in Figure 3-2.

---

[1]     Courant, R, D. Hilbert: "**Methoden der Mathematischen Physik.**" Band II, 2. Auflage, Springer-Verlag, Heidelberg, (1968).

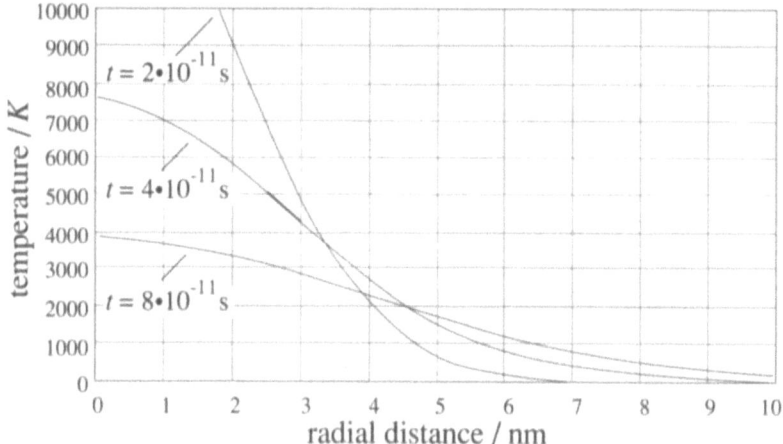

**Figure 3-2** *Time-space evolution of a thermal spike*, for example in glasses or in dielectric crystals. With increasing time the temperature maximum decreases rapidly while the total thermal energy is distributed over a cylindrical volume of increasing radius. Due to the cylindrical geometry, the damping of the outgoing thermal "wave" is high.

There exists a very rapid decrease of temperature with radial distance from the ion trajectory. It is possible to define an arbitrary track diameter very roughly by the maximum radial distance for which a specified temperature — sufficiently high to create defects — is exceeded during the course of the thermal-spike process. In the chosen case, the maximum diameter for which 1000 K are exceeded is roughly 7 nm, which is of the order of a typical experimental track radius.

**Thermally activated processes**

According to the thermal-spike concept, atoms or molecules along the ion trajectory are "activated". The activation process corresponds to the excitation of the atoms or molecules into metastable states with energies above the ground state (Figure-3-3).

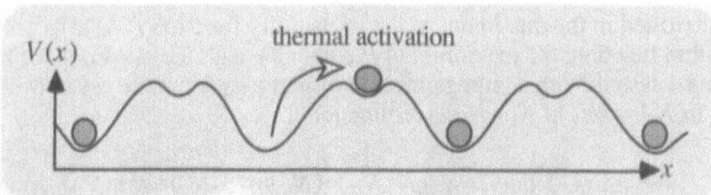

**Figure 3-3** *Thermal activation* of a lattice atom from its ground state into a metastable state. If the system is rapidly "quenched" from the activation temperature to a lower temperature, a certain fraction of the atoms will remain "frozen" in the excited states.

The activated sites or defects are prone to subsequent reactions. They are ultimately responsible for formation of the latent track and its preferential development. In order to derive the number of defects along the ion trajectory we have to know the time-temperature history of the volume element $d\tau$ at distance $r$ from the ion trajectory and the probability for its thermal excitation by the spike process, which is given by the Boltzmann factor, derived in the following.

**Boltzmann statistics for harmonic oscillators.** Our first goal is to determine the probability distribution of a large number $N$ of independent, distinguishable particles over $m$ energy states with energies $\varepsilon_i$ [1]. The number of particles in each energy state $\varepsilon_i$ is termed $N_i$. For example, if $m = 2$ and $N = 2$ we obtain Figure 3-4 with the three distinct macro states $\{N_0, N_1\}$: $\{2,0\}$, $\{1,1\}$, $\{0,2\}$ having the probabilities (weights) of 0.25, 0.5, and 0.25, and total energies of 0, $\varepsilon$, and $2\varepsilon$, respectively.

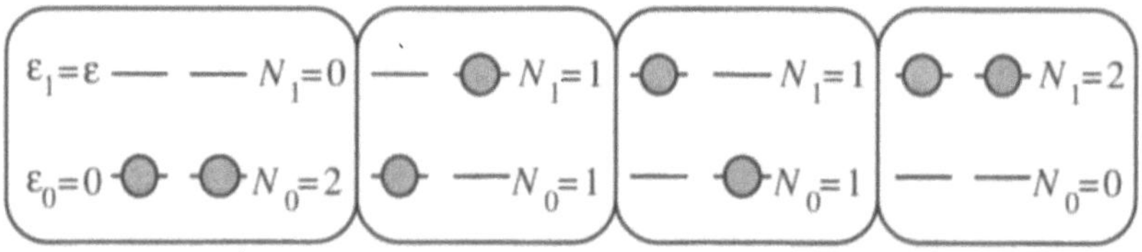

**Figure 3-4** *Determination of a-priority probability* distribution for two independent, distinguishable particles with two energy states $\varepsilon_0$ and $\varepsilon_1$. The number of particles in these energy states is denoted by $N_1$ and by $N_2$.

The a-priori probability $P$ of the macro state $\{N_0, N_1, ..., N_{m-1}\}$, corresponds to the stochastical distribution of $N$ particles in $m$ boxes of equal size:

$$P = \frac{N!}{N_0! \, N_1! \, ... \, N_{m-1}!} \tag{3.10}$$

and is a function of the $m$ variables $N_i$. These variables, however, are not independent of each other but have to obey the boundary conditions that the total number of particles is $N$ and that the total energy of the system is $E_{tot}$, respectively:

$$\sum_{i=0}^{m-1} N_i = N \, , \qquad \text{or} \qquad \left( \sum_{i=0}^{m-1} N_i - N \right) = 0 \, , \quad \text{and} \tag{3.11}$$

$$\sum_{i=0}^{m-1} N_i \, \varepsilon_i = E_{tot} \, , \qquad \text{or} \qquad \left( \sum_{i=0}^{m-1} N_i \, \varepsilon_i - E_{tot} \right) = 0 \, . \tag{3.12}$$

We are interested in the maximum of the probability function $P$. Due to the monotony of the logarithm function, the maximum of $P$ coincides with the maximum of $\ln P$. Since we are treating a system with a large number $N$ of particles we make use of Stirling's approximation, $\ln N_i! \approx (N_i \ln N_i - N_i)$, leading to:

$$\ln P = \ln N! - \sum_{i=0}^{m-1} \ln N_i! \approx N \ln N - N - \sum_{i=0}^{m-1} \left( N_i \ln N_i - N_i \right) = N \ln N - \sum_{i=0}^{m-1} N_i \ln N_i \, . \tag{3.13}$$

The maximum of the probability function is obtained by the condition $\delta P = 0$, or, alternatively by the condition $\delta \ln P = 0$, where $\delta$ is the variation operator:

$$\delta \ln P = - \delta \sum_{i=0}^{m-1} N_i \ln N_i = - \sum_{i=0}^{m-1} (\delta N_i) \ln N_i - \sum_{i=0}^{m-1} \delta N_i \overset{!}{=} 0 \tag{3.14}$$

[1]     Wedler, G.: "**Lehrbuch der Physikalischen Chemie**", VCH Verlagsgesellschaft, D-6940 Weinheim, pp. 84-92, and pp. 168-170, (1985).

From the eqs.(3.11) and (3.12) we obtain the additional conditions:

$$\delta N = \sum_{i=1}^{m-1} \delta N_i = 0 \quad \text{and} \tag{3.15}$$

$$\delta E_{tot} = \sum_{i=0}^{m-1} \varepsilon_i \, \delta N_i = 0 \; . \tag{3.16}$$

The eqs. (3.14) to (3.16) can be compactified into one single equation, using the method of Lagrange's multiplicators. For this purpose we multiply eq. (3.15) and (3.16) with arbitrary constants $\lambda$ and $\mu$, respectively, and subtract the result from eq. (3.14):

$$\sum_{i=0}^{m-1} \delta N_i \left( 1 + \ln N_i + \lambda + \mu \, \varepsilon_i \right) = 0 \; . \tag{3.17}$$

Due to the two boundary conditions $\delta N = \Sigma \delta N_i = 0$, and $\delta E_{tot} = \Sigma(\varepsilon_i \, \delta N_i) = 0$ all but two of the variations $\delta N_i$ are free variations. For these $(m - 2)$ variations the factor in brackets must vanish. For the remaining two variations which are pre-defined by the boundary conditions, the two terms within the brackets must vanish due to the free choice of $\lambda$ and $\mu$. Therefore all $m$ terms $(1 + \ln N_i + \lambda + \mu \, \varepsilon_i)$ vanish, which enables to determine the number $N_i$ of particles in the state i from $(1 + \ln N_i + \lambda + \mu \, \varepsilon_i) = 0$:

$$N_i = e^{-(1+\lambda)-\mu\varepsilon_i} \equiv \alpha \, e^{-\mu\varepsilon_i}, \quad \text{with } \alpha = \frac{N}{\sum\limits_{i=0}^{m-1} e^{-\mu\varepsilon_i}} \quad \text{and} \quad \mu = \frac{N}{E_{tot}} \; . \tag{3.18}$$

The factor $\alpha$ is determined from the boundary condition $\Sigma N_i = N$. The factor $\mu = N / E_{tot}$ corresponds to the inverse of the average particle energy and is determined from the boundary condition $\Sigma(\varepsilon_i \, N_i) = E_{tot}$. We are now specializing onto a system of harmonic oscillators for which $\varepsilon_i \equiv i \, \varepsilon \equiv E_i$. In addition we assume the classical condition that the average energy $E_{tot} / N$ of the particles or oscillators is proportional to the temperature, according to the equation $E_{tot} / N = k \, T$, where $k$ is the Boltzmann constant. We obtain the Boltzmann distribution for harmonic oscillators, which decays exponentially with the excitation energy $E_i$ of the particles:

$$N_i = N \frac{e^{-\frac{i\,\varepsilon}{kT}}}{\sum\limits_{i=0}^{m-1} e^{-\frac{i\,\varepsilon}{kT}}} \; , \quad \text{or} \quad N(E_i) = N \frac{e^{-\frac{E_i}{kT}}}{\sum\limits_{i=0}^{m-1} e^{-\frac{E_i}{kT}}} \; . \tag{3.19}$$

It describes the number of particles in a specific excitation state as function of $E_i$ and $T$.

The number of particles with energies larger or equal to $E_i$ can be determined from eq. (3.19) by summing up all numbers $N_i$ with indices larger or equal to $i$. After splitting off the multiplicative factor $N_i$ in front of a remaining sum, one obtains the frequently used Boltzmann formula:

$$n(E_i) = N \, e^{-\frac{E_i}{kT}} \; , \quad \text{or} \quad p(E_i) \equiv \frac{n(E_i)}{N} = e^{-\frac{E_i}{kT}} \; , \tag{3.20}$$

where $n(E_i)$ is a new function describing the number particles with energies overshooting the state energy $E_i$ and $p(E_i) = \exp[-E_i / (k T)]$ is the "Boltzmann factor".

The fraction of particles with sufficient kinetic energies to overcome a reaction barrier determines the reaction rate in many chemical reactions. The function $n(E_i)$ should not be confounded with $N_i$, the number of particles with exactly the energy $E_i$ according to eq. (3.19). According to eq. (3.20) the relative number or probability $p(E_i)$ of particles with energies overshooting the energy $E_i$ is equal to the "Boltzmann factor".

The eq. (3.20) is often displayed in the form of an "Arrhenius-plot" (Figure 3-5) which displays the logarithm of the observed effect as function of the inverse temperature $1 / T$, whereby the excitation energy $E_i$, given in eV, is used as parameter. Many track-formation and track-annealing studies [1], [2], [3] can be reduced to such a diagram.

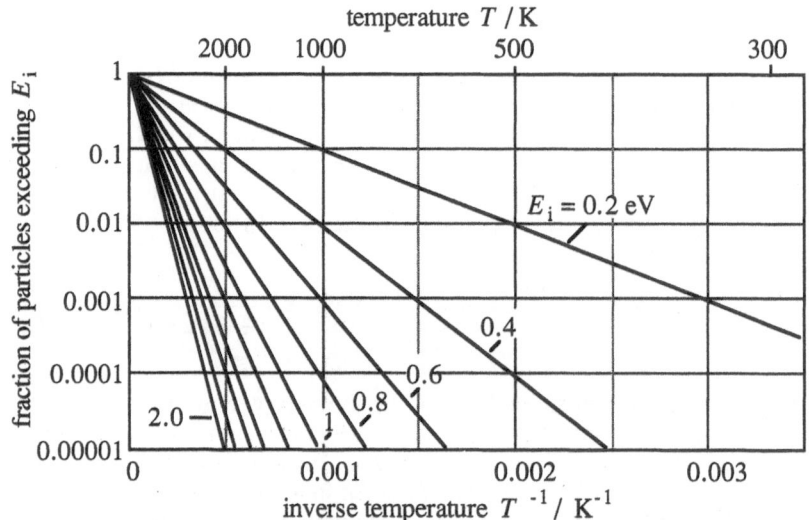

**Figure 3-5**  *Arrhenius-plot* of Boltzmann factor $\exp[- E_i / (k T)]$, describing the fraction of particles with thermal energies exceeding $E_i$.

**Thermal excitation and reaction rates.** The Boltzmann factor $\exp[-E_i / (k T)]$ corresponds classically to the number of particles with sufficient kinetic energy to surmount a potential-barrier of height $E_i$ and may settle down at a higher or lower potential energy level with respect to their original location. Particles which are "pumped up" thermally into a higher energy state have a high probability to return back into their ground state if the high temperature endures or if the temperature is gradually lowered over a prolonged time. If the high temperature endures only for a short time, the higher states have only a small chance to return to the ground state, due to the potential-energy threshold associated with the high energy level. Rapid cooling will leave the particles in their

[1]    Märk, T.D., W. Ritter: "Radiation Damage and its Annealing in Sodium Silica Glass." Mat. Res. Soc. Symp. Proc. Vol. **84**, pp. 659-669, (1987)
[2]    Walder, G., T.D. Märk: "Annealing Kinetics of Radiation Damage in Artificial Obsidian Glass." Nuclear Instruments and Methods in Physics Research, **B32**, pp. 303-306, (1988).
[3]    Märk, T.D., G. Walder: "Annealing and Leaching Studies with Natural and Artificial Obsidian Glass." Mat. Res. Soc. Symp. Proc. Vol. **112**, pp. 693-701, (1988).

metastable states. This "shock-freezing of defects" in metastable state is, on one hand, the origin of the lattice defects forming the latent ion track. On the other hand, the created defects have an increased tendency to react chemically, due to the reduced barrier height between their metastable state and a final state of reduced energy (Figure 3-6).

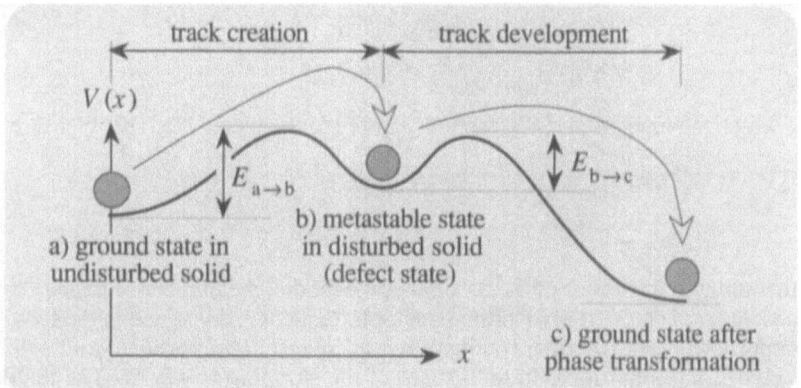

**Figure 3-6** *Principle of surmounting a reaction barrier* in track creation and development.

In first-order — or binary — chemical reactions, the Boltzmann factor $\exp[-E_i/(kT)]$ corresponds to the reaction rate $R$, in other words to the number of reactions per unit time and unit volume. In the formation of latent ion-tracks, the Boltzmann factor corresponds to the number of defects created per unit time and unit volume:

$$R = C\, n\, e^{-\frac{E_i}{kT}} \quad , \quad \text{where} \quad 0 \leq C \leq 1 \tag{3.21}$$

and $n$ is the number of particles per volume element, which are principally available for the specific reaction. The factor $C$ reflects the fact that not all particles possessing sufficient energy will necessarily surmount the barrier before dropping back into the pool of lower-energy particles. According to the classical point-of-view, the factor $C$ is proportional to the "knock-on frequency" or "trial-escape frequency", the oscillation frequency of the particle within its potential well, corresponding to the number of attempted escapes per unit time during which the particle tries to surmount the barrier.

Writing $R \equiv -dn/dt$, multiplying eq. (3.21) by $dt$, dividing by $n$, and integrating over $n$ or $t$, yields a successful formula ruling thermal annealing, $\ln(n/n_0) = -C\, t\, \exp[-E_i/(kT)]$ or $n = n_0 \exp(-\alpha\, t)$, where $\alpha = C \exp[-E_i/(kT)]$.

**Linear density of activated sites along the ion trajectory**. A thermal spike is a very rapid process. If we assume that the chance of the particles to return into the ground state, due to the very rapid shock freezing of the defect states, is negligible, we can calculate the number $\eta$ of defects created per unit track length. For this purpose we have to integrate the reaction rate $R$, in other words the number of defects formed per unit time and per volume element $d\tau = 2\pi\, r\, dr$, over time and space, according to:

$$\eta = \int_0^\infty dt \int_0^\infty d\tau\, R(t, \tau) = \int_0^\infty dt \int_0^\infty 2\pi\, r\, dr\; C\, n\, e^{-\frac{E_i}{kT(r,\, t)}} , \tag{3.22}$$

where $T(r, t)$ corresponds to the time history of the spatial distribution of the thermal spike, according to eq. (3.9). For solving the integral (3.22), the substitutions $t^* = [(4\pi\,\kappa_0\,E_i) / (S(E)\,k\,)]\,t$ and $\ln\xi = [\,c_{v,\,0} / (4\,\kappa_0)\,]\,[\,r^2 / t\,]$ are introduced, yielding — without derivation — the following transformed integral [1]:

$$\eta = \frac{C\,n\,k^2}{4\pi\,\kappa_0\,c_{v,\,0}}\,\left(\frac{S(E)}{E_i}\right)^2 \int_1^\infty \frac{d\xi}{\xi} \int_0^\infty dt^*\,t^*\,e^{-t^*\,\xi}, \tag{3.23}$$

of which the integral part corresponds simply to the factor 1/2. This yields the final result:

$$\eta = \frac{C\,n\,k^2}{8\pi\,\kappa_0\,c_{v,\,0}}\,\left(\frac{S(E)}{E_i}\right)^2. \tag{3.24}$$

According to this result of the thermal-spike model, the number of defects created per unit track length is proportional to the knock-on frequency, contained in the factor $C$, inversely proportional to the thermal conductivity $\kappa_0$, inversely proportional to the thermal heat capacity $c_{v,\,0}$, proportional to the square of the stopping power $S(E)$ of the projectile-ion, and finally inversely proportional to the threshold energy $E_i$ required for excitation of the particles into the metastable defect state.

A numerical evaluation of eq. (3.24) yields too low linear defect densities if normal values of $\kappa_0$ and $c_{v,\,0}$ are used. Therefore one has to assume that the thermal conductivity $\kappa_0$ as well as the specific heat $c_{v,\,0}$ of the material are drastically reduced, reflecting the small fraction of particles which participate in the collision cascade. A more advanced treatment is indicated in reference [2].

### 3.1.4    Resulting primary defects

In the following a short introduction into the defect creation in solids [3] and their diffusion [4] is given with special consideration of their potential influence in the formation of latent tracks. Usually the local concentration of ion-induced defects is very high. The created point defects will interact in many possible ways to form ultimately an entity vaguely circumscribed as an "extended defect".

#### Defect stages

With increasing energy-input along the ion trajectory four stages of increasing disorder can be discerned (Figure 3-7).

a)    At low energy input or very far from the ion trajectory, the solid, represented here as a regular crystal lattice, may take up thermal energy in the form of lattice vibrations (phonons). The transferred energy ultimately is distributed over the whole solid whereby the local information, concerning the position where the thermal input occurred, is completely lost by "delocalization".

[1]    Vineyard, G.H.: "Thermal Spikes and Activated Processes", Radiation Effects, **29**, 245-248 (1976).
[2]    Chadderton, L.T.: "On the Anatomy of a Fission Fragment Track." in "**Solid State Nuclear Track Detectors.**" Proceedings of the 14th International Conference, Lahore, Pakistan, 2-6 April 1988, Pergamon Press, 11-29, (1988).
[3]    Johnson, R.A., A.N. Orlov (editors): "**Physics of Radiation Effects in Crystals.**" North Holland, Amsterdam, (1986).
[4]    Mrowec, S.: "**Defects and Diffusion in Solids. An Introduction.**" Elsevier Scientific Publishing Company, Amsterdam, 466 pp., (1980).

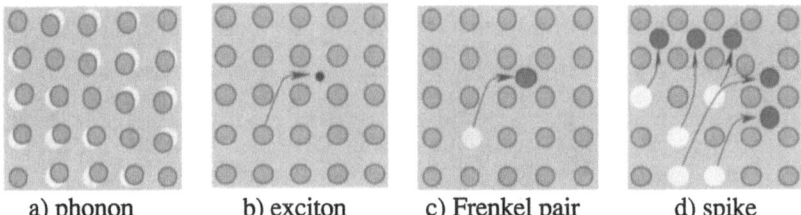

a) phonon         b) exciton       c) Frenkel pair      d) spike

**Figure 3-7** *Four defect stages with increasing disorder* related to increasing energy input or decreasing distance from the ion trajectory. (**a**) Generation of elastic waves, phonons, heat. (**b**) Electron excitation into localized "excitons" or dissociated electron-hole pairs. (**c**) Formation of an interstitial-vacancy pair. (**d**) Formation of a displacement spike consisting of a vacancy-rich core-zone surrounded by an interstitial-rich fringe-zone.

b)   At somewhat higher energy input, an electron can be expelled from a specific lattice site, forming an "exciton" or electron-hole pair together with its mother ion. Excitons may annihilate by the mutual recombination of the electron with the hole. The excited atom may react with a neighboring atom, leading to a permanent, locally confined defect. Alternatively the excitation energy may be conserved in the form of a chemical radical. With increasing distance from the mother ion (increasing excitation energy) the electron becomes more and more independent from its mother ion. It may induce a secondary reaction somewhere else.

c)   At medium energy input, an interstitial-vacancy pair or "Frenkel pair" can be created, whereby the considered atom is removed from its original site and squeezed-in between atoms on regular lattice sites. The interstitial and the vacancy is called a point defect. Point defects may wander about and annihilate mutually or aggregate to larger defects. Due to the different properties of interstitials and vacancies there exists an asymmetric behavior leading to a tendency that always one type of defects or defect aggregates remains as a majority defect.

d)   At high energy-input, a large number of defects are created as consequence of an atomic collision-cascade, forming a strongly disturbed quasi-amorphous zone, which is characteristic for the latent track.

**Self-trapped exciton mechanism.** A model case for the transfer of electronic excitation energy into atomic movement is provided by the "self-trapped exciton" mechanism found in alkali halides, alkaline earth fluorides and silicon dioxide (Figure 3-8) [1], [2], [3], [4]. Relaxation of this defect type leads to the expulsion of an interstitial along a

[1]   Itoh, N.: "Defect Formation in Insulators Under Dense Electronic Excitation." in "**Symposium on Swift Heavy Ions in Matter.**" Caen, May 16-19, 1989, to be published as special volume of Radiation Effects and Defects in Solids (editor: J. Biersack).

[2]   Tanimura, K., N. Itoh: "Mechanisms of Atomic Processes Induced by Electronic Excitation in Solids." Nucl. Instrum. and Methods in Physics Res., **B33**, pp. 815-819, (1988).

[3]   Itoh, N., T. Nakayama: "Electronic Excitation Mechanism of Sputtering and Track Formation by Energetic Ions in the Electronic Stopping Regime." Nuc. Instrum. and Methods in Physics Res., **B13**, pp. 550-555, (1986).

[4]   Itoh, N., K. Tanimura: "Radiation Effects in Ionic Solids." Radiation Effects, **98**, pp. 269-287, (1986).

{110} direction. The model provides thus — besides the Coulomb explosion — another efficient mechanism for the energy transfer from the electronic to the atomic collision cascade.

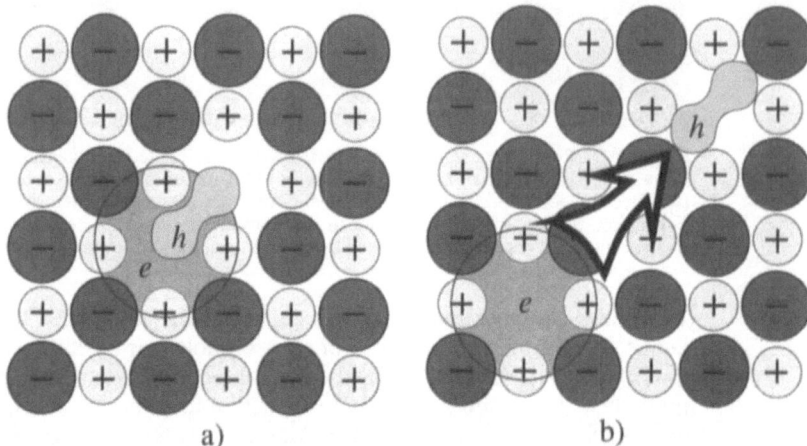

a)                                                              b)

**Figure 3-8** *Self-trapped exciton mechanism* converting electronic excitation into atomic displacement [1]. (**a**) The excitation of the electron *e* increases its orbital volume, thus reducing the binding energy between the negatively charged halogen core (the missing "-") and its positively charged metal ion neighbors (+). The system reorganizes to a metastable dumbbell configuration consisting of a molecular ion including a hole (*h*) with two halogen atoms. This is the so-called "self-trapped exciton". (**b**) The metastable system decays by expelling the dumbbell configuration in a {110} lattice direction. The latter configuration is a Frenkel pair consisting of a neutral vacancy (a so-called *F*-center) and a neutral interstitial (a so-called *H*-center), whereby one of the two dumbbell atoms represents the interstitial.

## 3.1.5    Diffusion and relaxation of defects

### Attractive potentials between defects

The short-range attraction between a vacancy and an interstitial is obvious from the required energy input to generate such a "Frenkel" pair. Another comprehension is to consider the lattice distortion around each defect which corresponds to a reservoir of elastic energy. Since the interstitial pushes its neighbors further away, while the vacancy attracts its neighbors to a closer distance, the associated lattice misfits will compensate each other increasingly with decreasing distance between the defect pair. The attraction may ultimately lead to the annihilation of the defect pair, a process which is associated with a complete recovery of the invested defect-creation energy (Figure 3-9).

With increasing distance between the interstitial and the vacancy, the process of annihilation is increasingly hindered. Thereby the increasing number of other lattice atoms which are coming in-between represents an increasing number of potential barriers which have to be overcome via thermal excitation. Further away, the attraction becomes negligible in comparison with the forces due to the nearest neighbors. Therefore at greater distance the interaction of the vacancy and the interstitial with other lattice defects becomes increasingly dominant.

[1]    Itoh, N., K. Tanimura: "Formation of Interstitial-Vacancy Pairs by Electronic Excitation in Pure Ionic Crystals." J. Phys. Chem. Solids, in print. (ca. 1989).

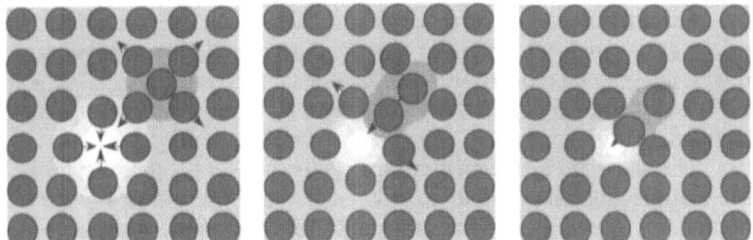

**Figure 3-9** *Attraction between vacancy and interstitial.* Final steps of vacancy-interstitial annihilation, reflecting the attractive potential between a vacancy and an interstitial in the decreasing lattice misfit. Alternatively, reading the figure from right to left, we observe the initial steps of a thermal vacancy-interstitial creation.

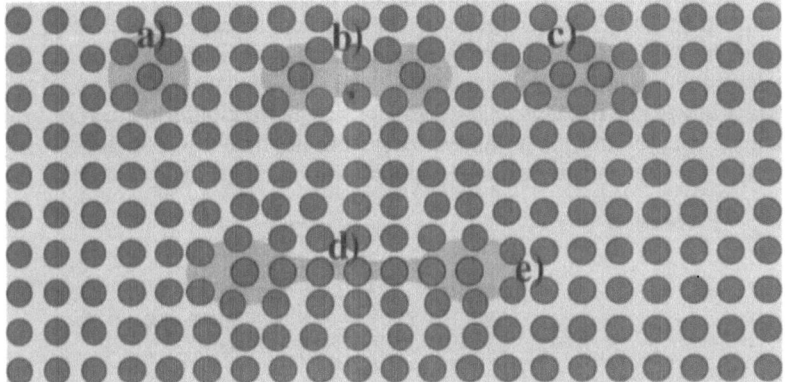

**Figure 3-10** *Attraction between interstitials.* (a) Discrete single interstitial with relaxed nearest neighbors. (b) Slightly separated pair of interstitials. (c) Fully aggregated pair of interstitials. (d) Nearly undisturbed central zone of a new crystal plane or aggregated multi-interstitial. (e) Slightly disturbed fringe zone of the new crystal plane providing still some attractive potential for approaching interstitials due to its increased atomic distances. The fringe zone around such an interstitial lattice "platelet" corresponds to a line dislocation.

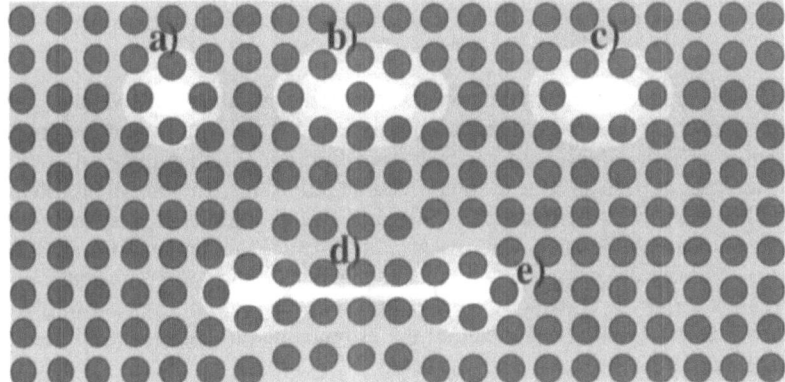

**Figure 3-11** *Attraction between vacancies.* (a) Discrete single vacancy with relaxed nearest neighbors. (b) Slightly separated pair of vacancies. (c) Fully aggregated pair of vacancies. (d) Nearly undisturbed central zone of a removed crystal plane or collapsed multi-vacancy. (e) Slightly disturbed fringe zone of the collapsed multi-vacancy still providing some attractive potential for approaching vacancies due to its decreased atomic distances. The fringe-zone around such a vacancy-platelet corresponds to a line-dislocation.

The attraction between a pair of interstitials becomes obvious if we realize that their agglomerates represent the initial stage of a new crystal plane, a process which is associated with the gradual return to an ordered crystal lattice (Figure 3-10). In this way, the "invested" defect energy is gradually dissipated with each merging process and transformed into heat. This represents a process of thermal annealing.

The same attractive mechanism occurs between a pair of vacancies, with the difference that the gradual return to an ordered crystal corresponds here to the removal of a crystal plane by the collapsing multivacancy and not to the addition of a new crystal plane (Figure 3-11). Again, the invested defect energy is thermally dissipated step by step with each merging process.

### Defect migration

Figure 3-12 shows the main mechanism of point-defect migration in solids, here an interstitial. The interstitial squeezes through a ring of neighboring atoms. The migration corresponds to the thermally activated diffusion of the defect in the periodic potential along a channeling direction of the crystal grid. An atom in its ground energy state on a regular lattice site has always to surmount a higher potential-barrier for its displacement than a vacancy or an interstitial resulting from its displacement (Figure 3-6. This is the reason for the dramatically increased mobility of defects in crystals.

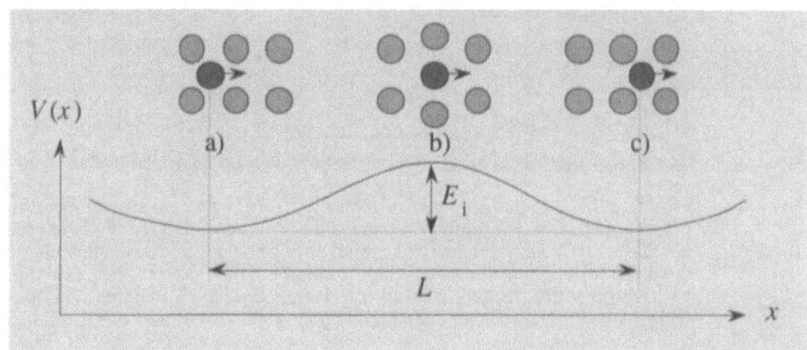

**Figures 3-12** *Migration of an interstitial* corresponding to the thermally activated diffusion — or tunneling — in a periodic potential.

Due to the asymmetry of the attractive and the repulsive part of the potential-energy curve between two atoms within the crystal, the required activation energy $E_i$ for migration of an interstitial and a vacancy are different, yielding quite different mobilities of the two types of defects. Due to their different mobilities, the interstitials and the vacancies do anneal at different temperatures. Therefore at a given temperature, one type of point defect can be still abundant — the "majority defect type" — whereas the other type of defect — the "minority defect type" — may have practically vanished. This fact is reflected in the decreased density in the core of latent tracks, corresponding to a surplus of vacancies of the order of 5 - 10 % [1], [2]. Due to the absence of the counterpart defect, the majority defects, here the vacancies, are "stabilized".

[1]    Albrecht, D., P. Armbruster, M. Roth, R. Spohr: "Small Angle Neutron Scattering Observations from Oriented Latent Nuclear Tracks." Radiation Effects, **65**, 145-148 (1982).

[2]    Albrecht, D., P. Armbruster, R. Spohr, M. Roth, K. Schaupert, H. Stuhrmann: "Investigation of Heavy Ion Produced Defect Structures in Insulators by Small Angle Scattering." Applied Physics, **A37**, 37-46 (1985)

In the vicinity of a defect there exists a long-distance gradient of the periodic potential along the channeling directions of the crystal grid. The gradient can be attractive or repulsive (Figure 3-13). Intuitively, if we consider only the widening of the grid around a vacancy and the compression of the grid around an interstitial, the long-distance term of the mutual potential between an interstitial and a vacancy should be weakly attractive, whereas the long-distance term of the mutual potential between two interstitials or two vacancies should be repulsive.

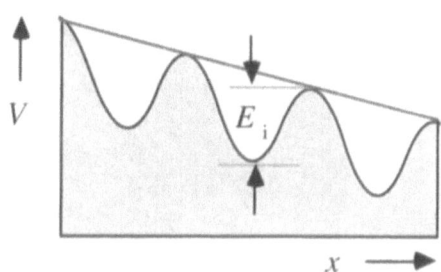

**Figure 3-13** *Long-distance interaction between defects.* The gradual variation of the periodic potential in the vicinity of defects leads to attraction or repulsion, depending on the defect type.

### Defect-hierarchy and defect sinks

With increasing "dimensionality of the defects" the available space for adapting other defects increases. In this sense, zero-dimensional defects are equivalent to point defects, one-dimensional defects to dislocation lines. The aggregation of point-defects (interstitials or vacancies) yields platelets of additional or missing lattice planes surrounded by rings of increased mismatch serving as defect sinks. Simultaneously such a "dislocation line" provides a zone of increased mobility and thus can be conceived as a linear conductor representing a path of enhanced defect-diffusion — similar to a metal wire conducting electric current. Such a "dislocation line", here the one-dimensional defect-ring surrounding an interstitial platelet, is shown in Figure 3-14.

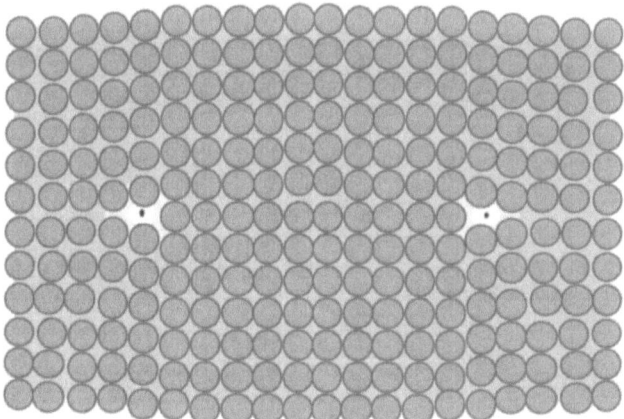

**Figure 3-14** *Dislocation line* surrounding an inserted interstitial platelet.

Real crystals consist of a mosaic of many small crystal grains with slightly different orientations. The grain boundaries correspond to a two-dimensionally extended defect-zone of increased mismatch, which can be conceived as a quasi-liquid zone of increased mobility and defect affinity (Figure 3-15). This "liquid model" of the grain boundary is reflected in the concept that the liquid state corresponds to a crystal with extremely many defects. Transferring this concept to the latent track, it consists of an amorphous — or glass — phase.

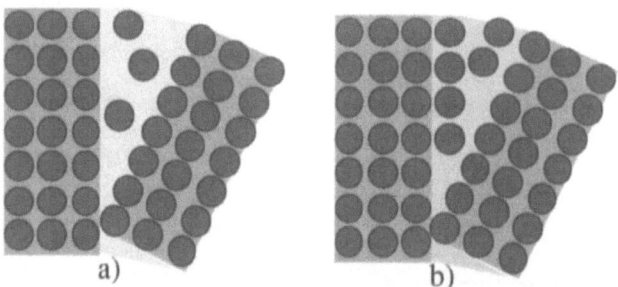

a)                                                                  b)

**Figure 3-15** *Formation of a grain boundary.* (**a**) During solidification, the competition between two crystal planes of different orientation leads to different affinities between the atoms in the solution and the different crystal planes. (**b**) A quasi-stable grain boundary has been established, corresponding still to a quasi-liquid zone of increased mismatch and mobility which can serve as a sink for defects.

### Defect-hierarchy and defect-annealing

Grain boundaries can be conceived as two-dimensional defect-sinks into which the one-dimensional dislocations conduct their surplus of defects which they in turn have collected from the bulk of the material. This corresponds to a very efficient mechanism for the expulsion of defects from the bulk material and for the recovery of the undisturbed crystal structure (Figure 3-16).

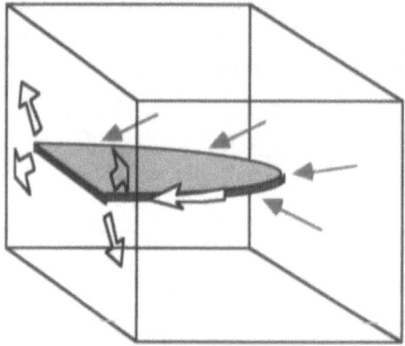

**Figure 3-16** *Excretion of point defects* from a disturbed crystal grain to the grain boundary. The shown dislocation line represents a channel of increased mobility along which the defects can migrate to the grain boundary.

At high defect-densities three-dimensional or "extended defects" [1] are

---

[1]     Dartyge, E., J.P. Duraud, Y. Langevin, M. Maurette: "New Model of Nuclear Particle Tracks in Dielectric Materials." Phys. Rev. **B 23**, 5213, (1981).

formed, corresponding to a quasi-amorphous zone. The latent track generated along the projectile trajectory consists of such extended defects forming a new phase which is embedded into a more or less undisturbed environment (Figure 3-17). At high defect densities the extended defects form a contingent zone of transformed material along the ion trajectory (Figure 3-18).

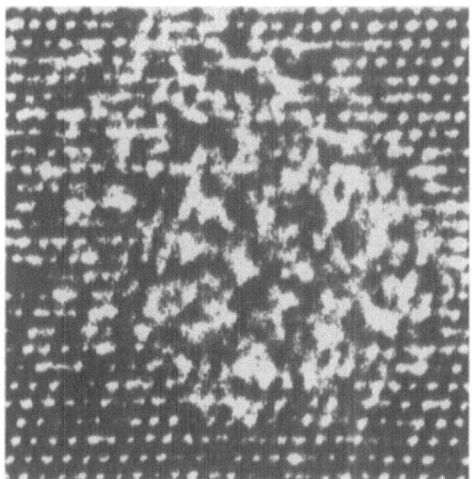

**Figure 3-17** *Displacement spike* due to the impact of a 100 keV bismuth ion onto a silicon monocrystal [1].

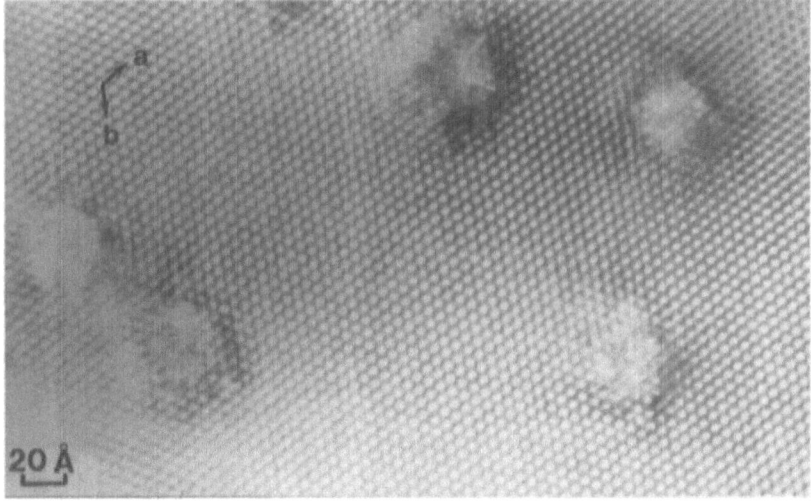

**Figure 3-18** *Latent track cross-sections* of 23 MeV/u xenon ions in $BaFe_{12}O_{19}$ [2].

[1]    Narayan, J, Oak Ridge National Laboratory, P.O. Box X, Oak Ridge, Tennessee, 37830/USA
[2]    Houpert, C., M. Hervieu, D. Groult, F. Studer, M. Toulemonde: "HREM Investigation of GeV Heavy Ion Latent Tracks in Ferrites." Nuclear Instruments and Methods in Physics Research **B32**, 393-396, (1988).

# 3.2    Track halo — electronic defects [1]

While the track core corresponds to a more or less densely populated aggregation of extended defects and has a diameter of about 10 nm, corresponding to the range of the atomic collision-cascade, the track halo corresponds to the radiation damage of the electronic collision-cascade, the diameter of which depends on the ion velocity and may reach about 1 μm. However, most of the electron energy is deposited quite close to the projectile trajectory.

Historically, electronic-damage track structure theory evolved from the embarrassment of the experimentalists who at a given dose (the deposited energy per unit mass) observed a decreased radiation response — or decreased relative biological effectiveness (*RBE*) — for heavier particles, such as neutrons or alpha particles, in comparison with light particles, such as high energy electrons, x rays, or gamma rays. The resulting description of the track halo is a successful synthesis of the features of the local radial dose distribution around the ion trajectory and the global radiation response of the used medium with respect to uniformly ionizing radiation, such as high energy electrons, x rays, or gamma rays.

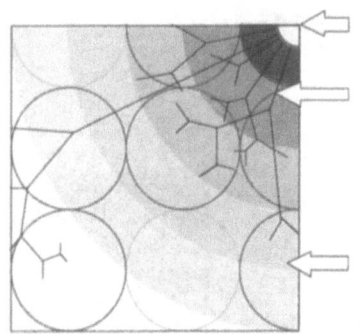

a) Ion trajectory, center of the electron emission.

b) Secondary-electron collision cascade, replaced by a smooth dose-distribution which is obtained by differentiation from an experimental energy-range relation for electrons.

c) Sensitive volume element, "grain", translates the deposited energy-dose into the observed effect.

**Figure 3-19** *Basic steps of electronic radiation damage model.* Quarter cross section through a track with the ion trajectory penetrating through the upper right corner, vertically with respect to the drawing plane. The target electrons are emitted mainly at 90° with respect to the ion trajectory. The available energy of the emitted electrons is spreading via collision cascades and is ultimately absorbed in the grains, indicated as circles. Dark circles are turned developable while light circles stay undevelopable.

The three basic steps used for determination of the electronic radiation damage are shown in Figure 3-19, whereby the emission of electrons is assumed to be at 90° to the ion trajectory. The main parameters of this semi-empiric model [2], [3], [4], are the inner cut-off radius $R_0$, the grain size $a_0$, the energy quantum $\Delta E_{hit}$ required to record a hit,

[1]    this section owes much to the work and fruitful criticism of J.W. Hansen, Forsoegsanlaeg Risoe, Postboks 49, DK-4000 Roskilde, Danmark.

[2]    Katz, R., S.C. Sharma, M. Homayoonfar: " The Structure of Particle Tracks." Topics of Radiation Dosimetry, supplement **1** (F.H. Attix, Editor), Academic Press, New York, pp. 317-383, (1972).

[3]    Katz, R.: "Track Structure Theory in Radiobiology and in Radiation Detection." Nuclear Track Detection, **2**, no.1, pp. 1-28, (1978).

[4]    Hansen, J.W.: "Experimental Investigation of the Suitability of the Track Structure Theory in Describing the Relative Effectiveness of High-LET Irradiation of Physical Radiation Detectors." Risø National Laboratory, DK-4000 Roskilde, Denmark; Risø-R-507, (1984).

and the "hittedness", defined as the minimum number $m$ of hits required to induce the desired radiation effect transforming the grain from an undevelopable state into a developable state [1].

The succeeding calculation of the electronic radiation damage and its observed effect involves in detail the following steps, whereby all calculations are performed per unit path length.

1.  Calculating the number of electrons emitted with a given energy $T$ per unit length of the projectile-ion trajectory.

2.  Calculating the number of electrons emitted with a given angle $\psi$ per unit length of the projectile-ion trajectory.

3.  Assuming that all electrons are emitted at 90° with respect to the ion path.

4.  Imposing an empiric energy-range relation $R(E)$ for electrons.

5.  Determining from this the radial energy-loss function $dT / dr$ for electrons.

6.  Calculating the radial dose distribution $D(r)$ of the emitted electrons.

7.  Calculating the integral energy-deposition $<D>$ per sensitive grain of radius $a_0$ and unit length at distance $r_0$ from the ion path.

8.  Translating the the integral energy-deposition $<D>$ per sensitive grain into an observable effect $W_m$.

### 3.2.1 Electron emission from ion trajectory

**Energy transfer from projectile-ion to target electrons**

For calculating the kinetic energy transferred from the projectile-ion to the much lighter target electrons we use eq. (2.89) in which we replace the projectile charge $q_1$ by $Z_{eff}\, e$, the target charge $q_2$ by the charge of the emitted electron $e$, the reduced mass $m_0 = m_1 m_2 / (m_1 + m_2) \approx m_2$ by the electron mass $m_e$, and obtain:

$$T = 2\frac{Z_{eff}^2 e^4}{m_e v^2}\frac{1}{p^2 + \left(\frac{b}{2}\right)^2} \quad , \text{ where } b \approx \frac{Z_{eff}\, e^2}{\frac{1}{2} m_e v^2} \quad . \tag{3.25}$$

According to this function $T(p)$, the fraction of scattering events within the impact parameter range between $p$ and $p + dp$ corresponds to the fraction of energy-loss events with a transferred energy between $T$ and $T - dT$ (Figure 3-20).

Since we are interested in calculating the differential cross section $d\sigma = 2\pi\, p\, dp$ we solve equation (3.25) for $p(T)$, determine $dp = (dp / dT)\, dT$ and obtain:

$$d\sigma = 2\pi\, p\, dp = -2\pi\frac{Z_{eff}^2 e^4}{m_e v^2}\frac{dT}{T^2} \quad . \tag{3.26}$$

---

[1]  The inner cut-off radius $R_0$ is considered as a constant and the grain size $a_0$ can be determined from the radiosensitivity with respect to uniformly ionizing radiation — such as gamma radiation — using target theory.

Since the differential cross section $d\sigma$ corresponds to the fraction of energy transfers occurring within the energy interval $(T, dT)$, we obtain the number $dn$ of electrons, emitted within the energy interval $(T, dT)$ and within the total solid angle per unit length of the ion trajectory, by multiplying $d\sigma$ with the number of electrons per volume element, $N_e = N Z_t = [\rho_t / (A_t u)] Z_t$, where $N$ is the number-density of the target atoms.

$$dn = d\sigma N_e = -2\pi N_e \frac{Z_{eff}^2 e^4}{m_e v^2} \frac{dT}{T^2} \; . \tag{3.27}$$

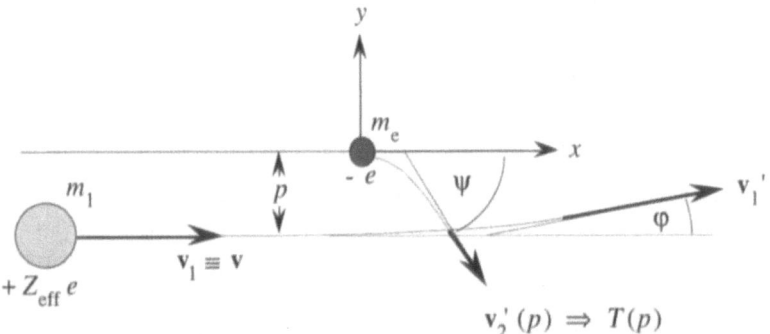

**Figure 3-20** *Electron emission from ion path* as observed in the laboratory coordinate system. With increasing impact parameter $p$ the transferred energy $T$ decreases, according to eq. (3.25), and the emission angle $\psi$ increases, according to eq. (3.28). At the same time the number $dn$ of emitted electrons increases according to eq. (3.29).

**Emission angle of the ejected electrons**

Since the radial range of the emitted electrons depends on the emission angle $\psi$ of the electrons in the laboratory system, we have to determine the emission angle $\psi$ as function of the energy transfer $T$. According to eq. (2.87) we have a relation connecting the energy transfer $T$ with the c.m. angle $\vartheta$ and according to eq. (2.68) a relation connecting the c.m. angle $\vartheta$ with the laboratory angle $\psi$. This yields $T = T[\vartheta(\psi)]$:

$$T = 2 m_e v^2 \sin^2\left(\frac{\vartheta}{2}\right) = 2 m_e v^2 \sin^2\left(\frac{\pi}{2} - \psi\right) = 2 m_e v^2 \cos^2 \psi \; . \tag{3.28}$$

Differentiation with respect to $\psi$ yields $dT$. Inserting $T(\psi)$ from eq. (3.28) and $dT$, obtained by differentiation of the same eq. (3.28) with respect to $\psi$ into eq. (3.27) yields the number $dn$ of emitted electrons as function of the laboratory angle $\psi$, which is Rutherford's representation of the energy-transfer formula of eq. (3.27):

$$dn = d\sigma N_e = 2\pi N_e \left(\frac{Z_{eff} e^2}{m_e v^2}\right)^2 \frac{\sin \psi}{\cos^3 \psi} d\psi \; . \tag{3.29}$$

This function increases rapidly with increasing angle $\psi$ and has a singularity for $\psi = 90°$. Therefore the overwhelming majority of the electrons are emitted at 90° with respect to the projectile-ion trajectory. But, on the other hand, their transferred energy decreases rapidly toward 90°, according to eq. (3.28). However, the deposited energy as function of the angle corresponds to the number d$n$ of emitted electrons, according to eq. (3.29), multiplied by their corresponding energy, according to eq. (3.28). The product-function, which reflects the deposited energy as function of angle, increases as $\sin \psi / \cos \psi$ and has a singularity for $\psi = 90°$. This means that most of the energy-transfer occurs at 90° with respect to the projectile-ion path. In the following we assume therefore for simplicity, that all electrons are exclusively emitted at 90° with respect to the projectile-ion trajectory.

## 3.2.2    Secondary-electron collision-cascade

Another important simplification of the model concerns the treatment of the secondary-electron collision-cascade [1], [2]. It is replaced by a smooth radial energy distribution, which is derived from a very rough energy-range relation of electrons transmitted through a thin foil.

### Energy-range relation for electrons

The range $R$ of electrons of uniform energy $T$ in a medium of density $\rho$ is defined as the maximum thickness of a target for which transmitted electrons can still be observed. Due to the build-up of a secondary-electron collision-cascade — due to multiple scattering — it is not possible to speak about the range of one single electron, in contrast to the range of one single heavy ion which is rather well-defined. The electron range $R$ follows approximately the empiric law:

$$R(T) = 5.2 \bullet 10^{-4} \frac{T^{1.67}}{\rho} \equiv C\, T^{\alpha}, \text{ where } C = \frac{5.2 \bullet 10^{-4}}{\rho}, \alpha = 1.67 . \quad (3.30)$$

where the range $R$ is given in nm, the electron energy $T$ is given in eV, and the density $\rho$ is given in g/cm³ (Figure 3-21).

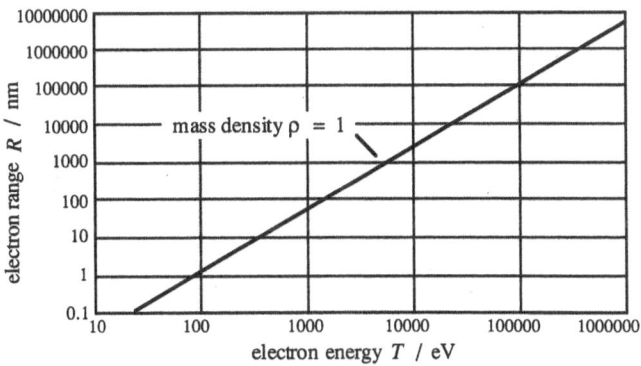

**Figure 3-21**   Rough energy-range relation for electrons in solid targets of density $\rho = 1$.

[1]    Schou, J.: "Transport Theory for Kinetic Emission of Secondary Electrons from Solids." Phys. Rev. B, **22**, No. 5, pp. 2141-2174, (1980).
[2]    Schou, J.: "Secondary Electron Emission from Solids by Electron and Proton Bombardment." Scanning Microscopy, **2**, pp. 607-632, (1988).

This very simple range-energy relation is approximately correct for electron energies above 1000 eV and becomes increasingly incorrect for electron energies below 1000 eV indicating one important limitation of the model and its potential improvement.

### Radial energy-distribution for electrons

According to the radial electronic damage model described here, the radial dose-distribution is determined roughly by a differentiation procedure from the energy-range relation, whereby averaged energies are considered. The procedure is as follows.

**Virtual radial energy-deposition for one electron.** It is possible to determine an average electron-energy $<T>$ as function of the target thickness $x$ or — adapted to our cylindrically symmetric problem — as function of the travelled distance $r$ from the projectile-ion trajectory, by multiplying the electron energy distribution at thickness $x$, $p(x, T)$, with their energy $T$ and integrating over all energies $T$ from zero to infinity:

$$<T(x)> = \frac{\int_0^\infty T p(x, T) \, dT}{\int_0^\infty p(x, T) \, dT} \quad ,$$

(3.31)

which is a function of the travelled distance $x$ and decreases monotonously with increasing $r$ and reaches zero for $r = R$ where $R$ is the maximum range of the electrons. The procedure for determining the average energy $<T>$ and the maximum range $R$ is symbolically represented in Figure 3-22.

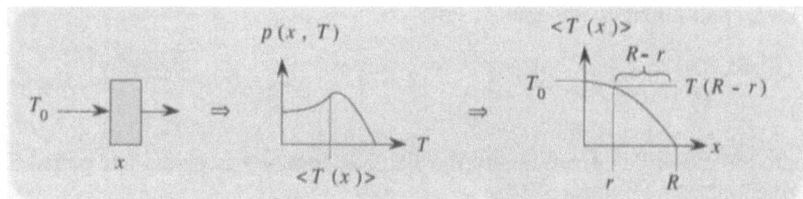

**Figure 3-22** Determination of average energy $<T(x)>$ and "virtual energy" $T(R - r)$.

In the following we are assuming that the range of the resulting electron-energy distribution at $x = r$ is identical to the range of a monoenergetic beam of electrons possessing virtually the discrete energy $T = <T(R - r)>$. Under this rather daring assumption that

$$T(x) \equiv <T(x)> \quad ,$$

(3.32)

which is only justified by the success of the model [1], [2] — we can invert the function $R(T)$ into the function $T(R)$ and omit the averaging further on. We obtain from eq. (3.30) by inversion:

$$T(R) = \left(\frac{R}{C}\right)^{\frac{1}{\alpha}} = 92.6 \, (\rho R)^{0.6} \quad .$$

(3.33)

[1]    Hansen, J.W., K.J. Olsen: "Experimental and Calculated Response of a Radiochromic Dye Film Dosimeter to High-LET Radiations." Radiation Research, **97**, pp. 1-15, (1984).
[2]    Hansen, J.W. K.J. Olsen:" Theoretical and Experimental Radiation Effectiveness of the Free Radical Dosimeter Alanine to Irradiation with Heavy Charged Particles. Radiation Research, **104**, pp. 15-27, (1985).

Electrons starting with energy $T$ from the axis (the projectile-ion trajectory) have the range $R(T)$ and will thus have the virtual energy $T(R - r)$ at distance $r$ from the projectile-ion trajectory (Figure 3-22).

$$T(R-r) = \left(\frac{R-r}{C}\right)^{\frac{1}{\alpha}} = \left(\frac{R}{C}\right)^{\frac{1}{\alpha}} \left(1-\frac{r}{R}\right)^{\frac{1}{\alpha}} = T(R)\left(1-\frac{r}{R}\right)^{\frac{1}{\alpha}}. \quad (3.34)$$

The deposited energy of the electrons emitted from the axis per unit path-length corresponds to the derivation of $T(R - r)$ with respect to the traversed distance $r$:

$$\frac{dT}{dr} = -T(R)\frac{1}{\alpha R}\left(1-\frac{r}{R}\right)^{\left(\frac{1}{\alpha}-1\right)}. \quad (3.35)$$

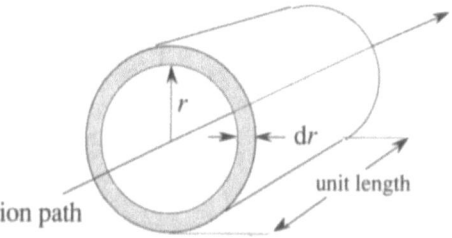

**Figure 3-23** Calculation of deposited energy within a cylindrical shell.

**Average dose-distribution of the emitted electrons.** For determining the deposited energy within a cylindrical shell of distance $r$ from the projectile-ion path (the axis) (Figure 3-23) — for all electrons emitted with kinetic energies $T_{min} \leq T \leq T_{max}$ — we have to multiply the fraction of electrons $dn$ emitted with kinetic energies in the interval $(T, T + dT)$, given according to eq. (3.27) with the energy $dT$, deposited within the radial interval $(r, dr)$ according to eq. (3.35) and integrate over all original energies $T$:

$$dE=\int_{T_{min}}^{T_{max}} dn(T)\ dT = 2\pi N_e \frac{Z_{eff}^2 e^4}{m_e v^2}\left(\int_{T_{min}}^{T_{max}} \frac{dT}{T^2}\frac{T}{\alpha R}\left(1-\frac{r}{R}\right)^{\left(\frac{1}{\alpha}-1\right)}\right)dr. \quad (3.36)$$

For distinction we have introduced here the term $dE$, corresponding to the total deposited energy within the cylindrical shell $(r, r + dr)$. In the integral, $T_{min}$ corresponds to the minimum energy transfer due to binding effects, and $T_{max} = 2\,m_e\,v^2$ corresponds to the maximum energy transfer in central collisions which is given by eq. (3.25) for zero impact parameter $p$. If we replace the energy $T$ in eq. (3.36) by $T(R) = (R/C)^{1/\alpha}$, according to eq. (3.33) we obtain:

$$dE = 2\pi N_e \frac{Z_{eff}^2 e^4}{m_e v^2}\frac{dr}{\alpha}\int_{R_{min}}^{R_{max}}\frac{1}{\alpha}\frac{dR}{R^2}\left(1-\frac{r}{R}\right)^{\left(\frac{1}{\alpha}-1\right)}, \quad \text{or} \quad (3.37)$$

$$dE = dE(Z_{eff}, v, r) = 2\pi N_e \frac{Z_{eff}^2 e^4}{m_e v^2}\frac{dr}{\alpha}\frac{1}{r}\left[\left(1-\frac{r}{R}\right)^{-\frac{1}{\alpha}}\right]_{R=R_{min}}^{R=R_{max}}. \quad (3.38)$$

In the integral the minimum range $R_{min} = C\,T_{min}{}^{\alpha}$ corresponds to the minimum energy of the emitted electrons and $R_{max} = C\,T_{max}{}^{\alpha}$ to their maximum energy, according to eq. (3.30). For distances $r > R_{min}$ the second term of the integral, $-[\,1 - r\,/\,R_{min}\,]^{1/\alpha}$, vanishes and is omitted further on due to the unrealistic smallness of $R_{min}$.

For example, if $Z_p = 36$ and $N_e = 3.4 \cdot 10^{23}$ we obtain the following radii $R_{min}$ as function of the specific energy $T_s$ of the projectile-ion: $R_{min}$ (0.1 MeV/u) $\approx$ $1 \cdot 10^{-4}$ nm, $R_{min}$ (1 MeV/u) $\approx$ $3 \cdot 10^{-6}$ nm, $R_{min}$ (10 MeV/u) $\approx 7 \cdot 10^{-8}$ nm, $R_{min}$ (100 MeV/u) $\approx 2 \cdot 10^{-9}$ nm. These radii are much smaller than the atomic dimensions which is an indication that the used, very rough range-energy relation for electrons according to eq. (3.30) gives too small ranges for small electron energies.

**Deposited dose.** The deposited dose $D$ within a cylindrical shell around the ion path of unit length at distance $r$ and of thickness $dr$ and outside of $R_{min}$ (Figure 3-23) is obtained by dividing $dE$ by the volume of this cylindrical shell:

$$D(Z_{eff}, v, r) = \frac{dE(Z_{eff}, v, r)}{2\pi\,r\,dr} = N_e \frac{Z_{eff}^2\,e^4}{m_e\,v^2}\,\frac{1}{\alpha\,r^2}\left(1 - \frac{r}{R_{max}}\right)^{\frac{1}{\alpha}}. \qquad (3.39)$$

For small radii the dose $D$ decays as $1\,/\,r^2$ and has a cut-off radius for $r = R_{max}$. This radial dose-distribution $D$ is represented in Figure 3-24 [1].

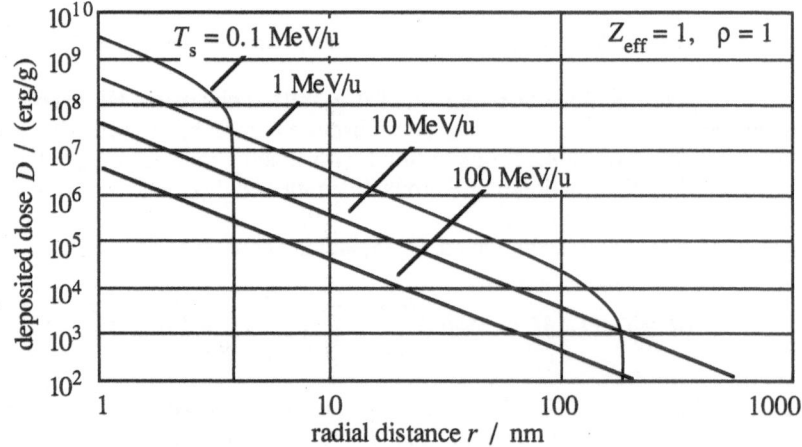

**Figure 3-24** *Radial dose distribution.* Deposited dose $D$ as function of distance $r$ from ion path for different specific energies $T_s$ of the ion, for $Z_{eff} = 1$, $\rho = 1$, $N_e = 2 \cdot 10^{23}$.

## 3.2.3    Translation of deposited energy into effect

The simplest assumption, that the observed effect is proportional to the deposited energy, does not reflect the observed behavior of track registration in general, as for example the phenomenon of a track-etch threshold, the development of silver grains in photographic emulsions, or the radiation effects occurring in physico-chemical media or in living cells [2]. A subdivision of the registration medium into sensitive volume elements

---

[1]    The SI unit of dose is gray (Gy). 1 Gy = 1 Joule / kg = $10^4$ erg / g.
[2]    Non-linearity is observed for any medium and any radiation when saturation is approached.

or grains and a non-linear translation of the average dose received by the grains into an observable effect is necessary to describe the experiments.

### Grain size

The described electronic radiation-damage model is originally [1] tailored to describe the registration of ionizing particles in photographic emulsions. Such a medium contains sensitive volumes or "grains" which have exactly two modes of response. They either are rendered developable or they stay undevelopable. We therefore assume that the recorded effect of the passage of an ion through the solid is due to the binary response of grains which are either becoming developable or nondeveolpable. The size of the grains thereby is used as fitting parameter to tailor the observed spatial resolution (Figure 3-25). Alternatively, the grain size can be determined from the experimental radiosensitivity of the medium with respect to uniformly ionizing radiation — such as gamma radiation — using target theory [2].

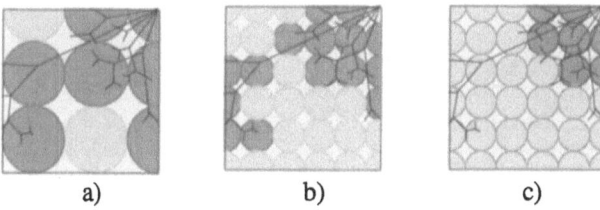

a)                    b)                    c)

**Figure 3-25** *Grain size, grain sensitivity, and resolution* of track recorders. Quarter cross sections through a track, as "seen" by different track recording media. The ion trajectory penetrates through the upper right corners of the picture frames, vertically with respect to the drawing plane. The available energy of the emitted electrons is spreading radially away from the ion path via collision cascades and is ultimately absorbed in the grains, indicated as circles. Only the dark circles are turned developable, whereas the light circles stay undevelopable. (**a**) "Large" grains are marginally adequate for observing the assumed electronic cascade. (**b**) "Small" grains are capable to improve the local resolution of the observation. (**c**) Decreasing the sensitivity of the grains reduces the overall size of the track, leading to an improved spatial definition of the track core by suppressing the fine details in the boundary zone of the electronic collision cascade. This suppression of detailed information by desensitization is a necessary precondition for high-resolution structural applications.

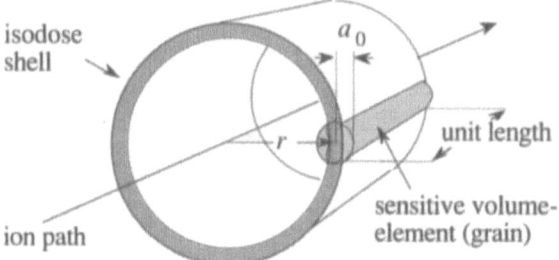

**Figure 3-26** Geometry for calculation of average grain-dose.

[1]    Katz, R., S.C. Sharma, M. Homayoonfar: "The Structure of Particle Tracks." Topics in Radiation Dosimetry, Supplement 1 (F.H. Attix, editor), Academic Press, New York, pp. 317-383, (1972).
[2]    Olsen, K.J., J.W. Hansen: "Experimental Data from Irradiation of Physical Detectors Disclose Weaknesses in Basic Assumptions of the δ Ray Theory of Track Structure." Radiation Protection Dosimetry, **13**, No. 1-4, pp. 219-222, (1985).

### Average grain-dose

We assume for simplicity sensitive elements or "grains" in the form of cylinders of unit length and radius $a_0$, aligned in the direction of the ion path, according to Figure 3-26. This choice of shape and orientation is only important close to the ion path.

For calculating the grain response for a grain located in the strongly varying radiation field in the vicintity of the ion path, an average grain-dose is calculated. The calculation is based on the geometry defined in Figure 3-27 in which the center of the grain is at the distance $r_0$ from the ion path or track center.

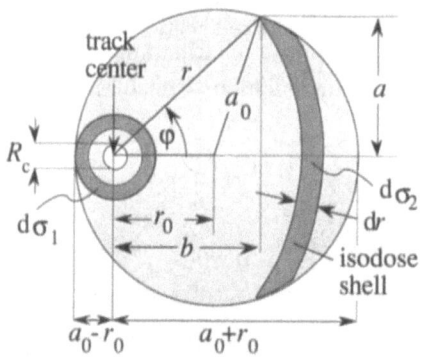

**Figure 3-27** *Calculation of average grain dose* [1]. If the ion trajectory passes through the grain, two types of iso-dose shells exist, closed cylindrical shells of volume $d\sigma_1$, and sections of cylindrical shells with the grain boundary of volume $d\sigma_2$.

The average grain-dose $<D>$ for a grain of radius $a_0$ located at distance $r_0$ from the track center is obtained by integrating the local dose over all volume elements $d\sigma$ within the grain and dividing by the grain volume $\pi a_0^2$:

$$<D> = <D(Z_{eff}, v, a_0, r_0)> = \frac{1}{\pi a_0^2} \int_{grain} D(Z_{eff}, v, r) \, d\sigma . \tag{3.40}$$

The area (or volume) $d\sigma$ depends on whether the corresponding iso-dose shell is contained completely within the grain or whether it is only partially contained within the grain, respectively:

$$d\sigma_1 = 2\pi r \, dr \quad \text{for } R_c < r \le |a_0 - r_0| , \quad \text{or} \tag{3.41}$$

$$d\sigma_2 = 2\varphi r \, dr \quad \text{for } |a_0 - r_0| < r \le a_0 + r_0 , \quad \text{respectively} \tag{3.42}$$

whereby $R_c$ is an inner cut-off radius which takes care that the integral does not diverge and the angle $\varphi$ is defined by the geometrical relation:

$$\tan^2 \varphi \equiv \frac{\sin^2 \varphi}{\cos^2 \varphi} = \frac{a^2}{b^2} = \frac{a_0^2 - (r \cos \varphi - r_0)^2}{r^2 \cos^2 \varphi} . \tag{3.43}$$

[1]     Hansen, J.W.: "Experimental Investigation of the Suitability of the Track Structure Theory in Describing the Relative Effectiveness of High-LET Irradiation of Physical Radiation Detectors." Risø National Laboratory, DK-4000 Roskilde, Denmark; Risø-R-507, (1984).

Applying the identity $\sin^2 \varphi + \cos^2 \varphi \equiv 1$ and solving eq. (3.43) for $\varphi$, we obtain:

$$\varphi = \arccos\left(\frac{r^2 + r_0^2 - a_0^2}{2\,r\,r_0}\right), \quad \text{for } r_0 + r > a_0 . \tag{3.44}$$

Inserting the dose $D(Z_{eff}, v, r)$ of eq. (3.39), into eq. (3.40), we obtain the average grain-dose (Figure 3-28):

$$<D> = \frac{1}{\pi a_0^2} N_e \frac{Z_{eff}^2 e^4}{m_e v^2} \frac{1}{\alpha}\left[\int_{R_c}^{a_0 - r_0}\left(1 - \frac{r}{R_{max}}\right)^{\frac{1}{\alpha}}\frac{d\sigma_1}{r^2} + \int_{a_0 - r_0}^{a_0 + r_0}\left(1 - \frac{r}{R_{max}}\right)^{\frac{1}{\alpha}}\frac{d\sigma_2}{r^2}\right] \tag{3.45}$$

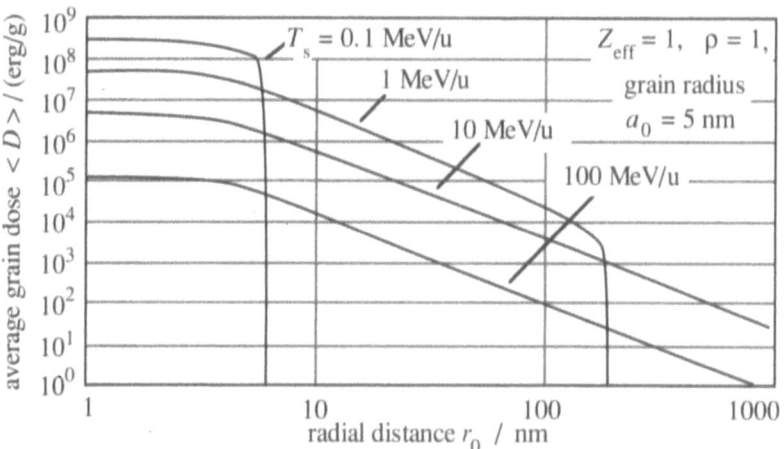

**Figure 3-28** *Radial distribution of grain-dose.* Average grain-dose as function of distance $r_0$ between ion path and grain center for different specific energies $T_s$ of the projectile-ion, for $Z_{eff} = 1$, $\rho = 1$, $N_e = 2 \cdot 10^{23}$, $a_0 = 5$ nm. The cut-off radius $R_c$ has been arbitrarily set to $R_c = 1$ nm.

According to eq. (3.45), the average grain-dose $<D> = <D(Z_{eff}, v, a_0, r_0)$ depends on the effective charge $Z_{eff}$ and velocity $v$ of the projectile-ion, on the grain size $a_0$, and on the grain distance $r_0$ from the ion path. In addition there exists the dependence on a rather arbitrary cut-off radius $R_c \gg R_{min}$, introduced to avoid the divergence of the first integral. In principle the cut-off radius $R_c$ should be chosen such, that the total energy deposited outside $R_c$ per unit path-length corresponds to the energy-loss $(dE / dx)$ of the projectile-ion [1]. For $Z_p = 36$, $N_e = 3.4 \cdot 10^{23}$ we obtain the following cut-off radii $R_c$ as function of the specific energy $T_s$ of the projectile-ion: $R_c$ (0.1 MeV/u) $\approx$ 6·10$^{-4}$ nm, $R_c$ (1 MeV/u) $\approx$ 5·10$^{-3}$ nm, $R_c$ (10 MeV/u) $\approx$ 1·10$^{-1}$ nm, $R_c$ (100 MeV/u) $\approx$ 3 nm. However, due to the chosen range-energy relation of electrons according to eq. (3.30) and due to the assumed emission of all electrons vertically to the projectile-ion trajectory, this

---

[1]    Another possibility is to use a fixed cut-off radius $R_c$, calculate the total energy deposited beyond $R_c$ by integration of eq. (3.45), and assign the remaining fraction of the energy-loss $(dE / dx)$ to radii smaller than $R_c$.

requirement obviously overstresses the model and gives much larger radii than $R_{min}$.
The average grain-dose for grains of radius $a_0 = 5$ nm and an arbitrary chosen cut-off
radius $R_c = 1$ nm is shown in Figure 3-28 as function of the distance $r_0$ of the grain
center from the track axis for different specific energies $T_s$. In comparison with
Figure 3-24, describing the dose $D$ the averaging procedure yields a $<D>$ which is flat-
tened in the innermost part of the distribution.

### Grain-response and target theory

The simplest assumption for describing the electronic radiation-effect is that
the grains respond proportional to the deposited dose. A linear response-function, however, does often not suffice to describe the radial radiation effect around an ion track. For
example, in the development of ion tracks one encounters a threshold phenomenon. Only
ion tracks with a sufficiently high energy-deposition density or energy loss ($dE / dx$) can
be developed. Tracks with insufficient energy-deposition density remain undeveloped.

In order to superpose this feature of nonlinearity to the electronic radiation-effect model, we use the model of target theory, in which the sensitive volume elements or
grains become either completely developable or stay completely undevelopable, depending
on the number of energy quanta or hits which are deposited within the grains.

More precisely,  according to target theory, a grain will become developable if
it obtains at least $m$ energy quanta $\Delta E_{hit}$ or "hits" and will stay undeveloped if there arrive
less than $m$ energy quanta or hits within the grain, shown symbolically in Figure 3-29.

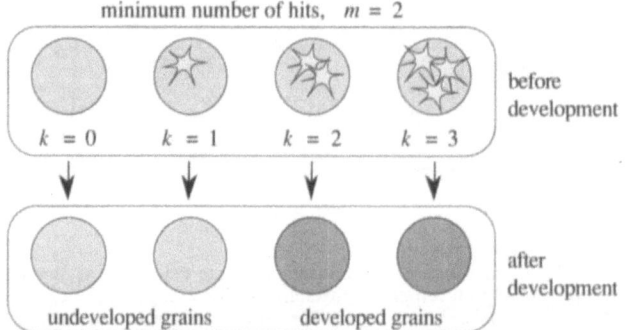

**Figure 3-29** *Basic assumption of grain response*: A grain will become developable if it is
hit by at least $m$ energy quanta $\Delta E_{hit}$, here shown for the case of $m = 2$.

The number $N$ of defect quanta per unit volume is given by the deposited energy per unit mass, the dose $<D>$, multiplied by the mass per unit volume, the density $\rho$,
divided by the size $\Delta E_{hit}$ of the energy quanta required for the accumulation of an effect:

$$N = \frac{<D> \rho}{\Delta E_{hit}} \tag{3.46}$$

The factor $\rho$ thereby just transforms the number of defect quanta per unit mass into the
number of defect quanta per unit volume.

The average number $P$ of energy quanta $\Delta E_{hit}$ deposited per grain corresponds to the number $N$ of energy quanta accumulated within the unit volume, multiplied
with the volume $V_{grain}$ of the grain:

$$P = N V_{grain} = \frac{<D> \rho}{\Delta E_{hit}} V_{grain} = \frac{<D> \rho}{\Delta E_{hit}} \frac{4\pi}{3} a_0^3 , \qquad (3.47)$$

While for a very large number of hits per unit volume, the average number of hits per grain, $P$, reflects the actual number $k$ of arriving hits in a specific grain with high precision, on the other extreme, fluctuations become dominant at small hit densities.

As shown in chapter "6.2 Stochastic track patterns" the probability $w_m$ for the arrival of exactly $m$ hits per grain is given by the Poisson formula:

$$w_m \equiv w_{k=m} = \frac{P^m}{m!} e^{-P} , \qquad (3.48)$$

where the index "$k = m$" indicates that the number of hits $k$ is exactly equal to the chosen number $m$ (Figure 3-30).

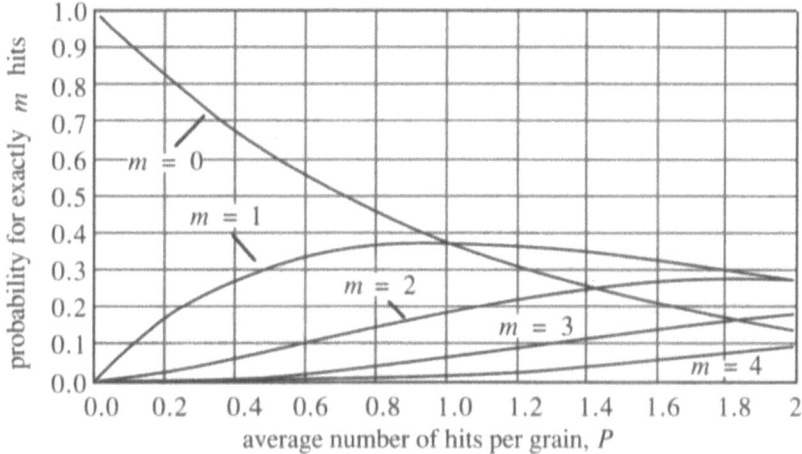

**Figure 3-30** Probability $w_m$ for the arrival of exactly $m$ hits in the selected grain as function of the average number $P$ of hits per grain.

The probability for the arrival of exactly $m$ hits is not sufficient to describe the grain response correctly since with increasing hit probability $P$ the number of exactly $m$ hits will decrease for small $m$. Therefore we need the cumulative probability $W_m$ to describe the arrival of at least $m$ hits, which increases monotonously with the deposited dose and is given by subtracting from all events the number of events with less than $m$ hits:

$$W_m \equiv w_{k \geq m} = 1 - \sum_{k=0}^{m-1} w_k , \qquad (3.49)$$

where the index "$k \geq m$" indicates that the number of arriving hits $k$ is at least equal or larger than the chosen number $m$ (Figure 3-31).

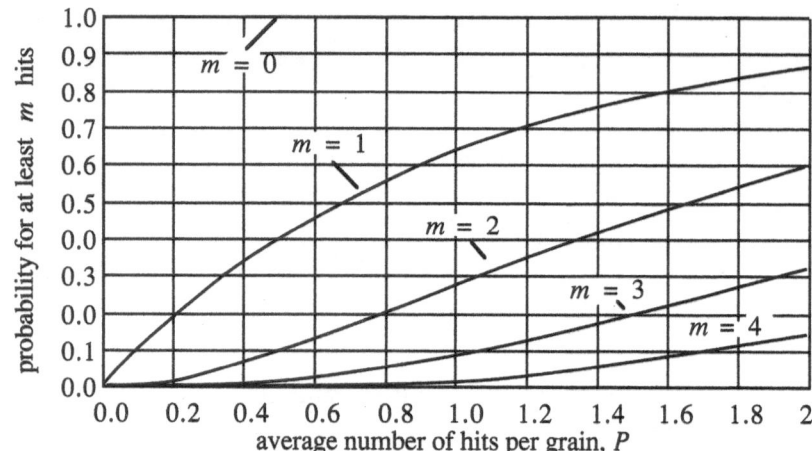

**Figure 3-31** Cumulative probability $W_m$ for the arrival of at least $m$ hits per grain as function of the average number $P$ of hits per grain.

### Radial effect-distribution

On the basis of the assumption, that the recorded effect of the passage of an ion through the solid is due to the binary response of grains which are either becoming developable if the number of hits exceeds a given number $m$ or stay quiescent if the number of hits is below this given number $m$, it is possible to fine-tune the response function of track-recording materials with high accuracy [1].

The average grain-response for grains of radius $a_0 = 5$ nm and an arbitrary chosen cut-off radius $R_c = 1$ nm is shown in Figure 3-32 as function of the distance $r_0$ of the grain center from the track axis for four different specific energies $T_s$ and two different minimum numbers $m$ of hits per grain required to render the grain developable. For this purpose, the average number $P$ of energy quanta $\Delta E_{hit}$ per volume element, according to eq. (3.47), is inserted into the probability $W_m$ for the arrival of at least $m$ energy quanta $\Delta E_{hit}$, according to eq. (3.49) and drawn as function of the distance between the grains and the projectile-ion trajectory. As quantum energy for the registration of the absorption-event as a hit $\Delta E_{hit} = 1$ eV was chosen.

We recognize from Figure 3-32 that for $T_s = 0.1$ MeV/u the grains are fully developable up to distances $r_0 = 7$ nm, for $m = 1$ and $m = 2$. This radius corresponds approximately to the maximum range $R_{max} = 4$ nm, folded with the grain radius $a_0 = 5$ nm. For $T_s = 1$ MeV/u the fraction of developable grains decreases somewhat smoother with increasing distance from the projectile trajectory. The slope becomes steeper with increasing minimum number $m$ of required hits. For $T_s = 10$ MeV/u only 45 % of the grains are developable at $r_0 = 1$ nm if $m = 1$ and only about 20% of the grains if $m = 2$.

[1]    Katz, R.: "Track Structure Theory in Radiobiology and in Radiation Detection." Nuclear Track Detection, **2**, no.1, pp. 1-28, (1978).

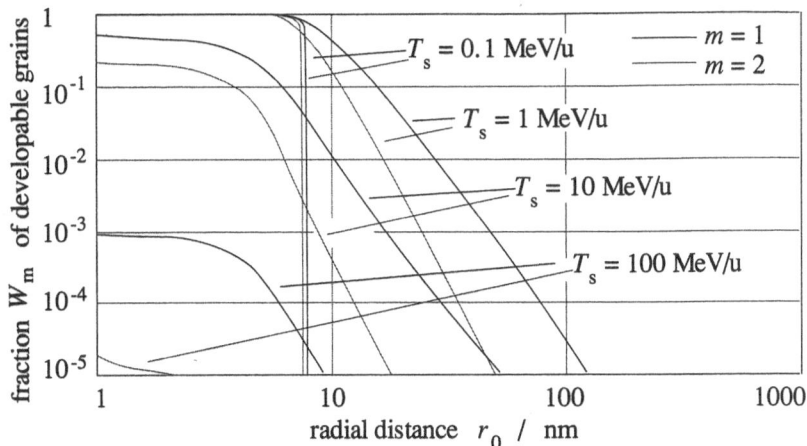

**Figure 3-32** *Radial grain-response* as function of distance $r_0$ between ion path and grain center for different specific energies $T_s$, for $Z_{eff} = 1$, $\rho = 1$, $N_e = 2 \cdot 10^{23}$, $a_0 = 5$ nm, $\Delta E_{hit} = 1$ eV. The cut-off radius $R_c$ has been arbitrarily set to $R_c = 1$ nm. The minimum number of hits required for grain development is $m$.

As obvious from Figure 3-32, the track radius for constant hittedness $m$ depends on the specific energy of the ion. If we follow a high energy ion on its path through the solid, the track radius $r$ — more precisely the radius corresponding to the zone of developable grains — increases along the ion path, reaches a maximum around — but not at — the Bragg peak and decreases rapidly towards the end of the track. Therefore, the same track radius $r$ is observed for two different specific energies — in general corresponding to two different values of the energy-loss function ($dE / dx$) — . Therefore the corresponding cross section $\sigma = \pi\, r^2$ is a double-valued function close to the Bragg peak, a phenomenon which is explained more detailed in the following section.

### Track cross section [1]

The activation or inactivation cross section $\sigma$ — in other words the probability for producing developable grains — depends strongly on the grain size and on the track structure only in that range where the density of energy deposition is greater than necessary to activate the grains. In the range of low energy-density, only the overall dose deposited in the grains is important. In this range of low values of linear energy transfer (LET), the observed effect increases with increasing energy deposition. — in other words the cross section curve versus LET forms a single valued linear function.

In the range of high energy deposition more energy than necessary to produce an effect can be transferred to the grain. This is called "overkill" effect. If the track would not have a finite diameter, the cross section curve would plateau at the size of the geometrical cross section $\pi\, a_0^2$ of the grain and stay constant even if the LET increases further. However, when the diameter of the track is comparable to the size of the grains, the grains can be activated by glancing collisions, in other words by collisions where only an outer part of the track mediated by the $\delta$ electrons hits the grains. Therefore, the activation cross section exceeds the geometrical cross section of the grains.

[1]    this section is due to G. Kraft, Gesellschaft für Schwerionenforschung (GSI) Darmstadt, Planckstr. 1, D-6100 Darmstadt.

At the end of the particle range, the track thins down. Consequently the δ ray contribution is diminished and the cross section is rapidly decreasing, according to a track diameter which is proportional to the velocity of the primary particles.

In Figure 3-33 the inactivation cross section for different grain sizes — different geometrical cross sections — are compared. For large grain sizes, the grains become more sensitive and the influence of the δ electrons — the production of hooks — becomes more pronounced.

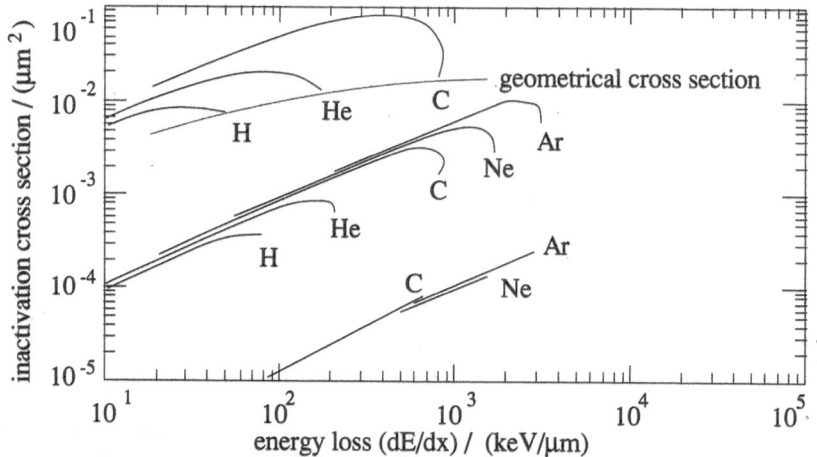

**Figure 3-33** *Inactivation cross section* for "small" grains as given by theory [1]. Assumed were three different grain sizes — around 0.1 μm², around 0.01 μm², and around 0.001 μm² — inactivated by various "light" ions.

The model of track structure has been also applied to radiological objects like living cells because the major contribution for the inactivation of DNA damage comes from the action of the electrons.

In Figure 3-34 the inactivation cross section — the cross section for the induction of critical damage in the DNA molecules — is given as a function of the LET for different cell lines. These cells — bacteria, yeast, and mammalian cells — exhibit different radiosensitivities and have different "grain" sizes — in other words different diameters of the cell nucleus or that areas where DNA is concentrated. According to these differences hooks are formed which are most pronounced for the most sensitive mammalian cells.

However, the track structure model provides only qualitative agreement with the measured data. For a detailed analysis, the biological "fine-structure" of the object has to be taken into account.

[1]    Katz, R., B. Ackerson, U. Homayoonfar, S.C. Sharma. Radiation Research **47**, pp. 402-425, (1971).

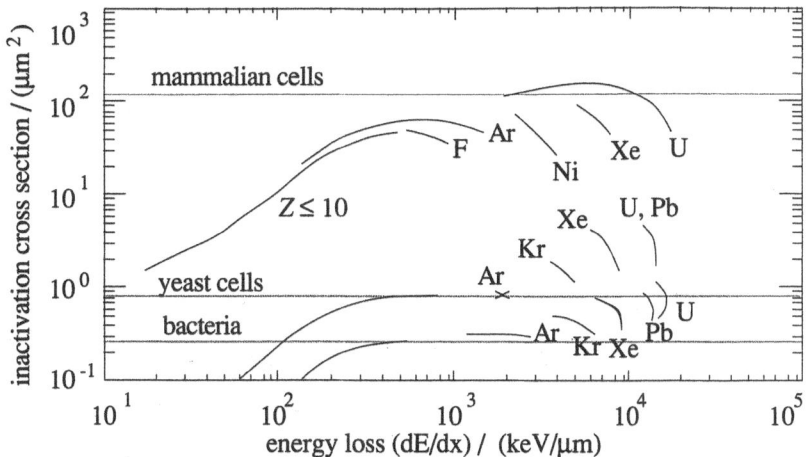

**Figure 3-34** *Inactivation cross section* for "large" grains as determined by experiment [1], [2]. Investigated were three different objects or grain sizes — bacteria around 0.3 $\mu m^2$, yeast cells around 0.8 $\mu m^2$, and mammalian cells around 100 $\mu m^2$ — inactivated by various "heavy" ions.

### Chemical equivalent of radiation damage

The hit energy $\Delta E_{hit}$ corresponds to a physical or chemical change of the molecules forming the grain and depends on the used medium. If a sufficient number $m$ of such energy quanta $\Delta E_{hit}$ is accumulated within the grain it is ultimately transformed into a developable grain.

### Long-term reactions

In the time interval between the defect creation and the ultimate development of the ion track the environmental conditions play an important role. While the process of atomic and electronic collision cascades is terminated about $10^{-12}$ s after the ion passage, the following steps of defect rearrangement, formation of radicals and chemical reactions may range between $10^{-12}$ s up to milliseconds, hours, or days. Most influential during this long time range is the storing temperature, and in the case of organic polymers the presence or absence of oxygen and water vapor.

[1]   Kraft, G.: "Radiobiological Effects of Very Heavy Ions: Inactivation, Inductions of Chromosome Aberrations and Strand Breaks." Nuclear Science Applications, **3**, pp. 1-28, (1987).
[2]   Kraft, G.: "Effects of LET, Fluence and Particle Energy on Inactivation, Chromosomal Aberrations and DNA Strand Breaks" in **Terrestrial Space Radiation and its Biological Effects**. P. D. McCormack, C. E. Swenberg, H. Bücker (editors), Plenum Press, pp. 163-184, (1988)

# 4 Development of ion tracks

During track development the damaged zone of the latent track is either transformed chemically or removed by an etchant, leading to the observable track, the end-product of all the preceding steps and ultimate goal of a structuring tool.

Track development corresponds to a phase transformation which starts preferentially along the latent track which serves as a sort of "condensation nucleus". There exist principally two different techniques for track development:

1.  Internal phase-transformations, characterized by short-range diffusion processes, in which the new material already exists within the bulk of the track recorder and transport distances are of the order of micrometers or less. Examples for internal phase transformations are the nucleation of a metallic phase in the silver halogen track-recorder [1], [2], [3], the development of ion tracks in photographic emulsions [4], [5] pre-soaked with the developer, and the condensation phenomena in cloud and bubble chambers in which the phase transformation of a supersaturated gas or liquid is triggered by ions.

2.  External phase transformations, characterized by long-range diffusion paths, in which the new material is transported over large distances to the condensation nucleus from outside the track recorder (often many micrometers). Examples for external phase transformations are track etching and development processes in which the track-recording material is immersed into an extraneous gaseous or liquid medium to reveal the ion tracks.

## 4.1 Nucleation of a new phase

Ion tracks can be conceived as a stochastically distributed, quasi-linear array of developable grains or droplets of a new phase which, at low damage densities, are more or less discretely distributed along the projectile-ion path and, at high radiation damage densities, form a contingent activated zone surrounded by the virgin material of the track recorder (Figure 4-1).

Classical examples for phase transformations are the condensation of water from air supersaturated with water vapor, such as in Wilson's cloud chamber, and the boiling of superheated liquid hydrogen, such as in Glaser's bubble chamber.

[1]     Slifkin: "The Photographic Latent Image." Physics Bulletin, **39**, No. 7, pp. 274-277, (1988).
[2]     Haase, G. E. Schopper, F. Granzer: "Solid State Nuclear Track Detectors. Track Forming, Stabilizing and Development Processes." Radiation Effects, **34**, p 25, (1977).
[3]     Haase, G., F. Zörgiebel, E. Schopper, F. Granzer, G. Henig, J.U. Schott, F. Wendnagel: "Silver Halide Crystals as Particle Track Detectors." Photog. Sci. Eng. **17**, 409, (1973).
[4]     James, T.H. (editor): "**The Theory of the Photographic Process.**" 4th edition, MacMillan Publ. Co. Inc., New York, 714 pp. (1977)
[5]     Granzer, F., E. Moisar: "Der photographische Elementarprozess in Silberhalogeniden." Physik in unserer Zeit, **12.2**, p. 36, (1981).

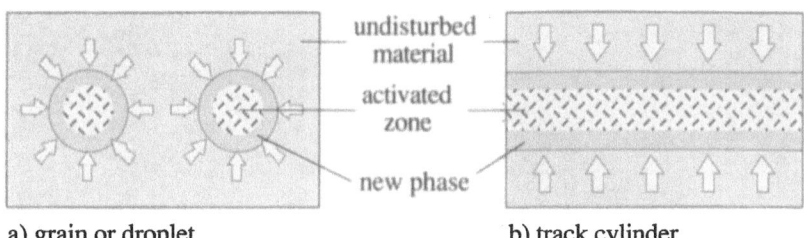

a) grain or droplet                                    b) track cylinder

**Figure 4-1** *Internal phase transformation*, triggered (**a**) by a discrete condensation nucleus, droplet or grain and (**b**) by a densely populated activated zone in the form of a cylinder centered about the projectile-ion path.

## 4.1.1   Origin of phases

### Stability of a planar interface

A phase is a homogeneous region of a system separated from the other parts of the system by an observable interface. The phase is stable if the formation of new interface area requires the input of energy

$$\Delta E = \gamma \Delta A > 0 \ , \tag{4.1}$$

where $\gamma$ is the interface energy per unit area and $\Delta A$ is the increase of interface area. For example, the deformation of the planar interface between water and oil increases its area and thus its interface energy and therefore is stable (Figure 4-2).

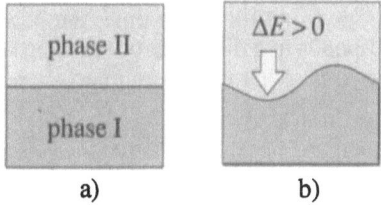

a)                          b)

**Figure 4-2** *Stability condition for a planar interface*. (**a**) Stable state. (**b**) The deformation of a stable planar interface requires the input of interface energy $\Delta E > 0$. If, on the contrary, the interface energy $\Delta E$ increases during such a deformation — which may be caused by some fluctuation — the interface will increase spontaneously and after some time a completely homogeneous state of the system will be reached.

## 4.1.2   Basic theory of interface energy

The mechanism of track development can be best understood as the growth of some activated zone or grain, which is always associated with the creation of new interface area. This problem is intimately related with the stability of grains or droplets during an internal phase transformation.

For obtaining a concept for the interface energy of grains or droplets we observe Figure 4-3 [1].

[1]     Walton, A.J.: "**Three Phases of Matter.**" Clarendon Press, Oxford, pp. 1-182, (1984).

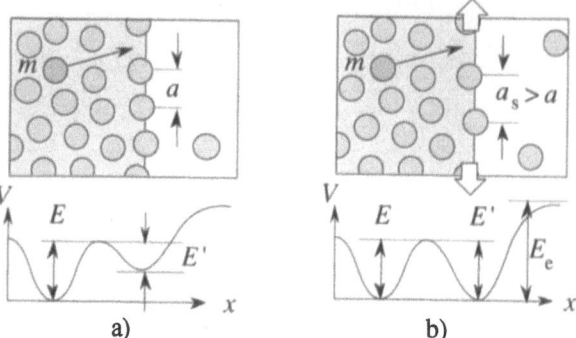

**Figure 4-3** *Origin of interface energy* in the case of a flat liquid/vapor interface. (a) Unrelaxed state with homogeneous density and reduced potential-well depth $E'$ close to the interface. (b) Relaxed state with reduced interface density, increased interface energy, and increased potential-well depth, $E' = E$.

In the unrelaxed state (Figure 4-3 a) the interface contains the same number-density of molecules as the bulk volume of the material and the interatomic distances $a$ are identical. Since the marked molecule $m$ has fewer neighbors on the interface than in the bulk, this results in a reduced diffusion barrier $E' < E$ at the interface as compared with the bulk volume. Therefore the probability of the molecule of escaping back from the interface into the bulk is larger than the opposite process. This leads to a gradual reduction of the number-density of the molecules present in the interface associated with a gradual increase of the potential-well depth at the interface.

If ultimately the relaxed state corresponding to the thermal equilibrium is reached (Figure 4-3 b), the depth of the potential-well of the interface layer of molecules corresponds to that in the bulk volume, $E' = E$. At the same time, the increased intermolecular distance $a_s$ of the interface corresponds to a dilatation of the interface. The two-dimensional dilatation of the interface layer is associated with an interface energy $\gamma$ measured per unit area of the interface:

$$\gamma \equiv \text{interface energy} = \frac{\text{stored energy}}{\text{interface area}} \;, \text{ or} \tag{4.2}$$

if we realize an experiment in which work has to be done against a force which proportional to the length of a surface, we obtain — using the same term and the same numerical value of $\gamma$ — the interface tension

$$\gamma \equiv \text{interface tension} = \frac{\text{effective force}}{\text{interface length}} \;. \tag{4.3}$$

**Vapor pressure over a flat interface**

The vapor pressure of a flat water interface in thermal equilibrium with its vapor can be determined by equating the rate of evaporation $R_e$ (the index "e" pointing to "evaporation") with the rate of condensation $R_c$ (the index "c" pointing to "condensation").

The frequency of evaporation $v_e$ for one selected molecule on the interface is given by the frequency of attempted escapes or "trial" escapes $v_{trial}$ multiplied by the probability $p(E_e)$ of the molecule for having sufficient energy to overcome the evaporation barrier of height $E_e$ (Figure 4-3 b):

$$v_e = v_{trial} \, p(E_e) \; . \tag{4.4}$$

The frequency of trial escapes $v_{trial}$ of the molecule is given by its average velocity $<u>$ and the characteristic interatomic distance $a_s$ of the interface, according to:

$$v_{trial} = \frac{<u>}{a_s} \; , \quad \text{where } <u> = \sqrt{\frac{3k\,T}{m}} \tag{4.5}$$

corresponds to the thermal translation energy of the molecule $m <u>^2 / 2 = (3/2)\,k\,T$.

The probability $p(E)$ of having sufficient energy to overcome the barrier of height $E_e$ is given by the Boltzmann factor

$$p(E) = e^{-\frac{E_e}{k\,T}} \; . \tag{4.6}$$

Therefore the evaporation frequency of the chosen molecule is:

$$v_e = v_{trial} \, p(E_e) = \frac{1}{a_s} \sqrt{\frac{3k\,T}{m}} \; e^{-\frac{E_e}{k\,T}} \; . \tag{4.7}$$

The rate of evaporation $R_e$ is given by the frequency of evaporation $v_e$ of one molecule multiplied by the number of the molecules per unit interface area, $1/a_s^2$, available for this process in the interface:

$$R_e = \frac{v_e}{a_s^2} = \frac{1}{a_s^3} \sqrt{\frac{3k\,T}{m}} \; e^{-\frac{E_e}{k\,T}} \; . \tag{4.8}$$

On the other hand, the frequency of condensation $v_c$ of a selected molecule in the vapor phase depends on much simpler rules than the frequency of evaporation $v_e$, since the condensation is not hampered by a potential barrier which has to be overcome by thermal activation. The frequency of condensation $v_c$ of a selected molecule is simply given by the frequency of arrivals $v_{arrival}$ — which corresponds to 1/6 of all molecules times the average velocity $<u>$ of the molecule — multiplied by a "sticking factor" $\vartheta$ which can be considered as an experimental fitting parameter with a magnitude between zero and one:

$$v_c = v_{arrival} \, \vartheta = \frac{1}{6} <u> \vartheta \; , \text{ where } <u> = \sqrt{\frac{3k\,T}{m}} \text{ and } 0 < \vartheta < 1 \; . \tag{4.9}$$

The rate of condensation $R_c$ is given by the frequency of condensation of one molecule, multiplied by the number of the molecules $n_v$ per unit volume available for this process in the vapor phase:

$$R_c = v_c \, n_v = \frac{1}{6} \sqrt{\frac{3k\,T}{m}} \; \vartheta \, n_v \; . \tag{4.10}$$

In thermal equilibrium, the evaporation and condensation rates are equal:

$$R_e = R_c \; \Rightarrow \; n_v = \frac{6}{\vartheta} \, n_s \, e^{-\frac{E_e}{k\,T}} \; , \text{ where } n_s = \frac{1}{a_s^3} \tag{4.11}$$

is the number-density of molecules in the interface.

In order to determine the saturation vapor pressure $P_0$ from the number-density $n_v$ of the molecules in the gas phase we remember the gas equation $P\, V_m = R\, T$ of an ideal gas, where $V_m = N_A\, v_m$ is the molar volume, $N_A$ is the Avogadro number and $v_m = 1 / n_v$ is the volume of one molecule. The gas equation may be applied with some precaution to a vapor and reads with respect to a unit volume of gas:

$$P_0 = n_v\, k\, T = \frac{6\, n_s}{\vartheta}\, k\, T\ e^{-\frac{E_e}{k\, T}} = A\ e^{-\frac{E_e}{k\, T}}\,, \qquad (4.12)$$

where $A$ is considered as a slowly varying constant and the quantity $P_0$ represents the saturation vapor pressure of a flat interface in contrast to the quantity $P$ — introduced in the section "stability of a grain or droplet" — representing the saturation vapor pressure of a spherical grain or droplet.

### Vapor pressure over a curved interface

For calculating the pressure difference $\Delta P$ existing between the interior and the outside of a spherical grain or droplet we observe Figure 4-4 [1]. If the pressure difference $\Delta P = f / A$, sustained by the external force $f$ on the piston of surface area $A$ is too small, the system will give off stored interface energy by contraction of the interface and thereby move the piston to the left. If the pressure difference is too large, the external force will move the piston to the right, the sphere will expand and energy will be stored on its interface in the form of interface energy. A labile equilibrium condition is reached if an infinitesimal external work done on the piston corresponds exactly to the infinitesimal increase of the stored potential energy on the interface. Since the stored interface energy per unit area is $\gamma$, this requirement leads to:

$$\Delta P\ 4\pi\, r^2\, dr = \gamma\, 8\,\pi\, r\ dr \ \Rightarrow\ \Delta P = \frac{2\,\gamma}{r}\,. \qquad (4.13)$$

Thereby the pressure is always higher on the concave side of an interface than on its convex side.

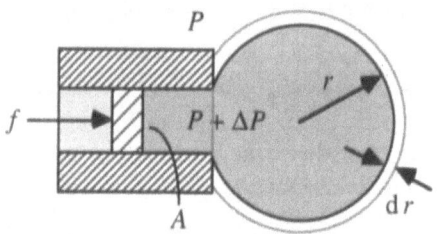

**Figure 4-4** *Equilibrium condition for a grain or droplet.* Pressure rise $\Delta P = f / A$ inside a spherical grain or droplet. Equilibrium exists if an infinitesimal external work done on the system corresponds to the infinitesimal increase of the potential energy of the system, stored in the form of interface energy.

[1]     Adamson, A.W.: "**Physical Chemistry of Surfaces.**" John Wiley & Sons, New York, pp. 1-698 ,
        p. 373, (1976).

## 4.1.3    Condition for grain growth

### Stability of a grain or droplet

The growth or shrinking of a grain or droplet of given size is intimately related with the exchange of matter between the the external and the internal phase through its interface. In the case of a water droplet we can obtain a concept for its stability with respect to growth or collapse by considering the saturation vapor pressure of a droplet of given size [1]. Already intuitively one anticipates that the vapor pressure over a droplet should be larger than over a flat interface since the number of nearest neighbors retaining a given molecule in the condensed phase and thus the evaporation energy per molecule $E_e$ is reduced for a curved interface, which according to eq. (4.12) leads to an increased saturation pressure.

In order to calculate the energy necessary for evaporation of one molecule from a droplet of given size $r$, we calculate first the total energy $E_{e, total}(r)$ necessary for evaporation of the whole droplet of radius $r$ which is the difference of the energy necessary to remove all molecules from a potential-well of depth $E_e$ to the vacuum and the energy recovered from the disintegration of the interface during complete evaporation:

$$E_{e, total}(r) = \frac{\frac{4\pi}{3} r^3}{v_m} E_e - 4\pi r^2 \gamma , \qquad (4.14)$$

where the factor $(4\pi / 3) r^3 / v_m$ corresponds to the number of molecules contained within the droplet. From this eq. (4.14) we obtain the energy necessary for evaporation of one molecule $E_e(r)$ from a droplet of radius $r$ by differentiation with respect to $r$ and multiplication with the change in radius $\delta r$ corresponding to the removal of that molecule:

$$E_e(r) = \frac{\partial E_{e, total}}{\partial r} \delta r = \left( \frac{4\pi r^2}{v_m} E_e - 8\pi r \gamma \right) \delta r , \qquad (4.15)$$

where $\delta r$ is calculated from the evaporation of one molecule of volume $v_m$ from the droplet which reduces the original droplet radius $r$ by an amount $\delta r$:

$$v_m = \frac{4\pi}{3} r^3 - \frac{4\pi}{3} (r - \delta r)^3 \approx 4\pi r^2 \delta r \implies \delta r = \frac{v_m}{4\pi r^2} \qquad (4.16)$$

Inserting $\delta r$ from eq. (4.16) into eq. (4.15) yields

$$E_e(r) = E_e - \frac{2\gamma v_m}{r} . \qquad (4.17)$$

If we insert the reduced evaporation energy $E_e(r)$ of a droplet of radius $r$ according to eq. (4.17) on the right side of eq. (4.12) instead of $E_e$, we obtain the saturation vapor pressure $P(r)$ of a droplet of radius $r$:

$$P(r) = A \, e^{-\frac{E_e}{kT} + \frac{2\gamma v_m}{rkT}} = P_0 \, e^{\frac{2\gamma v_m}{rkT}} . \qquad (4.18)$$

[1]    Walton, A.J.: "Three Phases of Matter." Clarendon Press, Oxford, pp. 1-182, (1984).

This means that in thermal equilibrium the vapor pressure $P(r)$ outside a droplet of size $r$ has to be larger by the factor exp $[2\,\gamma\,v_m\,/\,(r\,k\,T)]$ in comparison to the vapor pressure $P_0$ required for a flat interface.

For example, a water droplet with a radius of 1 nm, containing about 140 water molecules of volume $v_m = 18\,/\,6.0 \cdot 10^{23}$ cm$^3$ $= 3 \cdot 10^{-23}$ cm$^3$, will in thermal equilibrium at room temperature, $T = 293$ K, require a pressure ratio $P\,/\,P_0 = 2.9$, at which the interface energy of water is $\gamma = 72$ erg/cm$^2$, inserting the Boltzmann constant $k = 1.38 \cdot 10^{-16}$ erg/K.

### Nucleation barrier

Considering eq. (4.18), one finds for any vapor pressure $P$ and temperature $T$ a critical droplet radius $r_c$ which is stable under these conditions:

$$r_c = \frac{2\,\gamma\,v_m}{k\,T}\,\frac{1}{\ln\left(\dfrac{P}{P_0}\right)}\,.\tag{4.19}$$

Any smaller droplet will shrink to zero and any larger droplet will grow for ever if there exists no change in the external vapor pressure $P$ due to the evaporation or condensation. This phenomenon of instability of droplets with radii smaller than $r_c$ corresponds to existence of a "nucleation barrier" which has to be overcome during the creation of a grain or droplet. Only if the radius is beyond the barrier of the critical value $r_c$, the grain or droplet can grow. The mechanism is provided by "seeding" the homogeneous phase with suitable "condensation nuclei" capable to reduce the required saturation pressure and is one of the fundamental working principles in photographic process, in cloud and bubble chamber as well as in the ion track technique.

### Hindered phase transformation

A classical example for a phase transformation which energetically should take place, but is impossible without the presence of suitable condensation nuclei is the experiment of Marcellin-Berthelot in which purified water, in contact with a small amount of water vapor, is enclosed in a glass capillary and heated until the expanding water column has completely replaced the region containing water vapor (Figure 4-5). When cooled to the original temperature the water column still fills the complete volume of the capillary and has to be cooled considerably below the original temperature until finally a bubble of water vapor is explosively created. The large hysteresis of this process in the absence of suitable condensation nuclei points to their importance in phase transformations.

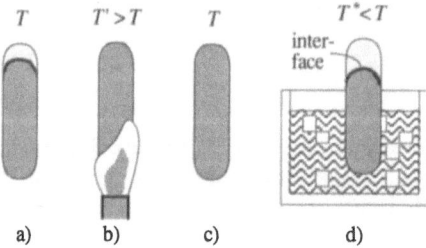

a)          b)          c)          d)

**Figure 4-5** *Necessity of condensation nuclei in phase transformations.* (a) Interface between water and its vapor at temperature $T$. (b) Vanishing interface at temperature $T' > T$. (c) After returning back to the original temperature $T$ no phase separation is observed. (d) Only after cooling to a considerably smaller temperature $T^* < T$ the interface is recreated during the sudden reappearance of a vapor bubble.

### 4.1.4 Formation of condensation nuclei

There exist mainly three mechanisms for creating condensation nuclei which are described in the following.

**Density fluctuations**

With increasing density of the aggregating molecules in the vapor phase the tendency of forming short-lived multimolecular aggregates increases. At very high supersaturation or pressures $P / P_0$ beyond the critical pressure, local density fluctuations can lead to the spontaneous creation of condensation nuclei. For estimating the order-of-magnitude of the density fluctuations within a given small volume we conduct the "Gedankenexperiment" of Figure 4-6 [1].

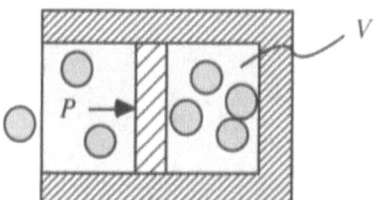

**Figure 4-6** *Density fluctuations* — the origin of condensation nuclei formation.

Let us separate a small volume $V$ containing $N$ molecules from the remaining system. Thermal fluctuations, providing the energy $\delta E = k T / 2$ are capable to compress the volume $V$ by $\delta V$. In this process, the pressure $P = N k T / V$ increases by an amount $\delta P = - N k T \, \delta V / V^2$. The thermal fluctuations perform the work:

$$\delta E = \frac{1}{2} k T = \frac{1}{2} \delta V \, \delta P = - \frac{1}{2} \delta V \frac{N k T}{V^2} \, \delta V \quad \text{or} \quad \left| \frac{\delta V}{V} \right| \approx \frac{1}{\sqrt{N}} \, . \qquad (4.20)$$

This means that for a small number $N$ of molecules enclosed in the volume $V$ the volume fluctuation $|\delta V / V|$ is large, which may rise the local density above the average density and lead to the spontaneous creation of condensation nuclei.

**Ionization**

Another possibility to increase the local density in a homogeneous phase such as water vapor above the critical value is the existence of an ion. Around the ion — due to the electrical polarization of its environment — an attractive force is acting on the water molecules which increases their tendency for aggregation [2].

**Defects**

A third possibility for the creation of condensation nuclei is the lattice misfit found in the environment of defects in solids which can start the aggregation of other defects around it. Defects in solids represent energy states above zero energy. By their agglomeration the crystal ameliorates its coherence and in this process the total energy of the system is decreased.

[1]   Adamson, A.W.: "**Physical Chemistry of Surfaces.**" John Wiley & Sons, New York, pp. 1-698 , p. 373, (1976).
[2]   Walton, A.J.: "**Three Phases of Matter.**" Clarendon Press, Oxford, pp. 1-182, (1984).

## 4.2 Track response function

Track response can be reduced to the response of a linear array of developable grains, stochastically distributed along the projectile trajectory in the track-recording solid. The creation of developable grains is intimately related to the energy-loss of the projectile-ion per unit path length and thus to the created ions along the path.

### Ionization and defect-density along the latent track

The ionization density $J$ corresponds to the number of ions $dN_{ion}$ formed along the projectile-ion path $dx$. It is roughly proportional to the energy-loss function ($- dE / dx$) according to eq. (2.148):

$$J \equiv \frac{dN_{ion}}{dx} = \frac{\left(-\frac{dE}{dx}\right)}{I} = \frac{4\pi}{I} \frac{Z_{eff}^2 \, e^4}{m_e v_p^2} \, N_t Z_t \, \ln\left(\frac{2 \, m_e v_p^2}{I}\right) , \qquad (4.21)$$

where $I$ is the average ionization potential of the target material, $Z_{eff}$ is the effective charge of the projectile-ion according to eq. (2.147), $e$ is the electron charge, $m_e$ is the electron mass, $v_p$ is the velocity of the projectile-ion, $N_t$ is the number of target atoms per volume element, and $Z_t$ is the nuclear charge of the target atoms.

Assuming as track-recorder a target in the form of an organic polymer of composition $C_n H_{2n}$, of density $\rho = 1$, and of average ionization potential $I = 5$ eV, we obtain a rough approximation of the ionization density for different ions on the basis of eq. (4.21), using the Bragg rule of eq. (2.153) for determining the energy-loss of this two-component compound material. The result is shown in Figure 4-7.

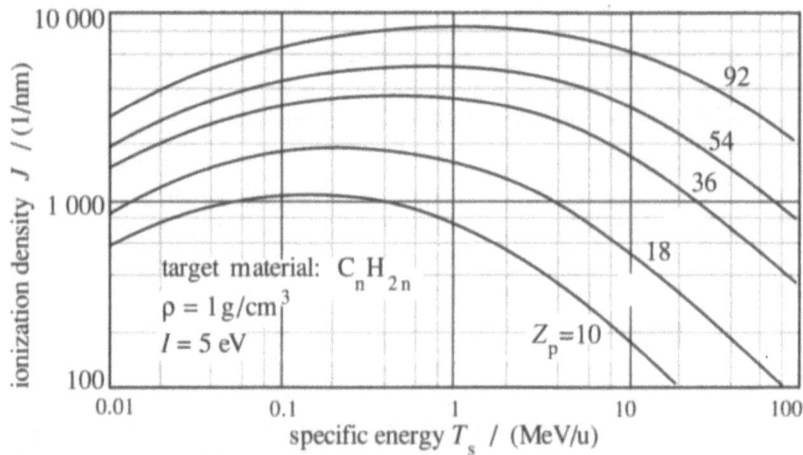

**Figure 4-7** *Linear ionization density J* — the number of target ions formed per unit path-length along the projectile-ion trajectory — in an organic polymer of composition $C_n H_{2n}$ and density $\rho = 1$, assuming an average ionization potential $I = 5$ eV, for different projectile-ions of nuclear charge $Z_p$.

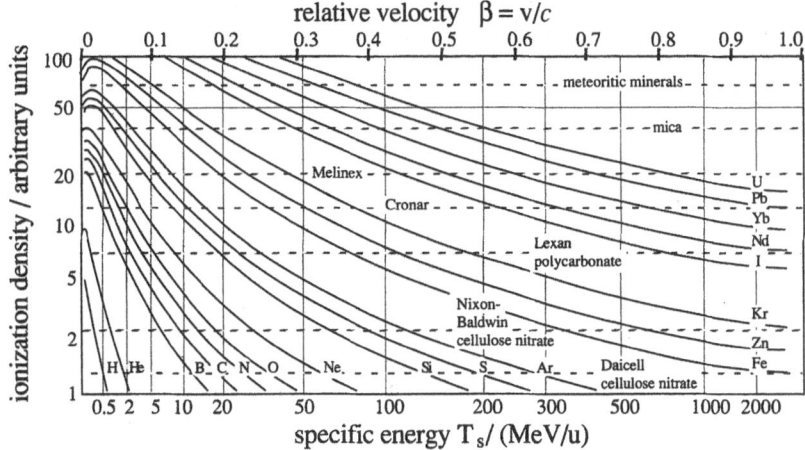

**Figure 4-8** *Linear ionization density and track-etch threshold* of several track recording materials [1].

**Linear density of defects**

The number of defects formed per unit path length — in other words, the linear defect-density $N_{defect}$ — is roughly proportional to the ionization density $J$:

$$N_{defect} = C\,J \quad , \tag{4.22}$$

where $C$ is a fitting-constant reflecting the sensitivity of the target material with respect to ionizing radiation.

Track-recording materials can be characterized by their threshold ionization densities, $N_{threshold}$, the minimum ionization density leading to developable tracks. In Figure 4-7 and in Figure 4-8 these values correspond to horizontal lines, characterizing regions of the ionization density curves which lead to developable tracks.

## 4.2.1    Track etch threshold

In photographic emulsions the developing agent is capable to diffuse through the gelatin matrix of the emulsion and approaches the irradiated silver halogen grains from all directions and over short diffusion paths. Therefore already very few developable grains embedded into an environment of undevelopable grains can be revealed in this type of internal development process. The developed grains form a quasi-linear array of stochastically distributed defects. At low defect-densities the developed grains are more or less discrete. With increasing defect-density the average distance between the developable grains decreases. Ultimately a contingent chain of developable grains is formed.

In contrast, a development process consisting of the removal of sensitized grains — in other words, an etching process — will only reveal a contingent chain of developable grains since the removal of all the preceding grains is required to remove a certain chosen grain (Figure 4-9). The necessity to remove a contingent chain of defects is

[1]    Fleischer, R.L.: "Nuclear Track Production in Solids." Prog. Mater. Sci. *X*, 97, (1981).

the basis for the observation of a track etch threshold (see also [1]). Thereby it is as-sumed, that the etch rate v in the grains is larger than the etch rate $g$ in the bulk material. For observing an etched track, always v has to be larger than $g$.

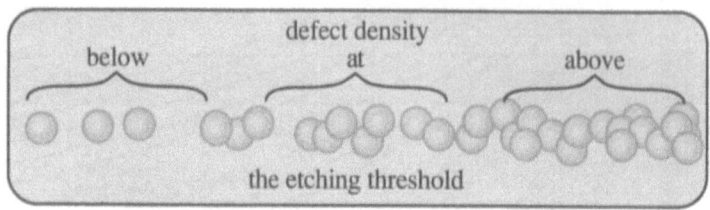

**Figure 4-9**  Basic gap model of track etch threshold.

For calculating the etch properties of a stochastical array of spherical defects located along a latent track, we replace it by a linear stochastical array of homogeneously developable cylindrical defects of identical length $D$ and stochastically varying gap sizes $G$, ideally aligned along the projectile-ion path with an average density $N_{defect} = <1/\lambda>$ (Figure 4-10) [2], [3]. The etch velocity within the cylinders is assumed to be $v_{defect}$. The etch velocity outside the cylinders is assumed to be $g < v_{defect}$.

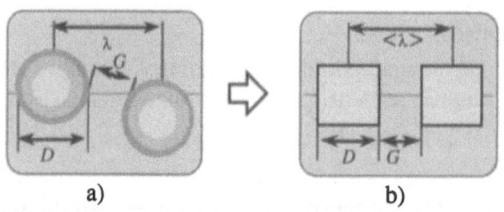

a)                                                     b)

**Figure 4-10**  *Simplified gap model* for track etching. The spherical defects (**a**) of diameter $D$ scattered around the ion track are replaced by a strictly linear array of homogeneously devel-opable cylinders (**b**) of identical length $D$, etch velocity $v_{defect}$ and linear density $N_{defect} = <1 / \lambda>$, separated by stochastically varying gaps of size $G$ and etch velocity $g < v_{defect}$, corresponding to the bulk etch-rate $g$ of the undisturbed material.

A regular array and a stochastical array of defects show quite different behav-ior. The effective track etch-rate v of a regular array of such cylinders starts at the bulk etch-rate $g$, increases linearly with increasing defect-density and attains its maximum, the track etch-rate $v = v_{defect}$, already at a linear defect-density $N_{defect} = 1/D$ for which all gaps are closed. The effective track etch-rate of a stochastical array reaches its maximum etch-rate $v = v_{defect}$ only at a much higher defect-density $N_{defect} \gg 1/D$. This is due the occurrence of gaps by chance even in quite densely populated stochastical defect-arrays.

A computer simulation of the average of 100 observations of the effective etch-rate v of 100 stochastically distributed defect-cylinders per unit track length for dif-ferent etch-ratios $v_{defect} / g$ and defect-densities $N_{defect}$ shows this delay in attaining the limiting etch-rate $v = v_{defect}$ with increasing defect-density $N_{defect}$ (Figure 4-11).

[1]    Lück, H.B.: "Mechanism of  Particle Track Etching in Polymeric Nuclear Track Detectors." Nucl. Instrum. Methods, **202**, 497, (1982).
[2]    Dartyge, E. J.P. Duraud, Y. Langevin, M. Maurette: "New Model of Nuclear Particle Tracks in Dielectric Minerals." Phys. Rev. **B 23**, 5213, (1981).
[3]    Bieth, C.: "Etude des caracteristiques parametres d'irradiations, de sensibilisation et d'attaque chimique des traces latentes", direct communication, GANIL, Boite Postale No. 5027, F-14021 Caen Cedex, France, (1987).

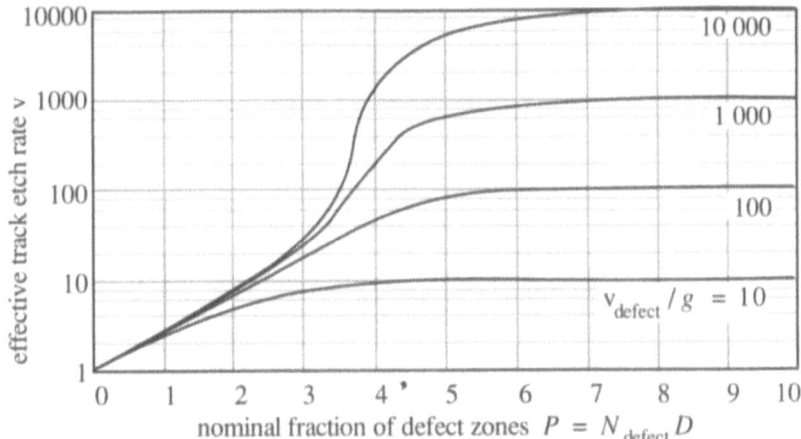

**Figure 4-11** *Track etch threshold according to gap model.* Effective track etch-rate v as function of the nominal fraction of defect zones *P* per unit track length for different ratios of the etch-rate $v_{defect}$ in defect zones and of the bulk etch-rate *g* in the undisturbed bulk material. Etch-rates are measured in unit length per unit time, for example in μm/h.

Only for very large defect-densities $N_{defect} \gg 1/D$ ultimately the characteristic defect etch-rate $v_{defect}$ is attained. This is the reason for the occurrence of a track etch threshold which enables track observation only beyond an ionization density that is characteristic for the radiation sensitivity of the used track-recording material.

In the computer simulation, the nominal fraction of defect zones *P* — in other words, the fraction of the projectile-ion path covered cumulatively by defects without counting overlap — is used as the independent variable, instead of the two variables, defect density $N_{defect}$ and defect size *D* which have the same proportional influence on the effective track etch-rate v. The defect fraction *P* is given by the product of the average defect density $N_{defect}$ and the characteristic defect size *D*:

$$P = N_{defect} \, D \quad . \tag{4.23}$$

## 4.2.2    Track sensitization and annealing

From Figure 4-11 it becomes obvious that in the vicinity of the steep rise of the effective track etch-rate v, already small changes of the defect diameter *D* have a large effect on the observed effective track etch rate v (Figure 4-12).

Track sensitization — corresponding in this model to a grain-growth — can be achieved in polymers for example by uv or oxygen treatment which increases the size of the dissolvable grains. On the other hand, track annealing — corresponding in this model to a shrinking of the grains — is generally achieved thermally [1], [2], [3] by rising

[1]    Märk, T.D., W. Ritter: "Radiation Damage and its Annealing in Sodium Silica Glass." Mat. Res. Soc. Symp. Proc. Vol. **84**, pp. 659-669, (1987)

[2]    Walder, G., T.D. Märk: "Annealing Kinetics of Radiation Damage in Artificial Obsidian Glass." Nuclear Instruments and Methods in Physics Research, **B32**, pp. 303-306, (1988).

[3]    Märk, T.D., G. Walder: "Annealing and Leaching Studies with Natural and Artificial Obsidian Glass." Mat. Res. Soc. Symp. Proc. Vol. **112**, pp. 693-701, (1988).

the temperature of the track recorder to a certain value for a certain time . In the case of polymers, already the irradiation conditions (air, rare gas, vacuum, temperature) have an eminent influence on the resulting track etch rate v.

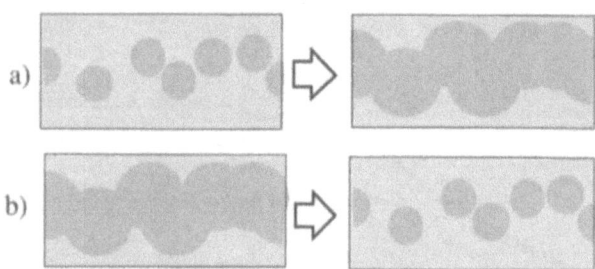

**Figure 4-12  Principle of track sensitization and track annealing,** according to grain model. (**a**) *Track sensitization* is achieved by a process that induces the growth of the developable grains, so that the overlap between neighboring tracks increases. (**b**) *Track annealing* is achieved by a process that reduces the size of the grains, so that the overlap between neighboring grains decreases.

# 4.3     Shape of etched tracks

## 4.3.1     Primary factors in track etching

The shape of the etched track depends mainly on the ratio of the track etch-rate v in the activated zone and of the etch-rate $g$ in the undamaged material (Figure 4-13). However, diffusion and convection play a very important role in short-range and long-range transport of the reaction products, respectively.

Primary factors in track etching are:

1.   The specific track-recording material used and the chosen etchant. The material should be as homogeneous as possible. The etchant should be a marginal solvent or a very weak etchant for the undisturbed virgin material.

2.   The defect density in the chemically activated zone along the ion path.

3.   The reaction rate in the boundary layer between the solid and the etchant which depends strongly on the chosen temperature and on the concentration of the etchant.

4.   The removal of the reaction products by diffusion and convection. Convection is very inefficient inside the track and very efficient on the recorder surface.

On the molecular level the track etch rate v is determined by the number of reactive sites in the form of chain ends or functional groups, by the so-called "free volume" in the vicinity of the reactive site which is accessible to the etchant during its approach to the reactive site, by the reaction rate of the chemical reaction, and by the required degradation of the polymer before it can be ultimately dissolved [1].

[1]     direct communication: Lück, H.B., Akademie der Wissenschaften, Zentralinstitut für Kernforschung, Postschließfach 19, DDR-8051 Dresden.

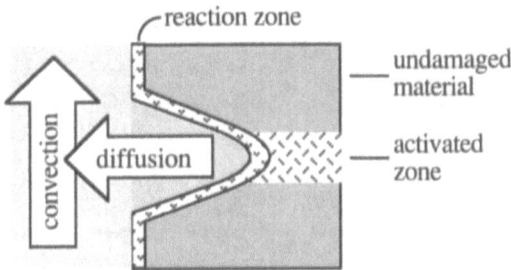

**Figure 4-13** *Primary factors in track etching.* The etching of a latent ion track depends on the density of the radiation damage along the ion track, on the reaction rate between the solid and the etchant, and on the removal of the etch products by diffusion and convection.

### Reaction zone

The thickness of the reaction zone represents a limiting factor for the possible reaction rate. With increasing thickness of the reaction zone, the diffusion of the etchant to the reaction sites and the back-diffusion of the reaction products into the bulk of the etchant are increasingly impeded. In polymers, the reaction zone consists of a thick layer of polymer chains interspersed with the etchant, whereby the length of the polymer chains — in other words the molecular weight of the polymer — decreases continuously from the bulk of the polymer to the etchant (Figure 4-14). Due to gradual decomposition of the polymer during etching the termination of an etch process by a flushing process is quite critical. Flushing will remove a fraction of the gel layer while another fraction of the gel layer coagulates again during flushing and may remain as a low-molecular weight layer on the surface of the track recorder and within the etched tracks.

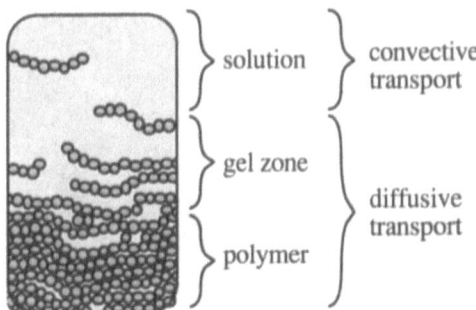

**Figure 4-14** *Reaction zone in a polymer.* The etchant penetrates into the polymer by diffusion and decomposes it gradually via chain scission. There exists a continuous transition of chain sizes between the virgin solid and the solution. Above the boundary layer, between the gel and the solution, long-range convection starts to become effective.

### Etch induction time

Due to the gradual decomposition of polymers in the etchant the removal of matter from the track recorder cannot begin immediately after the immersion of the track recorder into the etchant (Figure 4-15)

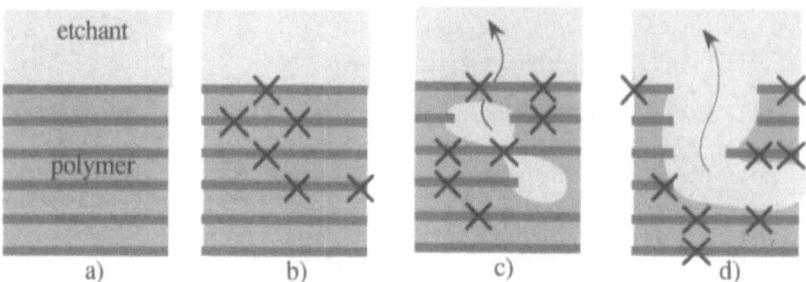

**Figure 4-15 Principal stages in polymer etching.** (a) *Virgin polymer chains.* (b) *Soaking stage,* characterized by the chemical or radiation-induced fragmentation of the molecular chains. During this stage the molecular fragments are still too large to be dissolved and thus cannot be transported away via diffusion. (c) *Digestion stage,* characterized by a decreasing mass density due to dissolving molecular fragments, mainly from the interior of the polymer body. During this stage, the molecular coherence and the planar surface of the track recorder are virtually preserved. (d) *Dissolution stage,* characterized by the observable change of the track recorder surface, a phenomenon characterized by the term "etch-induction time".

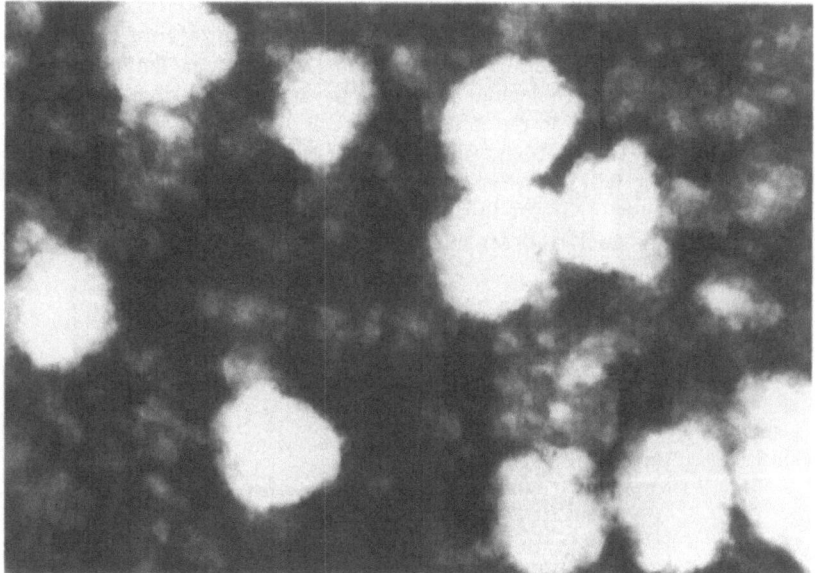

**Figure 4-16** *Resolution limit of a polymeric track recorder.* The washed-out diffuse shape of etched tracks in a 2 μm thick polycarbonate foil after prolonged etching — observed by high voltage transmission electron microscopy [1] — corresponds to an average density which gradually decreases from the bulk towards the surface. Correspondingly, the track boundary is not a sharp step function.

A chosen polymer chain undergoes a gradual decomposition via chain-breaking, reducing step by step its molecular weight until the fragments are small enough to be

[1]    Packard, R.E., J.P. Pekola, P.B. Price, R.N.R. Spohr, K.H. Westmacott, Zhu Yu-Qun: "Manufacture, Observation, and Test of Membranes with Locatable Single Pores." Rev.Sci.Instrum., 57 (8), 1654-1660 (1986)

dissolved and transported away. Only after this "digestion" reduced the length of the polymer chains at the interface of the track recorder to sufficiently small bits, capable to diffuse or dissolve in the etch solution, the track recorder will start losing substance into the etch solution. During the starting of the digestion stage and even in the beginning of the dissolution stage, the outer appearance of the track recorder will stay nearly unchanged. This phenomenon has been characterized by the term "etch-induction" or "etch-induction time". Concerning the two-step etch model see also reference [1].

The gradual decomposition of polymers is the reason for the build-up of a gel layer of undissolved, however, heavily decomposed material which in some cases may remain on the surface of the track recorder after the etch process is terminated [2]. This gel may be reduced in thickness by a suitable solvent.

After prolonged etching a track recording polymer may have a quite diffuse outer shape corresponding to the irregular build-up of microcavities reflecting material inhomogeneities (Figure 4-16).

## 4.3.2    Fick's first diffusion law

Diffusion transport is an important condition for the successful removal of matter in ion track etching. The following section is intended to give a rough idea about the parameters influencing diffusion transport processes. The diffusion of molecules in a viscous fluid is governed by laws very similar to the diffusion of heat. For deriving Fick's first law concerning the stationary diffusion of molecules in a concentration gradient we observe Figure 4-17.

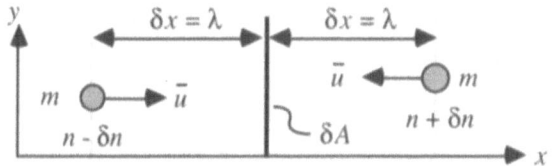

**Figure 4-17**   Deriving Fick's first law of diffusion.

The total number $\delta N$ of molecules crossing the interface $\delta A$ in $x$-direction per unit of time corresponds to the difference of the two particle currents crossing the interface from the left and from the right, respectively:

$$\delta N = \frac{1}{6}(n - \delta n)<u>\delta A - \frac{1}{6}(n - \delta n)<u>\delta A = -\frac{1}{3}\delta n<u>\delta A , \quad (4.24)$$

where the factor 1/6 corresponds to the fraction of molecules moving in the positive and in the negative $x$-direction, $n$ is the number-density of molecules, $<u>$ is the average velocity of the molecules, and $\delta A$ is the interface at which the particle flow is accounted for. Thereby the molecules are assumed to come from a distance $\delta x$ equal to the mean-free-path $\lambda$ of the molecules in the fluid.

[1]    Törber, G., W. Enge, R. Beaujean, G. Siegmon: "The Diffusion-Etch Model Part I: Proposal of a New Two Phase Track Developing Model." Supplement No.3 to the Journal Nuclear Tracks, pp. 307-310, (1982).
[2]    Enge, W. K. Grabisch, R. Beaujean, K.-P. Bartholomä: "Etching Behavior of a Cellulose Nitrate Plastic Detector Under Various Etching Conditions." Nuclear Instruments and Methods, **115**, pp. 263-270, (1974).

The particle current density $j$ corresponds to the ratio $\delta N / \delta A$:

$$j = + \frac{\delta N}{\delta A} = -\frac{1}{3} \delta n <u> = -\frac{1}{3} \delta n <u> \frac{\lambda}{\lambda} = -\frac{1}{3} \frac{\delta n}{\delta x} <u> \lambda , \text{ or} \tag{4.25}$$

$$j = -D \frac{\partial n}{\partial x} \quad \text{with} \quad D = \frac{1}{3} \lambda <u> , \text{ or in three dimensions} \tag{4.26}$$

$$j = -D \nabla n \quad \text{with} \quad D = \frac{1}{3} \lambda <u> . \tag{4.27}$$

### Cage model for diffusion in fluids

An atomistic interpretation of the diffusion in a fluid is obtained using the cage model (Figure 4-18) [1].

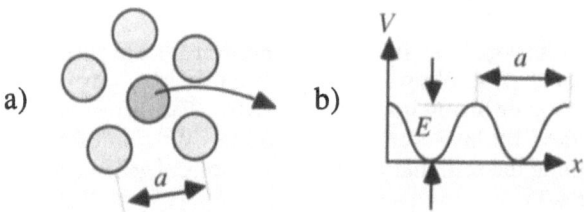

**Figure 4-18** *Cage model for diffusion processes in a fluid.* (a) Molecule surrounded by a "cage" of nearest neighbors. (b) Average potential in a channeling direction.

Similar to the diffusion of point defects in a solid, the migrating molecule in a fluid is surrounded by a potential wall which has to be overcome by thermal activation. The transition rate $\nu_{trans}$ for jumping from a given cage into the neighboring cage is determined by the product of the frequency of trial encounters $\nu_{trial}$ between the molecule and the wall and the probability $p(E)$ of the molecules of having sufficient energy to overcome the barrier of height $E$:

$$\nu_{trans} = \nu_{trial} \, p(E) . \tag{4.28}$$

The frequency of trial encounters $\nu_{trial}$ between the molecule and the wall is given by its average velocity $<u>$ and the characteristic size $a$ of the cage. The average velocity $<u>$ is determined from the average kinetic energy of translation which corresponds roughly to $kT / 2$ per degree of freedom and comprises 3 degrees of freedom:

$$\frac{m}{2} <u>^2 = \frac{3}{2} kT \quad \Rightarrow \quad <u> = \sqrt{\frac{3kT}{m}} . \tag{4.29}$$

This leads to the rate of trial encounters $\nu_{trial}$ between the molecule and its surrounding cage:

$$\nu_{trial} = \frac{<u>}{a} = \frac{1}{a} \sqrt{\frac{3kT}{m}} . \tag{4.30}$$

---

[1]     Walton, A.J.: "**Three Phases of Matter**." Clarendon Press, Oxford, pp. 1-182, (1984).

The probability $p(E)$ of the molecule having sufficient energy for passing over the potential barrier of height $E$ is given by the Boltzmann factor

$$p(E) = e^{-\frac{E}{kT}} .$$
(4.31)

The transition rate therefore becomes:

$$v_{trans} = v_{trial} \, p(E) = \frac{1}{a} \sqrt{\frac{3kT}{m}} \; e^{-\frac{E}{kT}} .$$
(4.32)

**Thermodynamic interpretation of diffusion constant**

For obtaining a thermodynamic interpretation of the diffusion constant $D = \lambda <u> / 3$ in accordance with the cage model, we assume that the mean-free-path $\lambda$ corresponds to the characteristic size $a$ of the cage and that the average velocity $<u>$ of the molecules used in Fick's first law corresponds to the average migration velocity $<u_{migration}> = a \, v_{trans}$. This yields

$$D \equiv \frac{1}{3} \lambda <u> = \frac{1}{3} a <u_{migration}> = \frac{1}{3} a^2 v_{trans} , \quad \text{or}$$
(4.33)

$$D = \frac{a}{3} \sqrt{\frac{3kT}{m}} \; e^{-\frac{E}{kT}} .$$
(4.34)

The main features of the diffusion constant $D$ are the dependence of the inverse square-root of the molecular mass $m$ and the exponential dependence on the inverse temperature $T$.

## 4.3.3    Calculation of track shapes

The evolution of track shape during etching can be simulated by considering the wave-front produced by an object moving with variable velocity $v(x)$ along the $x$-direction in a medium of constant phase velocity $g$ [1], [2]. Thereby the velocity $v(x)$ corresponds to the track etch rate which usually is a monotonous function of the energy deposited per unit track length — in other words of the energy loss function $(dE / dx)$. According to Huygen's construction the wave-front at the time $t + \Delta t$ is obtained from the wave-front at the time $t$ by drawing spheres with radius $g \, \Delta t$ around a sufficient number of points located on the wave-front at the time $t$ and determining their envelope which corresponds to the new wave front at the time $t + \Delta t$ (Figure 4-19).

For applying Huygen's construction two different domains of the solid have to be assumed, requiring different treatment, on one hand the three-dimensional volume of the undisturbed bulk material and on the other hand the quasi one-dimensional latent track, embedded in the undisturbed material as a cylindrical zone of nearly infinitesimal diameter with locally increased etch-rate.

[1]    Fleischer, R.L. , P.B. Price, R.T. Woods: "Nuclear Particle Track Identification in Inorganic Solids", Phys. Rev. **188**, 563, (1969).
[2]    Paretzke, H.G., E.V. Benton, R.P. Henke: "On Particle Track Evolution in Dielectric Track Detectors and Charge Identification Through Track Radius Measurement." Nuclear Instruments and Methods, **108**, 73-80, (1973).

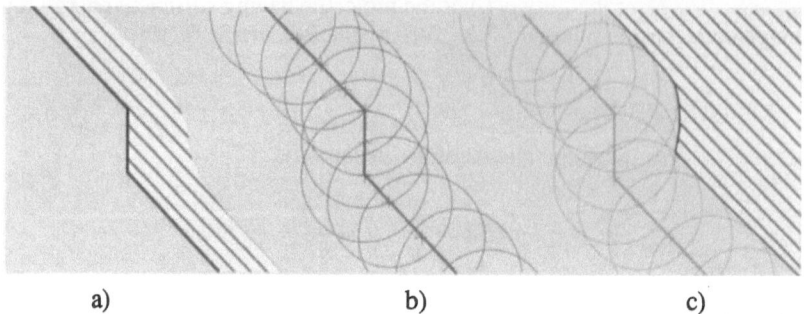

a)                              b)                              c)

**Figure 4-19** *Huygen's construction of a new wave-front.* (a) Original wave-front at time $t$. (b) Spreading elementary waves. (c) The wave-front at time $t + \Delta t$ is formed by the envelope of the elementary waves. — Only the envelope on one side of the original wave-front has to be considered.

From Figure 4-19 it is noted that concave parts of the etch figure are being rounded-off while convex parts of the of the etch figure are maintained quite long as sharp protrusions. This fact has far-reaching consequences in the generation of track structures. It is the reason for the very sharp edge of the entrance openings in single tracks and for the occurrence of very sharp pins in overlapping tracks.

### Track shape for constant track etch-rate

For constant track etch-rate, $v = \text{const} > g$, the result of Huygen's construction is a Mach cone (Figure 4-20). During the time interval $\Delta t$ the etch-front reaches the circumference of a sphere of radius $g \Delta t$ around point $P_1$ where $g$ is the bulk etch-rate. During the same time $\Delta t$, the etch-front proceeds along the track direction with the velocity $v > g$ and reaches the point $P_2$ located at distance $v \Delta t$ from $P_1$. The direction $\alpha$ of the etch-front, measured from the track axis, is given by

$$\cos \alpha = \frac{g \Delta t}{v \Delta t} = \frac{g}{v} \quad \text{or} \quad \alpha = \arccos \left( \frac{g}{v} \right) . \tag{4.35}$$

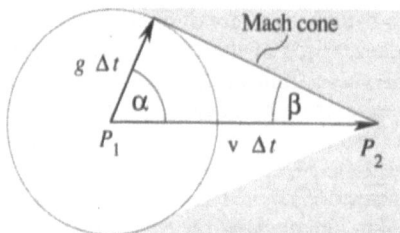

**Figure 4-20** *Formation of a Mach cone* for $v > g$. Direction $\alpha$ from the origin ot the elementary spherical wave to the etch-front and half-cone angle $\beta$ of the etch cone for constant track etch-rate $v > g$. For varying track etch-rate, $v = v(x)$, the angles $\alpha$ and $\beta$ become local functions of $v(x)$.

Assuming constant bulk etch-rate $g$, the angle $\alpha$ approaches 90° for $v \rightarrow \infty$ and 0° for $v \rightarrow g$. The half-cone angle $\beta$ is given by

$$\sin \beta = \frac{g}{v} \quad \text{or} \quad \beta = \arcsin \left( \frac{g}{v} \right) . \tag{4.36}$$

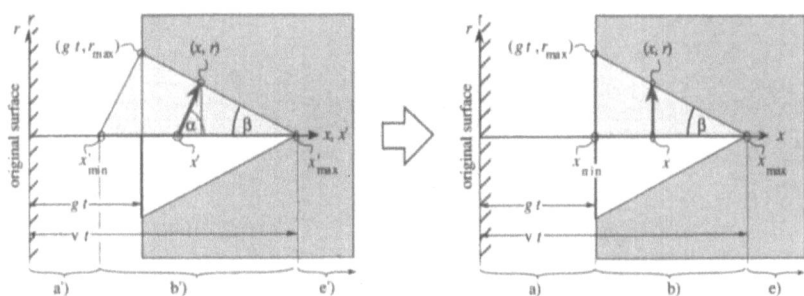

**Figure 4-21** *Calculation of track shape for constant track etch rate*, v = const, *at time* $t \leq$ range / v. **(left)** Geometry of intermediate-time solution, yielding $x$ and $r$ in parametric form as function of the time $t'$ at which the corresponding elementary "wave" started at the point $x'$ on the track axis. This solution corresponds to an oblique projection of the bottom side of the shaded triangle onto its top side. **(right)** Simplified geometry — using a distorted triangle — yielding directly $r = r(x, t)$.

The calculation of the track shape for constant track etch rate, v = const follows Figure 4-21 in which three zones are distinguished. During the etching, the tip of the etch cone reaches in succession the points $x'_{min}$, $x'$, and $x'_{max}$. The corresponding elementary "waves" reach at the time $t$ simultaneously the points $(g\,t, r_{max})$, $(x, r)$, and $(x'_{max}, 0)$, respectively. According to the simplified geometry of Figure 4-21 (right), the track radius $r = r(x, t)$ in the three different zones {(a), (b), (e)} is given by:

**(a)** $r = \infty$ for $0 \leq x < g\,t$ , $\hspace{6cm}$ (4.37)

**(b)** $r = (v\,t - x)\tan(\beta) = (v\,t - x)\tan\left[\arcsin\left(\frac{g}{v}\right)\right]$ for $g\,t \leq x < v\,t$, (4.38)

**(e)** $r = 0$ for $v\,t \leq x$ . $\hspace{6cm}$ (4.39)

**Rounding cone.** When the etch tip has reached the endpoint of the ion track — the ion range — a spherical section of the track starts to form (Figure 4-22).

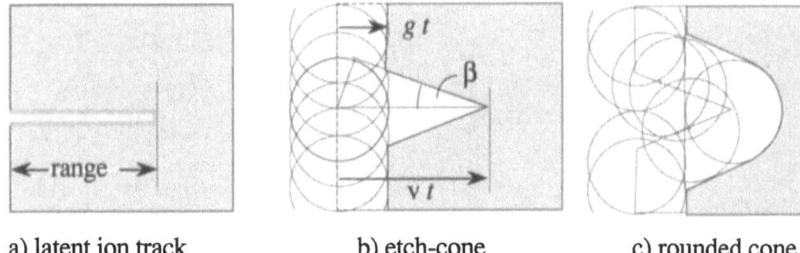

a) latent ion track $\hspace{2cm}$ b) etch-cone $\hspace{2cm}$ c) rounded cone

**Figure 4-22** *Shape of an etched track* with constant track etch-rate at an angle of incidence of 90° with respect to the interface plane. **(a)** Before etching. **(b)** The etchant has reached the end-point of the latent track. After this time the etching continues as an isotropic process. **(c)** After prolonged etching a combination of a conical section with a spherical section is obtained. Ultimately the conical section will be removed and only a spherical section remains.

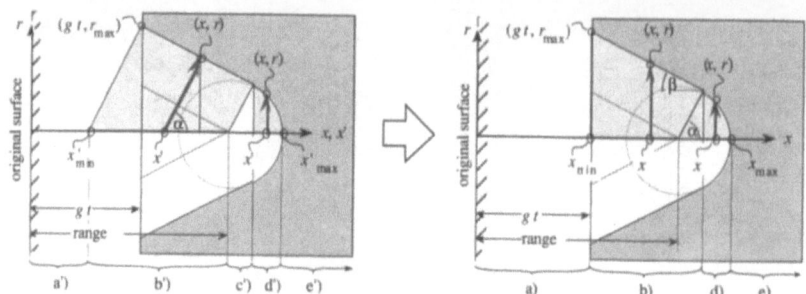

**Figure 4-23** *Calculation of track shape for constant track etch rate,* v = const at time *t* >
range / v. (**left**) Geometry of intermediate-time solution. (**right**) Simplified geometry, obtained by distortion of the shaded parallelogram and yielding directly *r* = *r*(*x*, *t*).

For describing *r* directly as function of *x* and *t* we transform the geometry of
the intermediate-time problem according to Figure 4-23 , in which four different zones —
{(a), (b), (d), (e)} — are distinguished :

(**a**)  $r = \infty$  for  $0 \le x < g\,t$ ,  $\hspace{4cm}$ (4.40)

(**b**)  $r = g\left(t - \dfrac{\text{range}}{v}\right) \sin(\alpha) + \left(\text{range} + g\left(t - \dfrac{\text{range}}{v}\right) \cos(\alpha) - x\right) \tan(\beta)$

$\hspace{1cm}$ for  $g\,t \le x < \text{range} + g\left(t - \dfrac{\text{range}}{v}\right) \cos(\alpha)$ ,  $\hspace{2cm}$ (4.41)

(**d**)  $r = \sqrt{\left[g\left(t - \dfrac{\text{range}}{v}\right)\right]^2 - \left[x - \text{range}\right]^2}$  for

$\hspace{1cm}$ $\text{range} + g\left(t - \dfrac{\text{range}}{v}\right) \cos(\alpha) \le x < \text{range} + g\left(t - \dfrac{\text{range}}{v}\right)$ ,  $\hspace{1cm}$ (4.42)

(**e**)  $r = 0$  for  $\text{range} + g\left(t - \dfrac{\text{range}}{v}\right) \le x$ .  $\hspace{3cm}$ (4.43)

The result of a track shape calculation for v = const is shown in Figure 4-24.

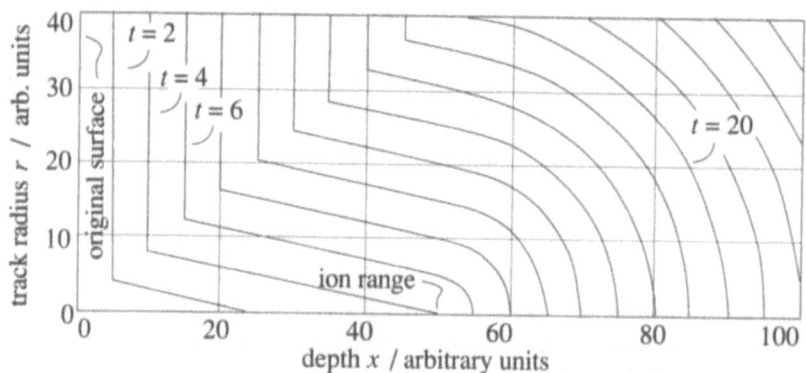

**Figure 4-24** *Track shape calculation for constant track etch rate* v *at different etch times*,
using arbitrary units — for example µm, hour, and µm / hour. General etch rate *g* = 1,
track etch rate v = 5, ion range = 50, time increment Δ*t* = 2.

### Track shape for varying track etch-rate

**Normal incidence.** For varying track etch-rate, $v = v(x)$, the angle $\alpha$ corresponds to the local track etch-rate $v(x)$ at the point $P_1$ and the angle $\beta$ at point $P_2$ corresponds to the half-angle of the infinitesimal etch-cone in the vicinity of the point $P_1$ during the passage of the etch-front through $P_1$. Typical shapes for increasing (Figure 4-26), constant, and decreasing etch-rate are shown in Figure 4-25.

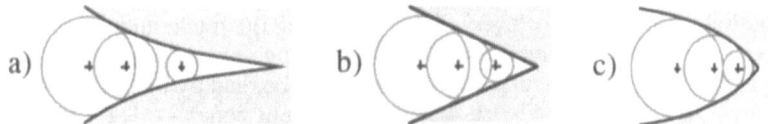

**Figure 4-25** *Typical etch figures* for (**a**) increasing, (**b**) constant, and (**c**) decreasing etch-rate v. The crosses mark the centers of the spherical "waves" emitted at equal time intervals.

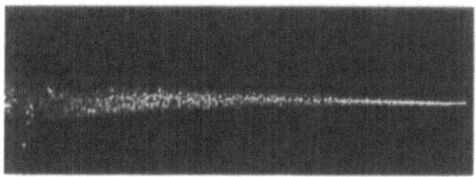

**Figure 4-26** Replica of cosmic ray track that ends in an Apollo helmet [1].

**Calculation of the shapes of etched tracks.** For calculating the shape of the etched track we observe Figure 4-27. During Huygen's construction we have first to distinguish between the two different domains of the solid which require different treatment, the bulk material and the the latent track material, embedded in the undisturbed material as a cylindrical zone of virtually infinitesimal diameter and increased etch-rate. In the following the exact description of the intermediate-time problem is given (Figure 4-27 left). The practical solution would follow again a simplified geometry (Figure 4-27 right).

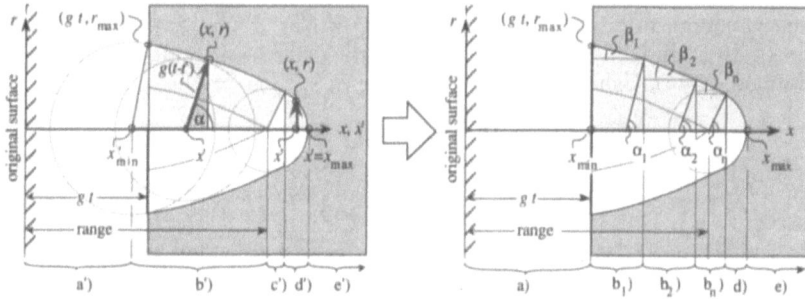

**Figure 4-27** *Determination of track shape for varying etch velocity* $v = v(x)$. (**left**) Geometry of intermediate-time solution. The etchant proceeds with the variable velocity $v(x)$ along the track direction, reaching the point $(x', 0)$ at the intermediate time $t'$. This point becomes the origin of a spherical etch-front, spreading with the constant velocity $g$, touching the actual boundary of the etched track at the point $(x, r)$ at the actual time $t$ after a time-delay of $(t - t')$. (**right**) Simplified geometry with subdivided zone (b) $\Rightarrow$ {(b$_1$), (b$_2$), ..., (b$_n$)}, suited for a direct approximative determination of $r = r(x, t)$.

[1]     Fleischer, R.L., General Electric Research and Development Center, K1-3C31, P.O. Box 8, Schenectady, N.Y. 12301, USA.

During etching, the etch front proceeds at first one-dimensionally along the latent track, starting from the origin. It reaches the point $(x', 0)$ at the intermediate time $t'$, given by the time-integral of d$t$ between 0 and $t'$:

$$t' = \int_0^{t'} dt = \int_0^{x'} \frac{dt}{dx} \, dx = \int_0^{x'} \frac{dx}{v(x)} \, . \tag{4.44}$$

According to eq. (4.44) for every value of $x'$ along the track, an intermediate time $t'$ can be determined. Thus a numeric list of corresponding values of $x'$ and $t'$ can be established from which $t'(x')$ or the inverse function $x'(t')$ can be read by interpolation.

According to Figure 4-27 five different zones — {(a'), (b'), (c'), (d'), (e')} — have to be distinguished in the calculation of the track shape:

(a')  $r(x) = \infty$  for  $0 \le x < g t$ , $\qquad\qquad$ (4.45)

(b') After reaching the point $(x', 0)$, the etch-front proceeds spherically and touches the actual boundary of the etched track in the point $(x, r)$ at the actual time $t$. The coordinates $(x, r)$ of this arbitrary point on the interface of the etched track at the actual time $t$ are obtained by observing the shaded triangle:

$$x = x(x', t) = x' + g \ (t - t') \ \cos\big(\alpha(x')\big), \tag{4.46}$$

$$r = r(x', t) = 0 + g \ (t - t') \ \sin\big(\alpha(x')\big) \, . \tag{4.47}$$

In these relations the difference $t - t'$ between the actual time $t$ and the intermediate time $t'$ corresponds to the retardation between the action taking place at the location $(x', 0)$ at time $t'$ and the arrival of the corresponding signal at the location $(x, t)$ at time $t$. This is in close analogy with the spreading of the retarded potential from a point-charge moving at a velocity larger than the speed of light. For evaluation of eqs. (4.46) and (4.47) at the actual time $t$ one chooses a series of values $x'$ between 0 and $x_{max}$ (to be defined below) and inserts them into the eqs. (4.46) and (4.47), using $\cos \alpha(x') = g \, / \, v(x')$ from eq. (4.35), $\sin \alpha = [1 - \cos^2 \alpha(x')]^{1/2}$, and determining the intermediate time $t'(x')$ from eq. (4.44). Using eq. (4.44), the value of $x_{max}$ is implicitly given by

$$t = \int_0^{x_{max}} \frac{dx}{v(x)} \, , \tag{4.48}$$

the actual time associated with the tip of the etch front, which is obtained from the already established numeric list of corresponding values of $x'$ and $t'$ by interpolation, choosing $t' = t$.

(c') Singular gap zone to be omitted in the calculation.
(d') Spherically rounding track tip, similar to eq. (4.42).
(e') Untouched bulk material, similar to eq. (4.43).

The result of a numerical calculation of track shapes for different etch times is shown in Figure 4-28. Thereby an ion range = 5 µm, a bulk etch-rate $g = 1$ µm/h, and a variable track etch-rate $v = 7 + 0.75 \, x$ µm/h was chosen which increases linearly along the latent track from 7 to 10 µm/h. Beyond the ion range the track etch-rate is replaced by the bulk etch-rate.

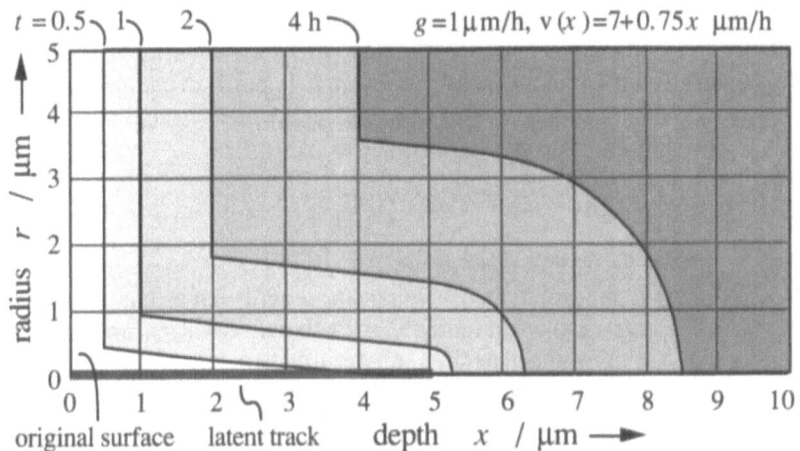

**Figure 4-28** Track shape obtained for ion range = 5 μm, bulk etch-rate $g$ = 1 μm/h, track etch-rate $v(x)$ = 7 + 0.75 $x$ μm/h, for four different times $t$ = 0.5, 1, 2, and 4 h. In the beginning an approximately conical track shape is obtained. At $t \approx$ 5 h approximately a half-sphere would be obtained. At still larger times the depth of the etch figure stays constant while its radius of curvature increases with the etch-rate $g$.

**Track length and track radius.** The main parameters of an etched track resulting from a normally incident ion are the track radius $R$, corresponding to the rim of the etch figure and the track length $L$ (Figure 4-29).

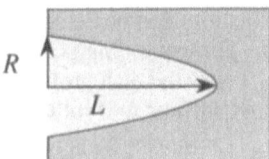

**Figure 4-29** *Primary track parameters*: Track radius $R$ and track length $L$.

For a normally incident track the track radius $R$ and the track length $L$ are determined in the following way.

On one hand, the rim of the etched track has the $x$ - coordinate $g\,t$, corresponding to the cut of the Huygen's envelope with the plane $x_{min} = g\,t$. Inserting $x = g\,t$ into eq. (4.46) $x_{min} = g\,t$ fulfils the condition:

$$g\,t = x'_{min} + g\,(t - t'_{min})\,\cos\left(\alpha\,(x'_{min})\right) \text{ or} \tag{4.49}$$

$$g\,t = \frac{x'_{min} - g\,t'(x'_{min})\,\cos\left(\alpha\,(x'_{min})\right)}{1 - \cos\left(\alpha\,(x'_{min})\right)} . \tag{4.50}$$

Realizing that $t'_{min} = t'(x'_{min})$ is a function of $x'$ according to eq. (4.44) we can therefore establish from eq. (4.50) a list of values $t\,(x'_{min})$ and obtain the inverse relation $x'_{min}(t)$ by interpolation. Inserting an arbitrary value $x'_{min}(t)$ into eq. (4.47) we obtain the value of the track radius $R$:

$$R = g\left(t - t'(x'_{min})\right) \sin \left(\alpha(x'_{min})\right) . \tag{4.51}$$

On the other hand, the vertex of the etched track has the $x$-coordinate $x_{max}(t)$, determined from the list of values corresponding to eq. (4.48). The track length $L$ corresponds to the difference between this coordinate $x_{max}(t)$ and the $x$-coordinate, $x_{min} = g\,t$, of the track rim:

$$L = x_{max} - x_{min} = x_{max}(t) - g\,t . \tag{4.52}$$

### Submicroscopic track shape [1]

Up to this point, track etching was described by a one-dimensional track etch rate $v(x)$. This approach is appropriate for track radii $R \gg r_{track}$, where $r_{track}$ is the radius of the latent track. For the generation of very fine structures with $R \approx r_{track}$ — comparable to the size of the latent track — it is necessary to consider also the radial dependence of the etch rate, that is $v(x) \to v(x, r)$, in order to determine the microscopic track etch profile.

Neglecting the linear dependence of the radiation damage on $x$ which varies slowly in comparison with its radial dependence — the submicroscopic track etch theory [2] assumes a purely radial dependence of the etch rate $v = v(r)$ and determines the envelope of a set of different etch fronts generated at arbitrary distances $r_0$ from the track axis. Therefore the etch profile is the envelope of the envelopes of the etch fronts generated for different distances $r_0$ from the track axis. This theory describes the etched track profile even in the case of an infinitesimally small track radius $R \gg r_{track}$ [3] for which a constant track etch rate $v = $ const can be assumed for simplicity.

The evolution of the track etch profile can be also analyzed using the least-time principle [4]. In this way, the least-time etch-trajectory can be determined from any arbitrary point $(0, r_0)$ of the the original surface to the corresponding point $(x_1, r_1)$ of the etch profile. The time $t$ necessary to reach a point $(x_1, r_1)$ of the etch profile is determined by starting at the arbitrary point $(0, r_0)$ of the original surface and following the corresponding trajectory $r(x)$:

$$t = \int_0^t dt = \int_0^{x_1} \frac{ds}{v[r(x)]} \equiv \int_0^{x_1} \frac{\sqrt{dx^2 + dr^2}}{v[r(x)]} \equiv \int_0^{x_1} f(x)\,dx , \tag{4.53}$$

$$\text{where } f(x) \equiv \frac{\sqrt{1 + \left(\frac{dr}{dx}\right)^2}}{v[r(x)]} \equiv \frac{\sqrt{1 + [r'(x)]^2}}{v[r(x)]} . \tag{4.54}$$

[1]  This section is due to Dr. Omar Bernaola, Department of Radiobiology / Atomic Energy Commission, Av. del Libertador 8250, 1429 Buenos Aires / Argentine.

[2]  Mazzei, R., O.A. Bernaola, G. Saint Martin, J.C. Bourdin, J.C. Grasso: "Submicroscopic Nuclear Track Kinetic Theory Applied to Initial Chemical Etching of Makrofol E. Nuclear Instruments and Methods, **B17**, pp. 275-279, (1986).

[3]  Mazzei, R., G. Saint Martin, O.A. Bernaola, J.C. Bourdin, J.C. Grasso: "Applications of the Submicroscopic Nuclear Track Kinetic Theory." Nuclear Instruments and Methods, **B34**, pp. 237-242, (1988).

[4]  Mazzei, R. J.C. Grasso, O.A. Bernaola, J.C. Bourdin, G. Saint Martin: "The Submicroscopic Track Kinetic Theory and the Variational Principle." Nuclear Instruments and Methods, **B34**, pp. 74-80, (1988).

Here $ds$ is the path element along the least-time trajectory, $v = v[\,r(x)\,]$ is the etch velocity, $r = r(x)$ is the least-time trajectory, and $r'$ is the derivation of $r$ with respect to $x$.

The used boundary conditions for the variational problem are that the arbitrary starting point $r(0) \equiv r_0$ can move freely on the original — unetched — surface and that the end point $r(x_1) = r_1$ is a fixed point on the etch profile. Applying the least-time principle to the etch time $t$, the corresponding Euler-Lagrange equations for the function $f(x)$ are (see for example reference [1]):

$$\frac{dr}{dx}\bigg|_{x=0} = 0 \quad \text{and} \tag{4.55}$$

$$\frac{d}{dx}\frac{\partial}{\partial r'}f = \frac{\partial f}{\partial r} \quad . \tag{4.56}$$

According to eq. (4.55), the least-time trajectories $r(x)$ are perpendicular to the original surface. From eq. (4.56) we obtain by insertion of $f\,[r(x)]$:

$$\frac{d^2r}{dx^2} + \left(1 + r'^2\right)\left(\frac{\partial}{\partial r}\ln(v)\right) = 0 \quad \text{or} \tag{4.57}$$

$$\frac{1}{2}\int_{r'(0)}^{r'(x)}\frac{d\left(r'^2\right)}{\left(1 + r'^2\right)} = -\int_0^x d\left(\ln(v)\right) \tag{4.58}$$

From eq. (4.55) and eq. (4.58) we obtain:

$$x = \int_{r_0}^{r}\frac{d\rho}{\sqrt{\left(\frac{v(r_0)}{v(\rho)}\right)^2 - 1}} \quad . \tag{4.59}$$

This defines the function $x(r)$, from which by numeric inversion the least-time trajectory $r(x)$ can be determined. For an arbitrary starting point $(0, r_0)$, eq. (4.59) corresponds to the least-time trajectory, and eq. (4.53) establishes the time necessary to reach the point $(x_1, r_1)$ along it.

Assuming a specific function $v(r)$, eq. (4.59) can be solved numerically for each given value of $r_0$. First, for each value of $r$, according to eq. (4.59), the point $(x, r)$ is determined. Second, the function $f(x)$ is evaluated and inserted in eq. (4.53) in order to determine the corresponding time $t$. The procedure is terminated when a value of $x$ is reached that satisfies eq. (4.53) for the given etch time $t$.

In this way, for a given etch time $t$, the last points of the least-time trajectories can be compared with the corresponding points of the experimental etch profile (Figure 4-30).

[1]     Bronstein, I.N., K.A. Semendjajew: "Taschenbuch der Mathematik." Verlag Harri Deutsch, Zürich, 1965, pp. 484-489, (1965).

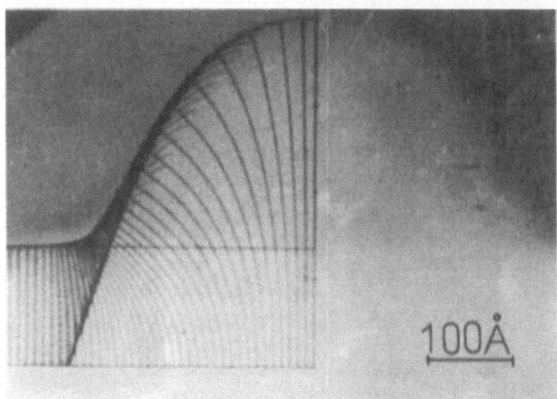

**Figure 4-30** Experimental track profile with overlayed least-time trajectories [1] starting perpendicularly to the horizontal original surface and ending perpendicularly on the etched track profile. The envelopes from the submicroscopic track etch theory are also shown.

**Oblique tracks.** If the track is not perpendicular but oblique to the surface normal, all preceding steps are identical, however, the planar cut performed at distance $g\,t$ from the surface of the track recorder has to be performed oblique with respect to the track. This cut is always parallel to the planar original surface of the track recorder.

## 4.3.4    Tracks in crystals

Track etching in crystals depends on the crystal structure (Figure 4-31), on crystal orientation (Figure 4-32), and on the specific etching conditions (Figure 4-33 and Figure 4-34). Quite complex structures can be obtained, depending on the chosen crystal, the angle of incidence, the chosen plane, and the etching medium (Figure 4-35).

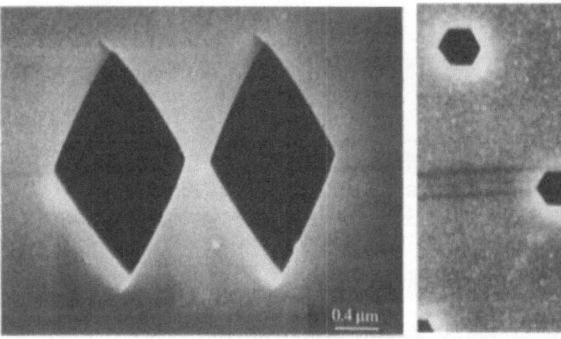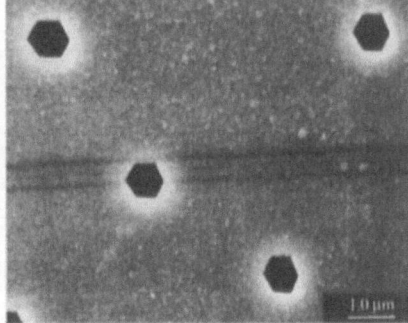

**Figure 4-31** *Crystal structure and track shape.* **(left)** Rhombic track cross-sections obtained in muscovite mica ($KAl_2Si_3AlO_{10}(OH,F)_2$), etched in 40% hydrofluoric acid. **(right)** Hexagonal track cross-sections obtained in iron garnet, etched 30 min in a mixture of 25% $HNO_3$ + 25% $CH_3COOH$ + 50% $H_2O$ at 70° C [2].

[1]    Dr. Omar Bernaola, Department of Radiobiology, Atomic Energy Commission, Av. del Libertador 8250, 1429 Buenos Aires, Argentine.

[2]    Krumme, J.P. I. Bartels, B. Strocka, K. Witter, Ch. Schmelzer, R. Spohr: "Pinning of 180° Bloch Walls at Etched Nuclear Tracks in LPE-Grown Iron Garnet Films." Applied Physics, **48**, 5191-5196 (1977).

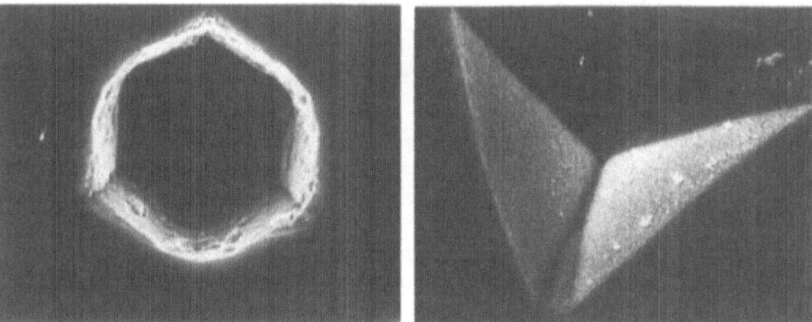

**Figure 4-32** *Crystal orientation and track shape.* Gd-Ga garnet, irradiated at 90° and 60° with respect to the surface plane, a {111} plane and etched few minutes in melted KOH [1].

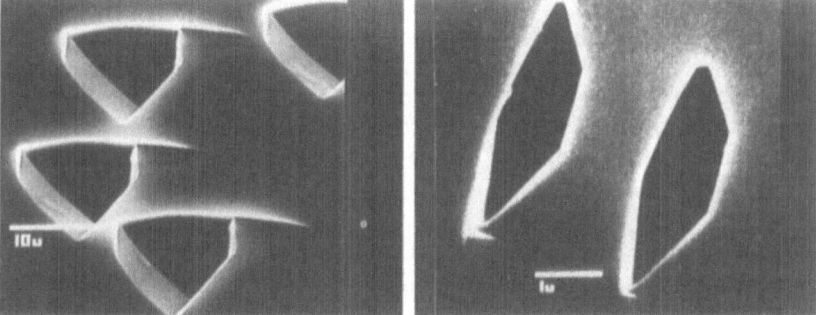

**Figure 4-33** *Etching medium and track shape.* Ion tracks in quartz oscillator foils [2]. (**left**) etched 15 h in 15 n NaOH at 133° C. (**right**) etched 10 h in 15 n KOH at 130° C.

**Figure 4-34** *Etching medium and track shape.* (**left**) Ion tracks in muscovite mica ($KAl_2Si_3AlO_{10}(OH,F)_2$), etched in HF, rounded by ion etching under vacuum conditions. (**right**) Ion tracks in muscovite mica, etched in HF, rounded by second etching in NaOH [3].

[1]     Thiel, K., Department for Nuclear Chemistry, Institute for Biochemistry, University of Cologne, D-5000 Köln.
[2]     Vater, P.: "Production and Applications of Nuclear Track Microfilters." Nuclear Tracks and Radiation Measurements, **15**, pp. 743-749, (1988).
[3]     Khan, H.A., N.A. Khan, R. Spohr: "Scanning Electron Microscope Analysis of Etch Pits Obtained in a Muscovite Mica Track Detector by Etching in Hydro-Fluoric Acid and Aqueous Solutions of NaOH and KOH." Nucl. Instrum Methods, **189**, 577, (1981).

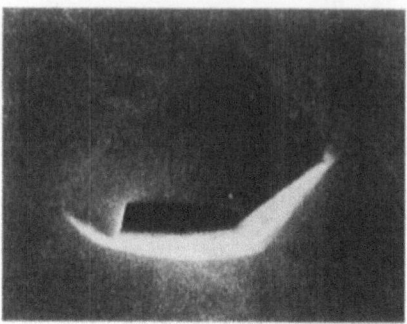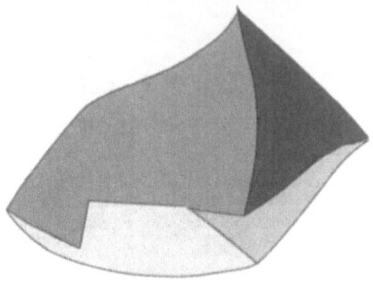

**Figure 4-35** *Complex track shape in Albite* (NaAlSi$_3$O$_8$) [1]. (**left**) Scanning electron micrograph. (**right**) Schematized observation.

## Prospects of channeled ion tracks

Channeling in crystals permits the generation of aligned ion tracks, related with fascinating structural features of the resulting track-etch technique [2] (Figure 4-36). A likely scheme to reveal channeled and suppress dechanneled tracks resides in their different lengths (Figure 4-37): first, a short pre-etching of the channeled latent tracks from the back side of the track recorder, second, the thermal annealing of the remaining latent tracks, and, third, the etching of the pre-etched tracks to the desired final diameter.

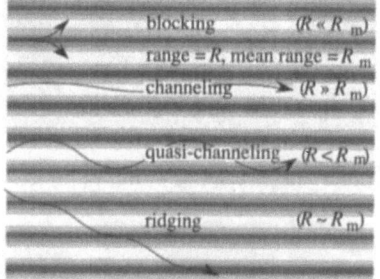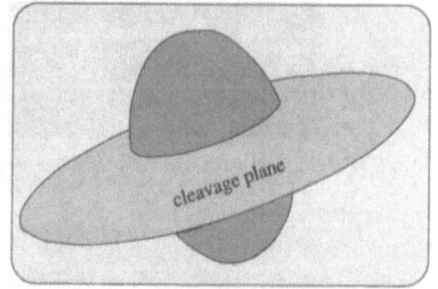

**Figure 4-36** *Principle of ion channeling in crystals* [3]. (**left**) Classes of ion trajectories. (**right**) Range-surface in layered crystal as function of track orientation to crystal plane .

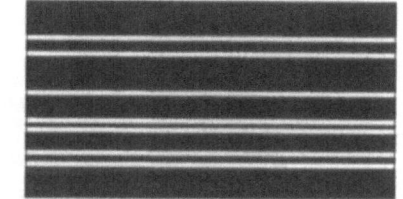

**Figure 4-37** *Channeled track generation in crystals.* (**a**) Latent tracks. (**b**) Resulting etched tracks after pre-etching from back-side, annealing, and final etching of remaining tracks.

[1] Thiel, K., Department for Nuclear Chemistry, Institute for Biochemistry, University of Cologne, D-5000 Köln.

[2] Chadderton, L.T., CSIRO, Research School of Physical Sciences, Institute of Advanced Studies, Australian National University, GPO Box 4, Canberra.

[3] Chadderton, L.T., S. Ghosh, A. Saxena, K.K. Dwivedi, J.S. Biersack, D.W. Fink: "On the Origin of Anomalous Ranges of 1.65 GeV $^{132}$Xe ions in Muscovite Mica." Nuclear Instruments and Methods, to be published (ca. 1989).

# 5      Observation of ion tracks

The use of ion tracks as a structuring tool depends very much on the availability of suitable observation techniques characterized according to the observed effects and recovered information:

1.      Physical effects: electron, nuclear, or radical densities, ionization, conductivity, diffusion, mechanical strength, thickness, various surface and volume effects.

2.      Recovered information: number of tracks, track diameter, track length and shape, density distribution of defects, individual tracks, track ensembles, various surface and volume properties.

### On-line and off-line observations

A possible classification of observational techniques is according to the time-delay between the irradiation and the observation. Observations performed during the irradiation at the ion accelerator are defined as "on-line" observations. Observations done after the irradiation — often at another location — are defined as "off-line" observations. On-line observations enable a direct feed-back of the gained information to the irradiation for adapting the irradiation parameters. Off-line observations are used on one hand to confirm the correctness of adjusted irradiation parameters and on the other hand to confirm physical changes of the irradiated samples after the irradiation took place. While on-line observations have to give almost instantaneous results, off-line observations can be performed without any haste often much more conveniently.

## 5.1      Microscopic observations

Microscopic observations have the advantage that they yield visual images of objects corresponding directly to our human intuition. The quantitative evaluation of microscopic images, however, requires often automated pattern recognition techniques which are still in their infancy.

### 5.1.1      Optical microscope

The generation of ion tracks of specified size and shape requires a microscopic control of the achieved results. The light-optical microscope (OM) provides adequate resolution down to or slightly below the wavelength of visible light of about 0.5 µm (Figure 5-1). Measuring microscopes enable to determine lateral dimensions of etched tracks with a precision of slightly less than 0.1 µm. Several techniques are available to enhance the microscopic contrast — for example by filling the etched tracks with a dye, by metal deposition within the etched tracks, or by metal evaporation on the surface of the etched track recorder. The main disadvantage of optical microscopy is its small depth-of-focus. Its great advantage its using ease requiring only very little preparation.

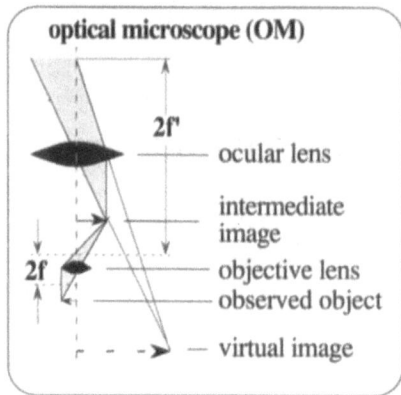

**Figure 5-1** Principle of optical microscope.

## 5.1.2  Scanning electron microscope

The scanning electron microscope (SEM) provides excellent shape contrast, resolution and depth-of-focus (Figure 5-2).

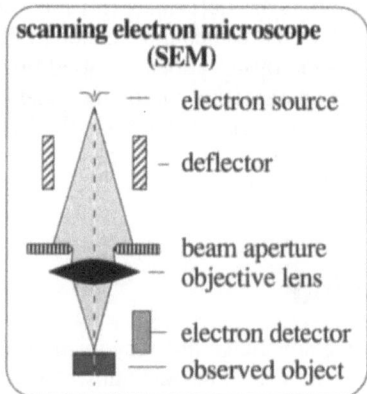

**Figure 5-2** Principle of scanning electron microscope (SEM).

The resolution of the SEM is about one order of magnitude better than the resolution of the optical microscope and is mainly limited by the diffusion-radius of back-scattered electrons from the depth of the sample which is around 0.01 μm or 10 nm. The main disadvantage of the SEM is its difficulty to observe structures deep inside etched tracks which requires microtomy or replica techniques. Due to the large interaction of the beam electrons with the electrons in the target, optically transparent samples appear opaque with respect to the scanning electron beam. The SEM observations require the deposition of a thin conductive metal film on the sample surface and have to be performed under vacuum conditions. During the drying of wet-etched tracks, very fine structures usually shrink or collapse, due to contractive force of surface tension. Critical-point drying is a useful technique to circumvent this problem.

### 5.1.3    Transmission electron microscope

For the high-resolution observation of latent ion tracks as well as of very fine etched tracks the transmission electron microscope (TEM) is an indispensable tool (Figure 5-3). For this purpose, however, very elaborate preparation techniques are required. The mainly used preparation techniques are thinning, vapor deposition, shading, and replication. The resolution of the TEM is again about one order of magnitude better than that of the SEM. It corresponds to 0.001 µm or 1 nm. There exist also instruments with a resolution limit of about 0.2 nm, capable to resolve individual atomic lattice planes.

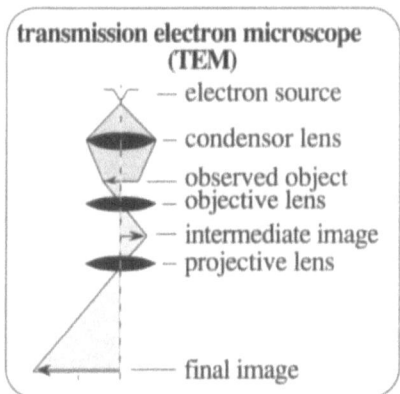

**Figure 5-3**  Principle of transmission electron microscope (TEM).

The interaction of electrons of a few eV with the atoms in a solid is so strong that they are completely absorbed after passing through only a few atomic layers. In order that an electron beam can penetrate very thin specimen of about 0.1 µm thickness, energies of the order of 100 keV are required. Specimen thinning is achieved by chemical or electrochemical etching, or by sputtering techniques. In high voltage electron microscopes with electron energies of the order of 1 MeV, specimens of about 1 µm thickness can still be observed. In all cases the image on the screen is a projection through the entire thickness of the specimen.

#### Observation of latent tracks

Magnified direct images of diffracted electron beams are possible and represent an important contrast-forming mechanism in the observation of tracks in crystals.

Under favorable conditions it is possible to resolve individual atomic planes in the transmission electron microscope. However, such high resolution (0.2 to 0.3 nm) is not required to detect latent tracks. Instead advantage is taken of the fact that in the neighborhood of a latent tracks the crystal is subject to internal stresses corresponding to a local deformation of the crystal planes.

When a single crystal is placed in the electron microscope, the electrons which pass through the specimen are diffracted into angles which are determined by the Bragg equation (5.2). For an arbitrary angle-of-incidence the produced image will be completely featureless. But if the specimen is rotated in such a way that a strong diffracted beam is generated, then the slight misorientation around latent tracks gives rise to a sharp change in image contrast and the latent tracks can be clearly observed. As the crystal is rotated, the contrast varies. However, the observed structure, corresponding to the misoriented zone around the latent track, appears much thicker than the real structure of the latent tracks which is of the order of 10 nm.

Transmission electron microscopy of ion damage in solids requires crystals sufficiently stable to the bombardment of the observing electron beam and depends on the diffraction contrast between lattice zones with slightly different grid sizes. The technique yields highly detailed pictures of individual tracks (Figure 3-17). However, it is difficult to extract quantitative data on the actual defect density from the obtained electron micrographs.

Without the art of using diffraction contrast imaging, latent ion tracks are practically inaccessible to the observation by transmission electron microscopy since the density difference between the ion track and the undisturbed material is too small .

## 5.2     Diffraction techniques

While optical methods enable the observation of individual tracks, diffraction techniques usually yield average observations of many tracks. The principal properties of diffraction techniques are shown in Figure 5-4.

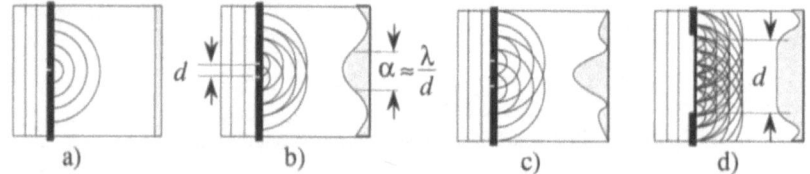

**Figure 5-4**  *Classification of interference phenomena.* (**a**) Spherical wave and corresponding intensity distribution. (**b**) and (**c**) Interference between two spherical waves and corresponding interference patterns for two object sizes or distances *d* between the spherical-wave centers. The diffraction angle α is proportional to the wave length λ and inversely proportional to the size *d* of the object. (**d**) Transition to classical optics with negligible diffraction.

The diffraction of a planar wave (a) at a very small object corresponds to a single spherical wave spreading with equal intensity distribution into the full solid angle. Spherical  waves originating from two small objects at distance $d$, (b) and (c), lead to an interference pattern with a maximum in forward direction and an angular width $\alpha \approx \lambda /d$ which is proportional to the wavelength $\lambda$ and inversely proportional to the distance $d$ of the objects. The same angular width is obtained for an extended object of total width or size $d$. For a large object with $d \gg \lambda$ the angular spread $\alpha$ is negligible in comparison with the lateral width of the beam which corresponds to the size $d$ of the object (d). This case corresponds to classical optics with a linear spreading of light "rays".

For an extended object the total intensity distribution is obtained from the quadrature of the total scattering amplitude. The total scattering amplitude is obtained by summing over the individual scattering amplitudes from infinitesimal volume-elements of the object. The scattering amplitude of an arbitrary volume-element is proportional to its size and scattering length, inversely proportional to the distance between the object and the detector, multiplied by a phase factor $\exp(i\ \varphi)$ which depends on its relative position to a phase reference point — the origin $O$ of the reference frame — which is arbitrary but kept fixed throughout the summation. Figure 5-5 shows the calculation of the phase difference $\varphi$ with respect to the origin $O$ by dividing the difference between the path-lengths

$s$ - $s_{ref}$ along the two respective rays by the wave length and multiplying the result by $2\pi$:

$$\varphi = \frac{2\pi\,(s\ -\ s_{ref})}{\lambda} = r\,(k\ -\ k')\ ,\quad \text{where}\ |k\ |=|k'|=\frac{2\pi}{\lambda}\ . \tag{5.1}$$

Here the wave vectors $k$ and $k'$ are introduced for convenience. They point in the direction of the primary and scattered beam and have in the case of elastic scattering — in other words, scattering without energy change — identical magnitudes $k = k' = 2\pi / \lambda$.

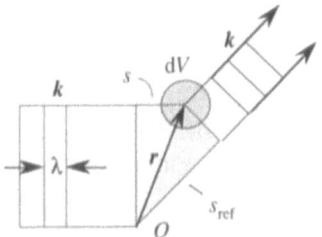

**Figure 5-5** *Calculation of phase difference between scattered and reference beam.* The phase difference corresponds to the difference of the path-lengths, marked in the shaded triangles as thick lines, divided by the wave length and multiplied by $2\pi$.

### 5.2.1    Basic principles

**Bragg diffraction**

The Bragg construction is used to determine the angles at which diffraction of radiation, such as x rays, electrons, or neutrons, occurs in crystals (Figure 5-6) [1].

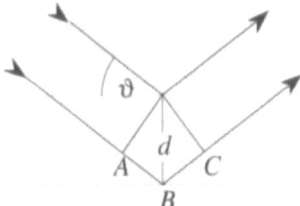

**Figure 5-6** *Derivation of Bragg's diffraction relation.* Constructive interference occurs if the phase difference of beams, reflected from two parallel neighboring planes of the crystal lattice, is an integer multiple $n$ of the wavelength $\lambda$.

Under the assumption of a specular reflection from the crystal planes, the reflected waves will interfere constructively, if their path difference $AB + BC$ corresponds to an integer number $n$ of wavelengths $\lambda$. Since $AB = BC = d \sin \vartheta$, where $d$ is the interplanar spacing, we obtain the Bragg condition

$$2\,d\ \sin \vartheta\ =\ n\ \lambda\ ,\ \text{which for small angles } \vartheta \text{ yields} \tag{5.2}$$

[1]  H.M. Rosenberg: "**The Solid State. An Introduction to the Physics of Crystals for Students of Physics, Materials Science, and Engineering.**" Oxford Physics Series. Oxford University Press, Oxford, pp. 1 - 274, (1975).

$$\vartheta = \frac{n\,\lambda}{2\,d} \quad \text{for } \vartheta \ll 1 \ . \tag{5.3}$$

Since always $\sin\vartheta \le 1$ we obtain as an upper limit for $\lambda$, at which an interference can still be observed, the condition

$$\lambda \le 2\,d\ . \tag{5.4}$$

Therefore no diffraction occurs for wavelengths $\lambda$ exceeding twice the interplanar spacing $d$, a condition which is necessary for small-angle scattering observations without the interference of signals from the crystal structure of the track recording material.

### De Broglie wavelength

The de Broglie wavelength $\lambda$ of a particle is given by $\lambda = h / p$, where $h$ is the Planck constant and $p$ is the momentum of the particle. The magnitude of the associated wave vector $k$ is $k = 2\pi / \lambda = (2\pi / h)\,p$. In terms of the non-relativistic kinetic energy $E$, $k = (2\pi / h)\,(2\,m\,E)^{1/2}$, and $E = [h / (2\pi)]^2\,k^2 / (2\,m)$. If $k$ is of the order of $10^8$ cm$^{-1}$ then $E$ will be about $6\bullet10^{-12}$ erg (4 eV) for electrons and about $3\bullet10^{-15}$ erg ($2\bullet10^{-3}$ eV) for neutrons.

In microscopic imaging techniques the de Broglie wave-length normally is smaller than the size of the observed object. In diffraction techniques, the de Broglie wave-length normally has about the magnitude of the observed object.

### Choice of the diffraction technique

Since the general form of the diffraction pattern will be the same whether x rays, electrons, or neutrons are used, one must decide which diffraction technique is most satisfactory for the various types of investigation.

In general x ray diffraction is the most convenient method and is by far the most widely used. X rays can be detected photographically or with an electronic counter, the latter enables the data to be recorded automatically in digital form for computer input. X rays are not absorbed very much by air. Therefore the specimen need not always be in an evacuated chamber. They do, however, have the disadvantage, that they do not interact very strongly with the lighter elements.

## 5.2.2    Small-angle scattering

Small-angle scattering enables the observation of large track ensembles with radiations for which as yet no optical microscopy exists, such as using x rays and neutrons. Also, large ensembles of etched tracks can be surveyed simultaneously using laser scattering. Or very small objects such as latent tracks embedded in thick volumes of undisturbed matter can be observed using small-angle x ray or neutron scattering. The applied wavelength is of the order of the object size or somewhat smaller. While x ray scattering corresponds to the electron density distribution of the target, which is always a positive quantity, neutron scattering corresponds to the scattering density distribution of the target nuclei, which may be positive or negative, depending on the interaction between the neutrons and the different target nuclei, associated with positive or negative values of the scattering lengths.

The small angle scattering of x rays or neutrons (Figure 5-7 and Figure 5-8) is capable to collect low-resolution information from many oriented ion tracks and yields quantitative data on the defect density in tracks even for thermally quite unstable organic polymers.

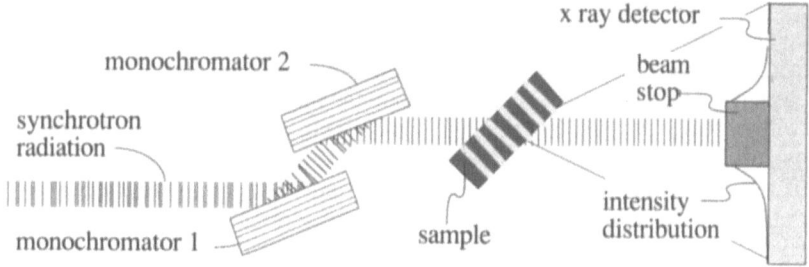

**Figure 5-7**  Principle of small-angle x ray scattering.

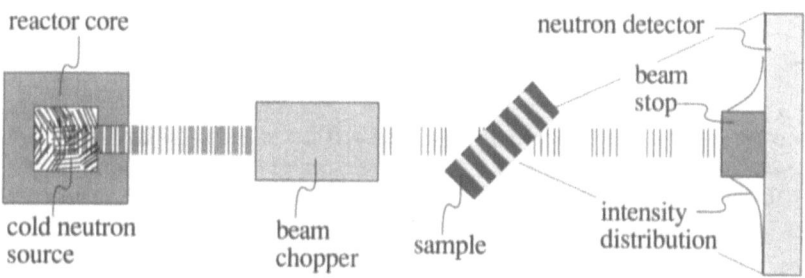

**Figure 5-8**  Principle of small-angle neutron scattering.

The main results of small-angle scattering experiments are the objective determination of a track radius $a$ and of a track density deficit — more explicitly, the 1/e radius $a$ and the maximum density decrease $(\Delta\rho / \rho)_{max}$ at the track axis with respect to the density $\rho_0$ of the undisturbed material — of an equivalent cylindrical track with a Gaussian radial density distribution [1], [2], [3]. The combined information of the track radius and the track density yields a quantitative picture of the ion track (Figure 5-9 ).

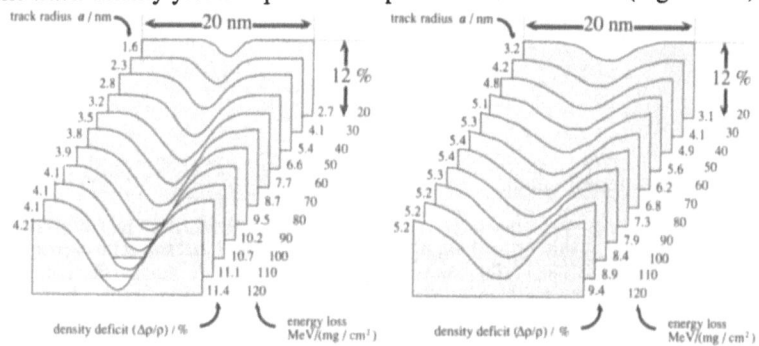

**Figure 5-9**  *Track structure* of muscovite mica (**left**) and polyethylene terephthalate (Hostaphan BN19, Hoechst AG, D-6200 Wiesbaden) (**right**) from small angle scattering.

[1]  Albrecht, D., P. Armbruster, M. Roth, R. Spohr: "Small Angle Neutron Scattering Observations from Oriented Latent Nuclear Tracks." Radiation Effects, 65, 145-148 (1982).

[2]  Albrecht, D., P. Armbruster, R. Spohr, M. Roth, K. Schaupert, H. Stuhrmann: "Investigation of Heavy Ion Produced Defect Structures in Insulators by Small Angle Scattering." Applied Physics, A37, 37-46 (1985).

[3]  Albrecht, D.: "Untersuchung der von schweren Ionen in Dielektrika erzeugten Defektstrukturen mittels Kleinwinkelstreuung." Dissertation, Technische Hochschule Darmstadt, D-6100 Darmstadt, GSI-Report 83-13, (1983).

The interaction of neutrons with most atomic nuclei is very weak and they penetrate most materials quite readily. Although the diffracted beams may not be very intense, they are, unlike electrons, able to escape from the sample very easily and can be detected with high efficiency. If we equate the minimum energy (maximum wavelength) for diffraction of about $3 \cdot 10^{-15}$ erg ($\lambda = 0.65$ nm), to the thermal energy $(3/2)\, k\, T$, then this corresponds to the mean thermal energy at a temperature of about 14.5 K. Neutron diffraction requires monoenergetic neutrons from a nuclear nuclear with a typical total flux of about $10^{14}$ neutrons per cm$^2$ per sec. After thermalization of the neutrons in a liquid hydrogen bath they escape from the reactor, are collimated and pass through a velocity selector, so that they are mono-energetic, before being used in neutron diffraction investigations at much reduced intensities of about $10^8$ monochromatic neutrons per s.

### 5.2.3    X ray topography

Lattice distortion used as contrast mechanism in TEM is also used in x ray topography. Changes in the intensity of the diffracted beam as it passes through or is reflected from different parts of the specimen can give an indication on the position and areal density of latent tracks. It is a useful method for examining ion tracks in nearly perfect crystals with low intrinsic dislocation densities.

# 5.3    Auxiliary techniques

### 5.3.1    Electron spin resonance

The formation and decay of chemical radicals formed along ion tracks can be selectively observed by the electron spin resonance technique. Chemical binding is associated with the mutual compensation of paired electron spins from the binding partners according to the scheme of Figure 5-10.

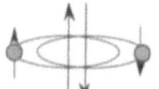    

a) saturated valence      b) radical electron

**Figure 5-10**  *Principle of electron spin resonance* (ESR) used for observing chemical radicals along freshly formed ion tracks. (**a**) The saturated valence-state corresponds to two mutually compensated orbital and magnetic momenta. (**b**) A radical state corresponds to an unpaired electron. Its interaction with the orbital magnetic moment leads in an external magnetic field to characteristic energy states.

Saturated valences correspond to mutually compensated orbital and magnetic momenta and therefore are little susceptible to an external magnetic field. However, chemically reactive radicals are characterized by free, unpaired electrons and are subject to a characteristic splitting of the electron binding energies in an external magnetic field. The corresponding transitions can be induced by a high frequency field. In addition to the external field, there exists an interaction with the magnetic field due to the spin of the atomic nucleus. This leads to a hyperfine splitting of the observed energy levels which enables the identification of the chemical radicals formed along ion tracks in solids by characteristic finger-prints.

## 5.3.2 Electrical observations

### On-line observations during the irradiation

**Counting ion impacts.** Due to the high energy of accelerated ions in comparison with the ionization energy of the target atoms in gaseous or solid matter, the on-line counting of individual ions is a relatively simple task as long as the count-rates do not exceed device-specific values between $10^3$ and $10^6$ counts per s. The counting of individual ion impacts on the target can be performed for example using solar cells, silicon surface-barrier detectors, ionization chambers, channeltrons, break-down counters, or charged-coupled devices. At counting rates and at total count numbers exceeding device-specific values it is necessary to reduce the ion current hitting the detectors by small apertures or by increasing the distance between the counter and the scattering target or aperture.

Most electronic particle counters are not position-sensitive detectors in the sense that they do not give information on the the the site of the ion impact on the detector surface. The break-down counter consists of an insulator deposited on a conducting substrate and covered by a thin metallic film, for example a sandwich consisting of a silicon substrate covered by an oxide layer and a deposited aluminum film. It represents a hybrid between an electronic detection system and a track recorder (Figure 5-11).

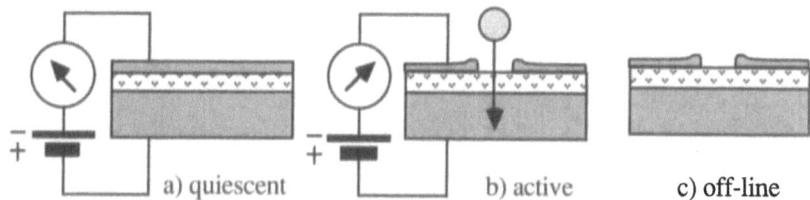

a) quiescent     b) active     c) off-line

**Figure 5-11** *Break-down counter,* for on-line counting of ion impacts during the irradiation as well as for off-line observation of ion tracks accumulated on the counter surface after the irradiation. (**a**) Quiescent state. (**b**) Ion impact associated with an electrical break-down and the evaporation of a small circular zone of the top metal film around the ion track. (**c**) Off-line counting of evaporated zones after the irradiation.

**Determination of total ion energy.** The working principles of ionization chambers, silicon surface-barrier detectors, and solar cells are very similar (Figure 5-12). The number $N$ of ion pairs created within the active volume of these detectors is proportional to the total energy $E$ of the projectile-ion divided by the average ionization potential $I$ for an atom in the detector:

$$N = \frac{E}{I} . \tag{5.5}$$

For thick targets, in other words if the projectile-ion is completely stopped in the ionizing medium the charge $-N\,e$ collected on the positive electrode is proportional to the total energy $E$ of the projectile-ion. Depending on the density of atoms within the detector, the required size of the detector has to be relatively large for gas-filled detectors and relatively small for solid-state detectors. The collection efficiency of ionization chambers is increased by increasing the electrode voltage. Above the "proportional counting" regime, however, discharge avalanches are created which can be used to increase the total collected charge by many orders of magnitude for easier counting. However, at the same time the proportionality between the number of collected ions and the energy of the pro-jectile-ion is lost.

Gas-filled ionization chambers are long-lived since they can be regenerated by exchanging their gas filling if their entrance window endures. In contrast, the life-time of solid-state detectors is much shorter since they accumulate radiation-induced defects in the form of tracks. Some solid state detectors can be annealed thermally in order to regenerate their capability to precisely determine ion energies.

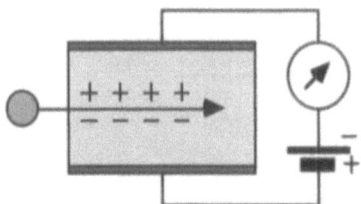

**Figure 5-12**  *Principle of (thick) ionization chamber* for measuring ion energies (*E*-detector) by completely absorbing the projectile-ion energy. At low collection voltages of the ioniza-tion chamber — in the proportional counting regime — the number *N* of collected ion pairs is proportional to the total energy *E* of the projectile-ion.

**Determination of ion energy-loss.** If the ionization chamber is thin in compari-son with the range of the projectile-ion within the ionizing medium (Figure 5-13), one obtains on the other hand a detector which measures the energy-loss ($dE / dx$) of the ion per unit path-length. The number *N* of formed ions thereby is proportional to the energy loss ($dE / dx$) per unit path-length, which according to eq. (2.145) is mainly a function of the effective charge $Z_{eff}$ of the projectile-ion and its energy *E*:

$$N = \text{const} \left(\frac{dE}{dx}\right) = \text{const} \frac{Z_{eff}^2}{E} \,, \tag{5.6}$$

where $Z_{eff}$ is the effective charge of the projectile-ion and *E* is its kinetic energy. Since the charge-collection efficiency of ionization chambers depends on the potential gradient, the collection efficiency can be increased in silicon surface-barrier detectors and in solar cells by reverse-bias operation. This increases the resistance and accordingly the potential gra-dient within the active zone of the detector. In nuclear physics experiments, a combination of such a thin ($dE / dx$)-detector with a (thick) *E*-detector in the form of an ionization chamber serves for simultaneously characterizing the energy and charge of emitted parti-cles created in nuclear reactions.

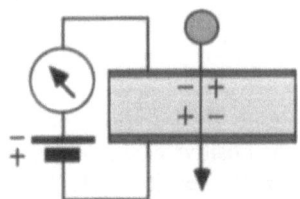

**Figure 5-13**  *Principle of thin ionization chamber* or ($dE / dx$)-detector, used for nonde-structive particle counting, for example in front of a track recorder.

### Off-line observations

**Spark counter.** Track counting with an optical microscope is a time-consuming procedure with a natural tendency toward subjective errors. To avoid these, one can make a simple track counting device out of any track-perforated foil by evaporation of two thin-film metal electrodes on both sides of the foil (Figure 5-14).

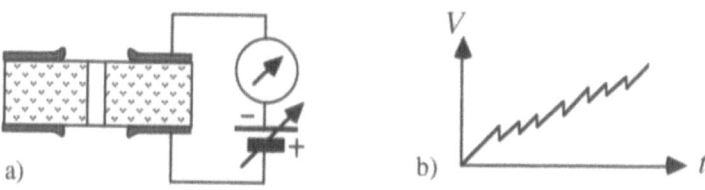

**Figure 5-14** *Principle of spark counter.* (**a**) Evaporation of thin film electrodes during registration of one pore. (**b**) Voltage-time characteristic during counting of many pores.

The track-perforated foil thereby serves as the dielectric of a self-healing capacitor. As the voltage over the capacitor is gradually increased, a succession of sparks occurs which can be counted electronically. Since each spark evaporates a well-defined circular area of the thin film electrodes, in this way a second spark cannot occur through the same etched hole and each hole is counted only once. A still simpler technique is to place the track-perforated foil between a conductive substrate and an aluminized foil.

**Conductivity cell.** Similar to the gas permeation through pores, the electrolytic conduction of electrical current through etched tracks is used to characterize their diameter [1], [2]. The individual track of length $L$ and cross-section $A = \pi\, r^2$ corresponds to a resistance $R = \rho\, L\, /\, A$, where $\rho$ is the specific resistivity of the electrolyte. The electrical current $I = dQ\, /\, dt$ trough $N$ cylinders of radius $r = d\, /\, 2$ and length $L$ is:

$$I = N\,\frac{U}{R} = N\,\frac{A\,U}{\rho\,L} = N\,\frac{\pi\,r^2\,U}{\rho\,L} = N\,\sigma\,\frac{\pi\,r^2\,U}{L}\,, \tag{5.7}$$

where $\sigma$ is the specific conductivity, $U$ is the applied voltage, and $\rho = 1\,/\,\sigma$ is the corresponding specific resistivity of the electrolyte.

## 5.3.3    Gas-permeation

The permeation of gas through etched pores of known diameter can be used for calibration of the pore diameters. In the laminar flow regime — valid under atmospheric pressure conditions at now too high flow rates — the mass current $I = dM\,/\,dt$ trough $N$ cylinders of diameter $r$ and length $L$ is given by

$$I = N\,\frac{\pi}{8\,\eta\,L}\,r^4\,\Delta P\,, \tag{5.8}$$

where $\eta$ is the viscosity of the transmitted gas, and $\Delta P$ is the pressure difference between the entrance and the exit of the channels.

[1]  DeBlois, R.W., C.P. Bean, R.K.A. Wesley: "Electrokinetic Measurements with Submicron Particles and Pores by the Resistive Pulse Technique." Colloid Interface Sci., **61**, 323 (1977).
[2]  Apel, P.Yu., G. Pretzsch: "Investigation of the Radial Pore-Etching Rate in a Plastic Track Detector as a Function of the Local Damage Density Around the ion Path." Nucl. Tracks Radiat. Meas. **11**, Nos 1/2, pp. 45-53, (1986).

### 5.3.4    Mechanical observations

#### Thickness of the etched sample

In isotropic materials, the pore diameter of etched tracks is directly related to the thickness of the material removed in the direction of the surface normal. If the track etch rate $v$ is large in comparison with the bulk etch rate $g$ — in other words for sufficiently large track etch-ratio $v / g$, the removed thickness $h - h'$ is directly related with the pore diameter $d$ (Figure 5-15). In this case the pore diameter $d$ corresponds roughly to the decrease in thickness $h - h'$ of the track recorder during etching. The observation can be performed either intermittently after interrupting or continuously during the etch process — for example by measuring the absorption of light or the electrical capacitance of the track recording material. This represents one of the simplest possibilities for an indirect observation of track sizes.

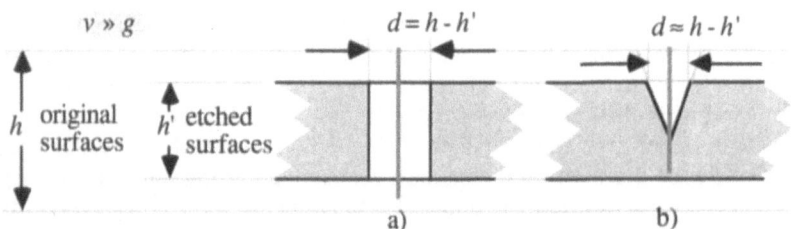

**Figure 5-15** *Thickness calibration of etched track diameters*, valid for track etch rates $v$ large in comparison with the bulk etch rate $g$. (**a**) For a perforated thin sample. (**b**) For a thick sample not penetrated by the ions or for a thin sample before perforation.

#### Double exposure technique

If a test zone of the sample is irradiated with a known areal dose $N$ of tracks, significantly higher than in the remaining sample, macroscopic property changes — such as light scattering or mechanical coherence — of the test zone can be used to interrupt the etching at a precisely defined single-pore size (Figure 5-16). In this way the observed effect is amplified, can be easily detected and used for calibration of the remaining sample. It can be used to interrupt etching at a precisely defined nominal porosity $P = N S$ in the test zone and thus a precisely defined pore area $S$ in the normal zone. The technique could be used for testing different etch recipes rapidly, however, it requires a precise control over the areal dose $N$ of the irradiation.

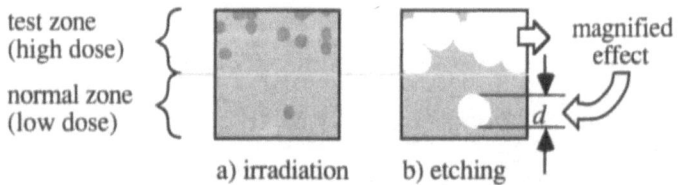

**Figure 5-16** *Double-exposure technique for track diameter calibration.* (**a**) Irradiated sample. (**b**) Etched sample around the time of a mechanical break-down of the test zone.

# 6 Resulting structures

## 6.1 Fundamental shapes of etched tracks

The ion track etch technique enables the generation of concrete shapes which can either be used individually in the form of single-ion tracks, or collectively in the form of track arrays composed of many tracks. In the following, a short overview on the possible shapes is given. The used techniques comprise mainly (1) variation of the track recording material, (2) variation of the etchant, and (3) partial or complete annealing.

### 6.1.1 Single-ion tracks

Due to the high local damage-density of ion tracks already one ion suffices to induce a submicroscopic change physically and chemically in the track recording material and render it susceptible to the development process. Single-ion tracks (Figure 6-1 to Figure 6-7) provide a number of shapes which can be employed individually or in the form of track patterns of many tracks. A practical realization is shown in Figure 6-8.

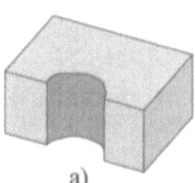 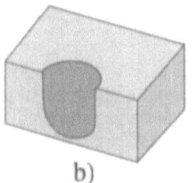 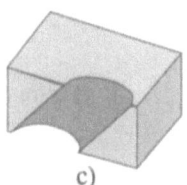

a)           b)           c)

**Figure 6-1** *Cylindrical track shapes* corresponding to high track etch-rates. (**a**) Cylindrical channel. (**b**) Spherically rounded cylinder just before perforation, obtained by, first, applying a short pre-etching, second, annealing the latent track after the etch-front has reached the required depth, and, third, etching to the desired track diameter. (**c**) Oblique cylinder.

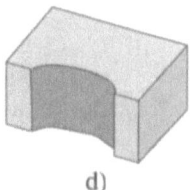 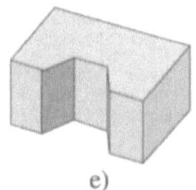 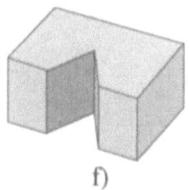

d)           e)           f)

**Figure 6-2** *Prismatic track shapes* in anisotropic track recorders at high damage densities or high track etch-rates. (**a**) Elliptical channel typical of one-dimensionally stretched polymer films. (**b**) Hexagonal channel typical of yttrium-iron garnet single crystals . (**c**) Rhombic channel typical of muscovite mica ($KAl_2Si_3AlO_{10}(OH,F)_2$) single crystals

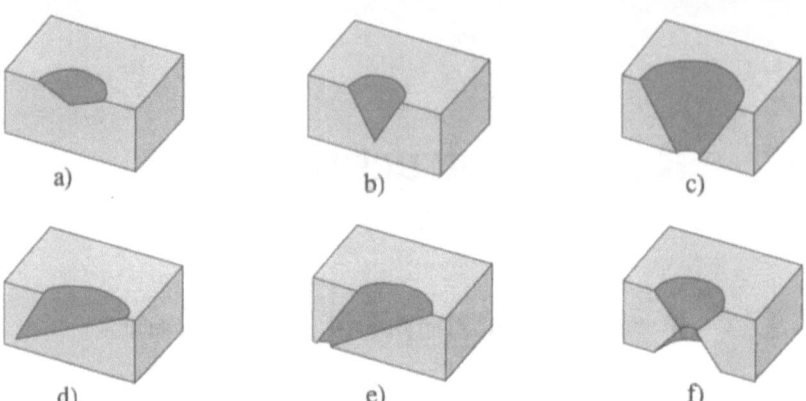

**Figure 6-3**  *Conical track shapes* obtained in isotropic track recorders — for example in glasses — for low damage densities or low track etch-rates. (**a**) Flat etch cone corresponding to very low track etch rate, for example a partially annealed track. (**b**) Somewhat steeper etch cone. (**c**) Conical perforation, etched only from top surface. (**d**) Oblique cone just before perforation occurs, etched only from top surface. (**e**) Same cone after perforation occurred. (**f**) Double cone, obtained by double-sided etching.

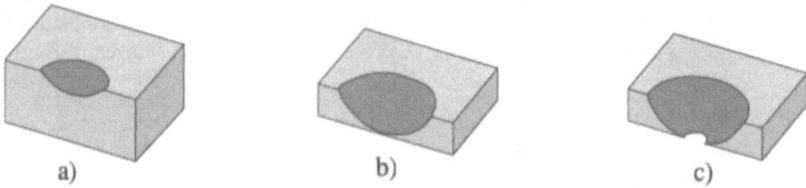

**Figure 6-4**  *Spherical tracks* obtained in isotropic track recorders at low to medium track etch-rates. (**a**) Shallow spherical zone obtained at low track-etch rate after the conical part of the etched track has been etched away. As limiting case, a half-sphere is obtained at high track-etch rates. (**b**) Somewhat deeper spherical zone, obtained at higher track etch-rates — just before the perforation occurs. (**c**) Same track just after the perforation occurred.

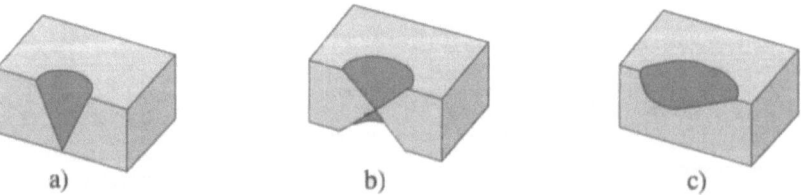

**Figure 6-5**  *Limiting etched track shapes*, obtained by interrupting the etch process after reaching certain conditions. (**a**) Etch cone just at the instant of perforation. (**b**) Double cone just at the instant perforation. (**c**) Spherically rounded etch cone just at the instant when the overhanging parts — the undercut — are removed.

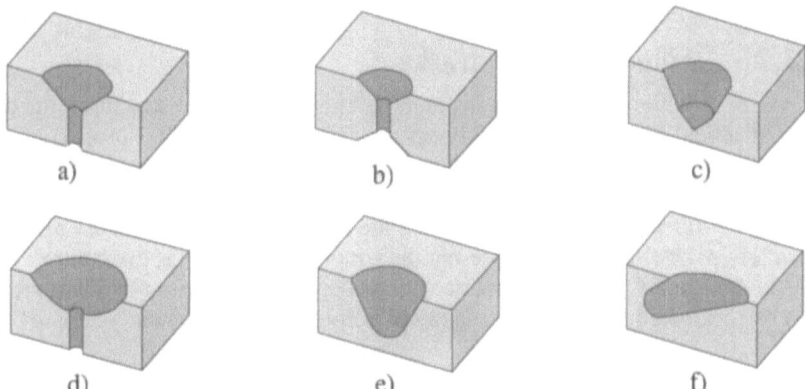

**Figure 6-6** *Two- and multi-step etch processes.* (**a**) Beveled cylinder — or funnel — obtained by, first, etching with an etchant of low track etch-rate — or low selectivity — from top surface and , second, etching with high track etch-rate from bottom surface. (**b**) Doubly beveled cylinder obtained by first etching with low track etch-rate and then with high track etch-rate. (**c**) Broken-angle cone obtained by, first, etching with medium track etch-rate, second, partial track annealing, and, third, continuing the etch process at reduced track-etch rate. (**d**) Spherical funnel obtained by, first, etching from bottom at high track etch-rate up to the required track length, second, etching from top down to the required track length, third, complete annealing of the latent track, fourth, etching from bottom to the desired track diameter, fifth, etching from top until the sphere and the cylinder meet. (**e**) Rounded cone, obtained by total annealing after reaching a certain depth of the track etch cone and continuing the etch process to the desired radius of the spherical zone. (**f**) Same as before, but oblique track.

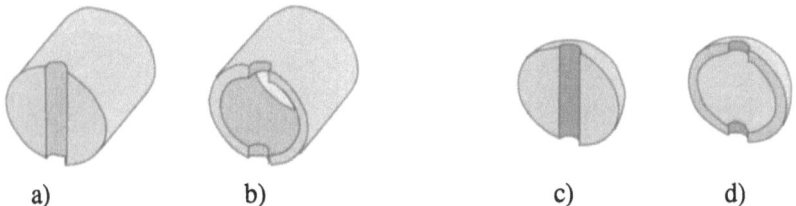

**Figure 6-7** *Track etching in cylindrically or spherically shaped track recorders.* The shown geometries correspond to aimed tracks through the axis or center of symmetry. (**a**) Perforated fiber. (**b**) Perforated hollow fibre. (**c**) Perforated sphere. (**d**) Perforated hollow sphere.

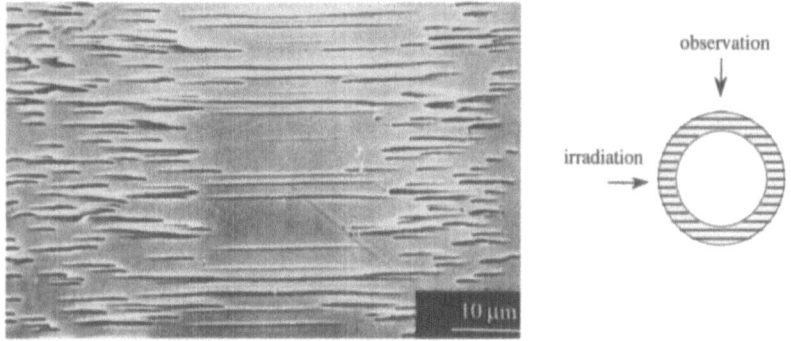

**Figure 6-8** Track-perforated polycarbonate tube [**1**].

[1]   Vetter, J., Gesellschaft für Schwerionenforschung (GSI), Planckstr. 1, D-6100 Darmstadt 11.

### 6.1.2    Non-overlapping tracks

While single-ion tracks represent highly localized structures, the overall effect of distributed ion tracks can be better described by the global properties of the solid, especially if the size of the tracks is smaller than the critical size of the observed phenomena.

Tracks can be distributed on the surface of the track recording material either regularly or stochastically. At present the simplest possibility is the generation of stochastically distributed tracks. At low track densities such stochastically distributed ion tracks can be considered as consisting mainly of non-overlapping tracks.

### 6.1.3    Overlapping tracks

With increasing track density more and more multiple tracks occur within a stochastical distribution of ion tracks on a given surface, ultimately replacing the matrix of the original track recording material completely. Multiple tracks display an infinite number of different shapes of unprecedented sharpness, thereby revealing a glimpse into the possibilities of future pattern generation with aimed microbeams. One example is shown in Figure 6-9

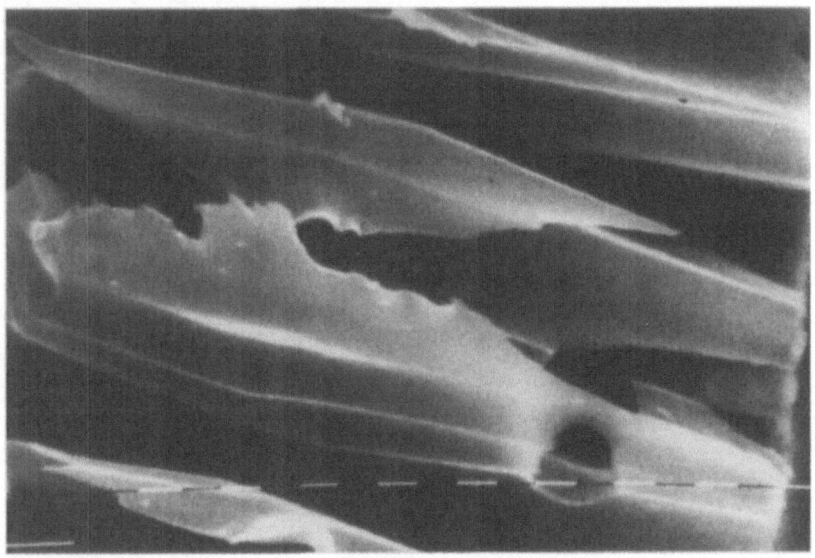

**Figure 6-9** *Overlapping track shapes* display a multitude of different forms and shapes, here very sharp, thin structures, due to overlapping double cones. Cross section trough 25 μm polyimide (Kapton, DuPont, CH-1200 Geneva), irradiated with $^{238}$U ions of 15 MeV/u, etched in NaOCl solution [1]. Bar unit = 1 μm.

Their main application at present is to imprint arbitrary structures in solids, for example in ion lithography.

[1]    Vetter, J., Gesellschaft für Schwerionenforschung (GSI), Planckstr. 1, D-6100 Darmstadt 11.

### 6.1.4 Further possibilities

#### Multi-layer track recorders

Conventional track recorders consist of a homogeneous material providing more or less constant bulk etch rate throughout the depth of the material. In contrast to these homogeneous materials, multi-layer track recorders consist of layers of different bulk etch rates and may be used in the future to generate a variety of unusual track shapes (Figure 6-10). In this way almost any arbitrary volume of rotation could be generated, comparable to the shapes obtained with the classical mechanical lathe. The miniature-sized "turned" bodies could open up new territory at the boundary of classical physical phenomena. A two-layer example is shown in Figure 6-11.

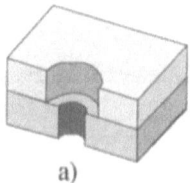 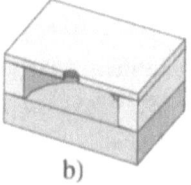 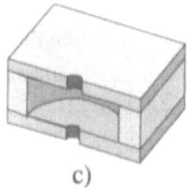

a) b) c)

**Figure 6-10** *Single tracks in two- or multi-layer track recorders* with different lateral etch-rates. (**a**) Stepped cylinder corresponding to two different lateral etch-rates. (**b**) Open cavity. (**c**) Cavity with entrance and exit opening.

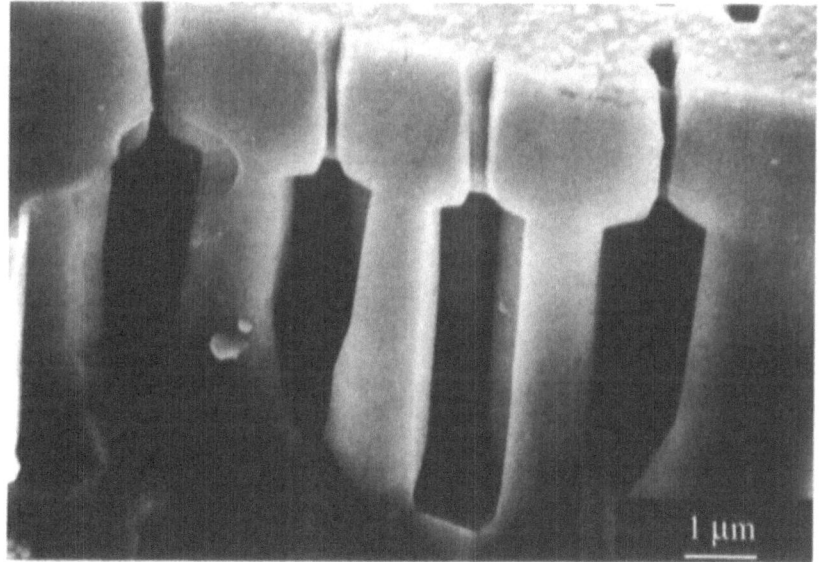

**Figure 6-11** *Open microcavities* in a double-layer track recorder — an epitaxially grown iron garnet film on top of a single-crystal garnet substrate — [1].

[1] Hansen, P., H. Heitmann, B. Strocka, R. Spohr: "Der Einfluß von Ionenstrahlen auf die magnetischen Eigenschaften von ferrimagnetischen Schichten." BMFT Bericht FB T83-048 Fachinformationszentrum Karlsruhe, D-7514 Eggenstein-Leopoldshafen 2, 1-180 (1983).

### Etch alternatives

Besides tack etching — representing a process that preferentially removes the activated zone along the ion trajectory — there exist alternative development processes corresponding to internal phase transitions which do not require the long range exchange of material with the environment. Examples are track grafting, metal precipitation, and the silver chloride system [1].

Using polymerizable — negative resist type — material one might be able in the future to replace in radiation dosimetry the optical observation of stochastically distributed etched tracks by a sequential electrical counting of track-equivalent aggregates using the resistive pulse technique [2]. In this way even composite materials with oriented fibers might be generated, whereby the fibers would consist of polymeric material aggregating along the latent ion tracks. This might enable ultralight structures which only nature is able to build by now.

### Replica techniques

The structuring possibilities of the ion track technique are further widened by using replication (Figure 6-12, Figure 6-13).

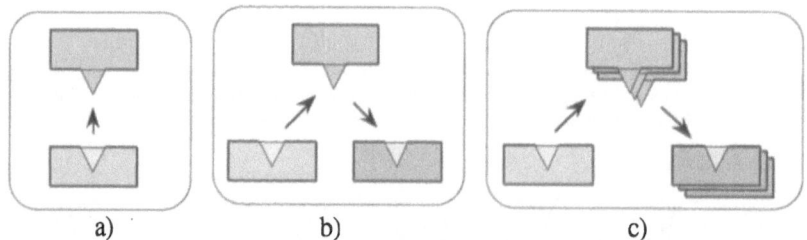

a)                              b)                              c)

**Figure 6-12** *Principal possibilities of replication techniques.* (**a**) Formation of the complementary shape. (**b**) Transfer of shapes to some other material. (**c**) Multiple reproduction of shapes.

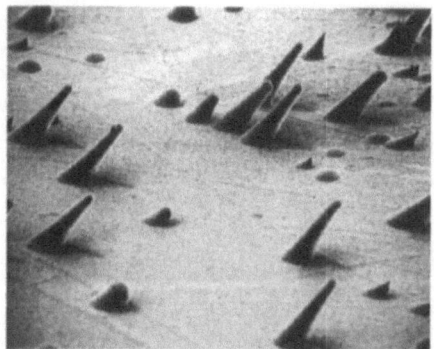

**Figure 6-13** Replica of argon ion tracks in Apollo helmet material [3].

[1]     Fischer, B.E., R. Spohr,: "Production and Use of Nuclear Tracks: Imprinting Structure on Solids." Rev.Mod.Phys., **55**, No.4, 907-948, (1983).
[2]     Tommasino, L.: "Future Developments in Etched Track Detectors for Neutron Dosimetry." Radiation Protection Dosimetry, **20**, No. 1/2, pp. 121-124, (1987).
[3]     Fleischer, R.L., General Electric Research and Development Center, K1-3C31, P.O. Box 8, Schenectady, N.Y. 12301, USA.

### Regular track patterns

The generation of regular track patterns of single ions requires a scanning ion microbeam. A necessary requirement for the scribing technique using single-tracks is the careful suppression of scattered particles by anti-scattering apertures since already single-ion tracks lead to an observable effect (Figure 6-14). This requirement can be circumvented by decreasing the sensitivity below the track etch threshold, such that only multiple tracks are revealed.

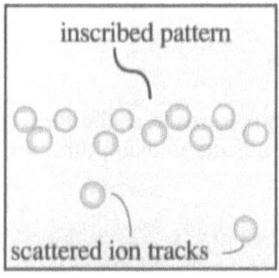
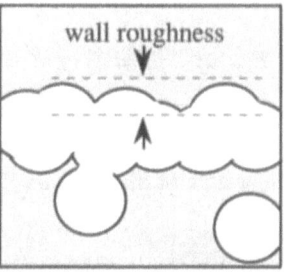

a) after inscription      b) after development

**Figure 6-14** *Influence of scattered ions on resolution of ion microbeams.* While conventional lithographies always use the volume effect of many particles, the ion track technique usually reveals even single-ion tracks and scattered ions have to be avoided. (**a**) Latent tracks after inscribing procedure. (**b**) Wall roughness after track development.

## 6.2 Stochastic track patterns

There exist two extreme types of track distributions, first, the regular track lattice which is analogous to a two-dimensional crystal lattice and requires for its generation an aimed ion beam together with the capability of detecting individual ion impacts and, second, stochastically distributed tracks, corresponding to the unrestricted irradiation of a target with a homogeneous ion beam (Figure 6-15).

a) regular track lattice      b) stochastical track distribution

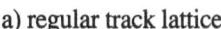
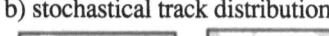

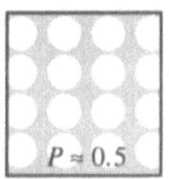

$P \approx 0.1$    $P \approx 0.5$     $P \approx 0.1$    $P \approx 0.5$

low porosity   high porosity     low porosity   high porosity

**Figure 6-15** *Regular track lattice and stochastical track distribution.* (**a**) The regular track lattice requires that the individual tracks are precisely located. Such a lattice has no overlap up to a porosity $P = \pi / 4 \approx 0.79$, but beyond this porosity value a thin perforated structure will disintegrate abruptly. (**b**) The stochastical track pattern possesses track overlap already at very low porosities. With increasing porosity it will change its properties more smoothly in comparison to the regular track lattice. Multiple tracks composed of more than one hit can have a strong influence on the properties of a stochastical track array.

While track-overlap for a regular quadratic array starts abruptly at a nominal porosity $P \approx 0.79$ — and for a regular close-packed lattice at $P \approx 0.91$ — track overlap of a stochastical pattern starts only gradually with increasing porosity.

## 6.2.1    Two-dimensional track overlap

Two-dimensional stochastical track patterns are primarily determined by the product of the number $N$ of ion hits per unit area — the areal density of hits — and the cross sectional area $F$ of the tracks:

$$P = N \, F \quad \Rightarrow \quad P = N \, \pi \, r^2 = N \, \pi \frac{d^2}{4} \quad , \tag{6.1}$$

where $P$ is called the "nominal porosity" and the cross-section of the track is assumed to be a circle (Figure 6-16). In this context a hit is considered as a point, while a track is considered as a circular area of diameter $d = 2 \, r$.

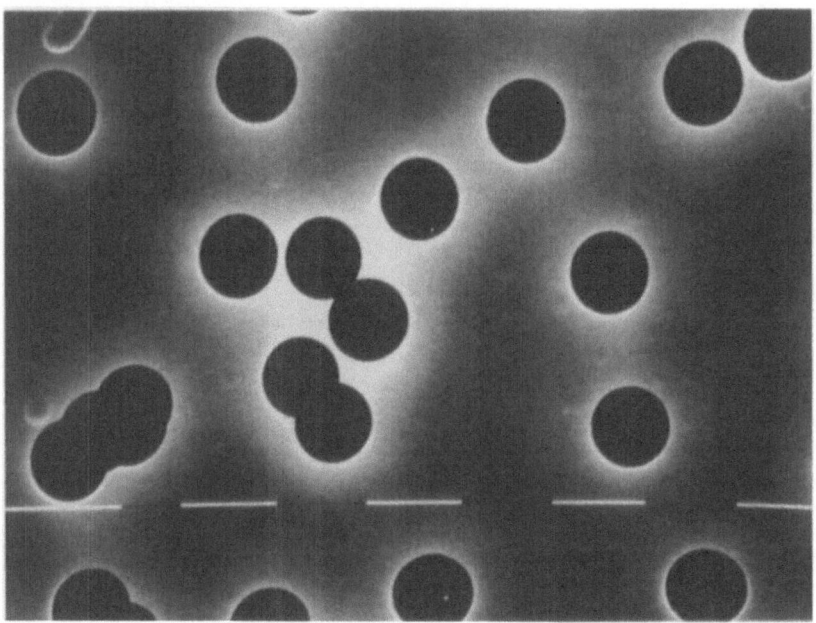

**Figure 6-16** *Stochastic distribution of etched ion tracks* in polycarbonate (Makrofol, Bayer AG, D-5090 Leverkusen). Pore diameter 8 µm. Bar unit = 10 µm.

The stochasticity of the track distribution can improve or deteriorate the overall properties of a specific track array. For example, stochasticity may have the advantage to suppress the interference between different tracks in scattering experiments. On the other hand, stochasticity may have the disadvantage to decrease the selectivity of a track filter. The main observable effect of stochasticity is the occurrence of multiple tracks. Therefore the probability of multiple tracks is of special interest in the use of stochastically distributed tracks and has been discussed by several authors [1], [2].

[1]    Armitage, P.: "An Overlap Problem Arising in Particle Counting." Biometrica, **36**, pp. 257-265, (1949).

The parameters of two-dimensional stochastical track arrays are defined in Figure 6-17, whereby the influence of the fringe zone decreases as $4\,d = 8\,r$ with decreasing track diameter $d = 2\,r$.

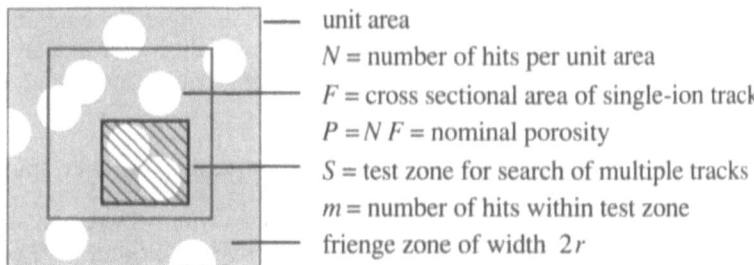

— unit area
$N$ = number of hits per unit area
$F$ = cross sectional area of single-ion track
$P = N\,F$ = nominal porosity
$S$ = test zone for search of multiple tracks
$m$ = number of hits within test zone
— frienge zone of width $2\,r$

**Figure 6-17**  *Parameters of two-dimensional stochastical track distributions.*

We will first determine the probability for the occurrence of exactly $m$ hits within the test zone of area $S$. A more difficult task is the determination of the probability of track overlap, due to the complicated geometry of the resulting shapes. A multi-pore of multiplicity $n$ is defined as a contingent track configuration due to $n$ hits. Double pores can be treated analytically. For higher multiplicitiies computer simulation is advantageous.

### Probability of *m*-hit event

The calculation of the probability of a certain number of ion impacts — or hits — within a chosen test zone $S$ is the basis for the calculation of the probabilities of multiple tracks. Assuming that it is possible to number the ion impacts — or hits — according to their time-of-arrival from $1 \le i \le N$, the probability for the $i$-th ion hitting within the test zone of surface $S$ corresponds to the ratio of the test area $S$ to the unit area, in other words to $S$. For determining of the probability of an $m$-hit event we observe Figure 6-18.

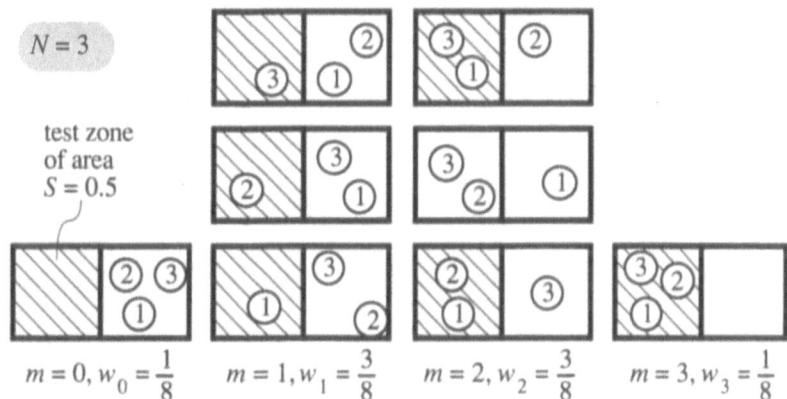

**Figure 6-18**  Probability to find $m$ out of 3 impacts in one half of an available area, for $0 \le m \le 3$. The problem is analogous to the probability to find $m$ out of 3 molecules in one half of the available space, for $0 \le m \le 3$.

[2]  Riedel, C., R. Spohr: "Statistical Properties of Etched Nuclear Tracks. I. Analytical Theory and Computer Simulation." Radiation Effects, **42**, 69, (1979).

The probability that an arbitrarily chosen projectile-ion of number $i$ arrives on the test zone $S$ is equal to the ratio between the test zone $S$ and the total area 1, in other words to $S$. The probability that the remaining N-1 particles arrive outside of the test zone $S$ is $(1 - S)^{(N-1)}$. The probability that the first ion arrives within the test zone $S$ and all other projectiles arrive outside the test zone $S$ corresponds therefore to the product of the above probabilities and is equal to $S \cdot (1-S)^{(N-1)}$. The probability $w_1$ that any one of the $N$ particles but no other particle arrives on the test zone is $N$ times as large as the probability of the first ion arriving on the test zone:

$$w_1 = N S (1 - S)^{(N-1)} = \text{probability for exactly one hit within } S \ . \qquad (6.2)$$

When determining the probability for exactly two hits within $S$ we have to take into consideration that the event in which particles $i$ and $k$ arrive within $S$ are physically equivalent to all events in which exactly any two of the $N$ particles arrive within $S$. The probability that particle the first and the second ion arrive within $S$ and all other particles do not arrive within $S$ is the product of the individual probabilities, $S \cdot S \cdot (1 - S)^{(N-2)}$. The probability that exactly two particles arrive within $S$ corresponds therefore to all combinations, "$N$ over 2" of two particles chosen from an ensemble of $N$ particles multiplied by the probability of the first and second ion arriving within $S$:

$$w_2 = \binom{N}{2} S^2 (1 - S)^{(N-2)} = \frac{N!}{2! \, (N-2)!} S^2 (1 - S)^{(N-2)} \, , \qquad (6.3)$$

which represents the probability for exactly two hits within $S$. This statement can be generalized to an arbitrary number $m < N$:

$$w_m = \binom{N}{m} S^m (1 - S)^{(N-m)} = \frac{N!}{m! \, (N-m)!} S^m (1 - S)^{(N-m)} \, , \qquad (6.4)$$

which is the binomial distribution giving the accurate probability for the occurrence of exactly $m$ hits within $S$. For large numbers $N$ and sufficiently small $m \ll N$ — in other words, for a sufficiently small test zone $S$ — one can write:

$$\binom{N}{m} = \frac{N!}{m! \, (N-m)!} = \frac{N(N-1)(N-2) \ldots (N-m+1)}{m!} \approx \frac{N^m}{m!} \quad \text{and} \qquad (6.5)$$

$$(1-S)^{(N-m)} = 1 + \binom{N}{1}(-S) + \binom{N}{2}(-S)^2 + \ldots + (-S)^{(N-m)} \quad \text{or} \qquad (6.6)$$

$$(1-S)^{(N-m)} \approx 1 - \frac{(N S)}{1!} + \frac{(N S)^2}{2!} - \frac{(N S)^3}{3!} + \ldots \approx e^{-N S} \, , \qquad (6.7)$$

whereby the $N S = P$ is the average number of hits within the test zone $S$. By inserting eqs. (6.5) and (6.7) into eq. (6.4) we therefore obtain the result:

$$w_m = \frac{(N S)^m}{m!} e^{-N S} \quad \Leftrightarrow \quad w_m = \frac{P^m}{m!} e^{-P} \, , \quad \text{for } P = N S \, , \qquad (6.8)$$

which is the Poisson distribution giving the approximate probability for the occurrence of exactly $m$ hits within $S$ for large $N$ and $m \ll N$. This condition is fulfilled in most cases of practical interest for ion track applications.

### Probability of multiple tracks

The calculation of the probability of $m$-hit events is rather straightforward due to the simple geometry of objects composed of points. In contrast hereto, the probability for the occurrence of multiple tracks of multiplicity $n$ is rather complex due to the complicated geometry of objects composed of circles. Multiple tracks of multiplicity $n$ consist of $n$ individual tracks forming a contingent shape.

It is important to realize that while for example $P_2$ is the fraction of hits belonging to double-track configurations, the actual fraction of counted double-tracks per unit area is only $P_2 / 2$, since double tracks are composed of two tracks each. In general the probability of the occurrence of a hit in a multi-hit configuration has to be divided by the proper multiplicity in order to obtain the number of multi-tracks of this multiplicity.

### Probability of single tracks

For determining the probability for the occurrence of single tracks which do not overlap with any other track we observe Figure 6-19.

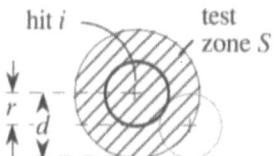

**Figure 6-19** *Probability of single tracks,* obtained by the probability that within a radius of $d = 2\,r$ around an arbitrary hit no other hit occurs.

We chose an arbitrary hit $i$ out of the ensemble of $N$ hits distributed over the unit area and ask for the probability that no other hit occurs within a circle around hit $i$ of a radius equal to the diameter of the individual tracks. In other words, we have to determine the probability $w_0$ for the occurrence of exactly no hits within the test zone $S$ of area $\pi\,d^2$. Since this holds for any arbitrary hit $i$ this represents the probability for any arbitrary hit leading to a single-track configuration — in other words, $w_0$ represents the probability $P_1$ for single tracks:

$$P_1 \equiv w_0 = \frac{(N\,S)^0}{0!}\, e^{-N\,S} = e^{-N\,\pi d^2} = e^{-4N\,\pi r^2} = e^{-4P}, \text{ with } P = N\,\pi\,r^2 \qquad (6.9)$$

### Probability of double tracks

**Rough estimate.** For obtaining a rough approximation of the double-track probability we refer again to the Figure 6-19 which can be used alternatively also to calculate the probability $w_1$ that within the test zone $S$ another hit occurs. According to eq. (6.8) this probability is given by:

$$P_2 \equiv w_1 = \frac{(N\,S)^1}{1!}\, e^{-N\,S} = 4P\, e^{-4P} \text{ with } P = N\,\pi\,r^2, \qquad (6.10)$$

where $F = \pi\,r^2$ is the cross sectional area of a single track and the nominal porosity $P = N\,F$ is defined according to eq. (6.1). This approximation (Figure 6-21 a) neglects the possibility of a track chain formed with a third hit outside of the test zone $S$.

**Excluding triple-tracks.** For a more accurate determination of the probability for the occurrence of double-track configurations we have to ensure that the second track does not overlap with a third track centered outside of the test zone $S$ and forming a track chain with the tracks $i$ and $k$ (Figure 6-20).

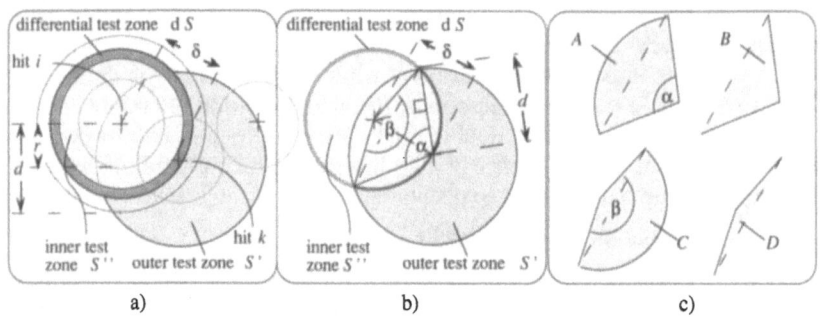

**Figure 6-20** *Probability of double-track configurations*, obtained by summing up the differential probabilities that exactly one hit $k$ occurs within the differential test zone $dS$ <u>and</u> that exactly no hit occurs within the outer <u>and</u> inner test zones $S'$ and $S''$, the sizes of which depend on the distance $\delta$ between hit $i$ and hit $k$. The upper boundary of the summation is the circle around hit $i$ with radius $d = 2r$. (**a**) General sketch of one possible track geometry. (**b**) Calculation of test-zone areas $S'$ and $S''$ involving the overlap zone between two circles of radius $d$ and $\delta$ at distance $\delta$. (**c**) Exploded view of overlap zone.

The probability $P_2$ that the chosen hit $i$ belongs to a track configuration consisting of exactly two tracks — in other words, the probability of double-hit events or the double-track probability — is given by summing (or integrating) up all probabilities that there occurs one hit $k$ within the differential test zone $dS$ — represented by the ring ($\delta$, $d\delta$) around hit $i$ — <u>and</u> no other hit within the test zones $S'$ <u>and</u> $S''$. For obtaining the differential probability $dP_2$ we chose an arbitrary hit $i$ and ask for the probability that exactly one other hit $k$ occurs within the differential test zone $dS$ of radius $\delta$ <u>and</u> that no third hit occurs within the test zones $S'$ <u>and</u> $S''$. In the calculation we have to distinguish two regions.

First region: For $\delta < d/2$ the circle with radius $d$ around hit $k$ overlaps completely with the test zone $S''$, since the smaller circle is completely inside the larger one. The areas of the test zones $S'$ and $S''$ therefore are given by :

$$S' = \pi d^2 - \pi \delta^2 \quad \text{and} \quad S'' = \pi \delta^2 , \quad \text{for} \quad \delta < \frac{d}{2} . \tag{6.11}$$

Second region: For $\delta \geq d/2$ the circles around hit $i$ and hit $k$ have only partial overlap and the calculation of $S'$ becomes more complicated. For determining the size of the test zone $S'$ as function of $\delta$ we observe Figure 6-20 b and c. The areas $A$, $B$, $C$, and $D$ are given by:

$$A = \frac{1}{2}\alpha d^2, B = d^2 \cos\left(\frac{\alpha}{2}\right)\sin\left(\frac{\alpha}{2}\right), C = \frac{1}{2}\beta \delta^2, D = \delta^2\cos\left(\frac{\beta}{2}\right)\sin\left(\frac{\beta}{2}\right), \tag{6.12}$$

where $\alpha = 2\arccos\left(\frac{d}{2\delta}\right)$ and $\beta = 2(\pi - \alpha)$. This yields $\tag{6.13}$

$$S' = \pi d^2 - (A - B) - (C - D) \quad \text{and} \quad S'' = \pi \delta^2 , \quad \text{for} \quad \delta \geq \frac{d}{2} . \tag{6.14}$$

The test zone $S' = S'(\delta)$ depends via $A$, $B$, and $\alpha$ in a complex way on the mutual distance $\delta$ between hit $i$ and $k$.

According to eq. (6.8), the probability $dw_1$ for the occurrence of one other hit $k$ within the differential test zone $dS$ — in other words inside the ring interval $(\delta, d\delta)$ around hit $i$ — is given by:

$$dw_1 = N \cdot 2\pi \, \delta \, d\delta \cdot e^{-N \cdot 2\pi \, \delta \, d\delta} \approx N \cdot 2\pi \, \delta \, d\delta \quad \text{for } d\delta \to 0. \tag{6.15}$$

According to eq. (6.8), the probability $w_0$ of finding no other hit within the test zones $S'(\delta)$ and $S''(\delta)$ is given by:

$$w_0 \equiv e^{-N(S'(\delta) + S''(\delta))} = e^{-N(S'(\delta) + \pi \delta^2)}, \tag{6.16}$$

where $S'(\delta)$ is given by eqs. (6.11) or (6.14), depending if $\delta < d/2$ or $\delta \geq d/2$, respectively.

The differential probability $dP_2$ that the chosen hit belongs to a double track configuration is given by the probability that there occurs one hit $k$ within the differential test zone $dS$ and no other hit within the test zones $S'$ and $S''$

$$dP_2 = dw_1 \cdot w_0 = N \, 2\pi \, \delta \, d\delta \cdot e^{-N(S'(\delta) + \pi \delta^2)}, \tag{6.17}$$

from which the probability of the hit $i$ belonging to a double-hit configuration is obtained by integration:

$$P_2 = \int_0^d dP_2 = 2\pi N \int_0^d e^{-N(S'(\delta) + \pi \delta^2)} \delta \, d\delta, \tag{6.18}$$

the numeric evaluation of which is shown in Figure 6-21 b. This evaluation corresponds to the approximate function:

$$P_2 = 4P \, e^{-4.74 P} \quad \text{where } P = N \pi r^2, \tag{6.19}$$

**Computer simulation.** The computer simulation gives much more accurate values for the double-track probability since it considers track configurations up to very high multiplicities which tend to reduce the number of double-track configurations. According to reference [1] the double-track probability $P_2$ is approximated by:

$$P_2 = 4P \, e^{-5.645 P} \quad \text{with } P = N \pi r^2, \tag{6.20}$$

which is shown in Figure 6-21 c.

[1]    Riedel, C. R. Spohr: "Correcting Overlapping Counts in Dose Calibration at High Event-Densities." Nuclear Tracks, **5**, 265-270, (1981).

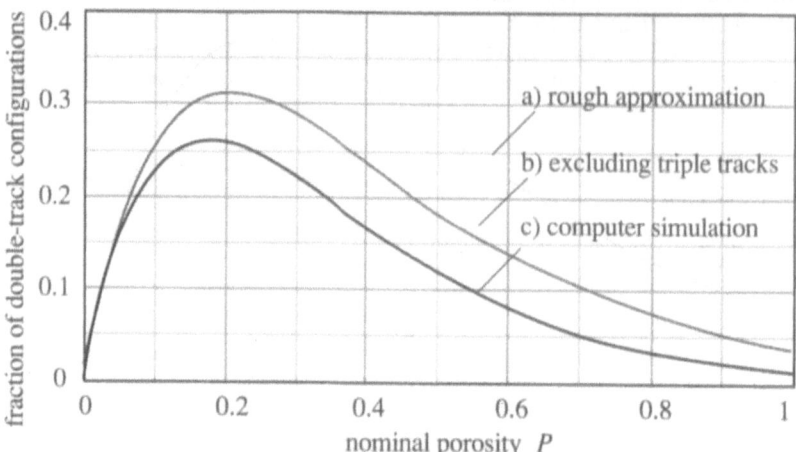

**Figure 6-21** *Double-track probability $P_2$* — the probability that an arbitrary hit $i$ belongs to a track configuration consisting of two hits. (**a**) Probability of another hit occurring within a circular test zone of radius $d = 2\,r$ around the hit $i$, neglecting the possibility of a third track outside this circle which overlaps with track $k$, according to eq. (6.1). (**b**) Probability of another hit within a circular test zone of radius $d = 2\,r$ around hit $i$, excluding the possibility of a third track outside this circle which overlaps with track $k$, according to eq. (6.18). (**c**) Limiting curve according to a computer simulation [1].

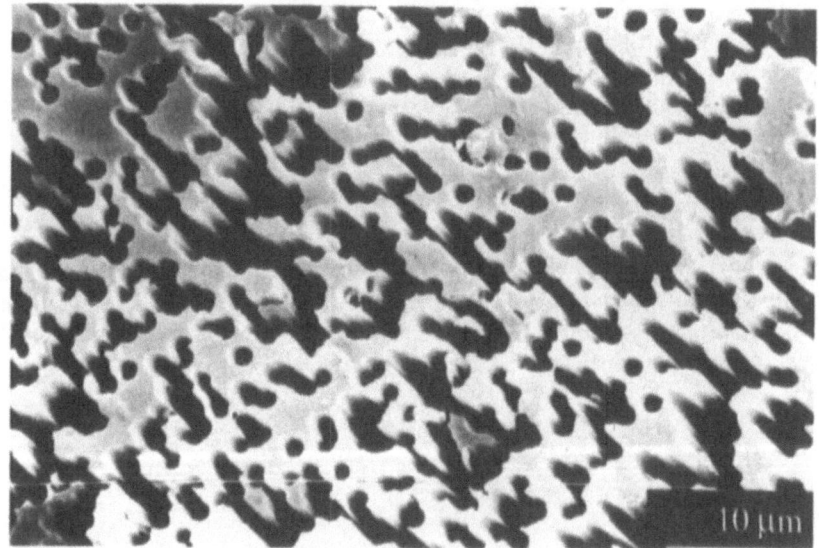

**Figure 6-22** *Disintegration limit*. Track filter at a high porosity, close to its mechanical collapse [2].

[1]     Riedel, C. R. Spohr: "Correcting Overlapping Counts in Dose Calibration at High Event-Densities." Nuclear Tracks, **5**, 265-270, (1981).

[2]     Vetter, J., Gesellschaft für Schwerionenforschung (GSI), Planckstr. 1, D-6100 Darmstadt 11.

### Effective porosity

Even for high nominal porosities $P$ there exists still a finite probability that there remains an island on the track recorder which contains no etched track. The "effective porosity" corresponds to the ratio of the porous surface area to the total area and is given by [1]:

$$P_{\text{eff}} = 1 - e^{-P} \, , \quad \text{where } P = N \, S \tag{6.21}$$

is the nominal porosity, $N$ is the areal density of hits, and $S$ is the area of a single track. This relation becomes important in ion lithography where arbitrary zones of the track recorder have to be removed completely. Figure 6-22 shows a perforated track structure close to its mechanical disintegration due to chains of overlapping tracks.

## 6.2.2 Three-dimensional track overlap

The main advantage of ion track filters is the very narrow pore diameter distribution of etched ion tracks. However, if the ion tracks are parallel, multi-pores can deteriorate the selectivity already at relatively small nominal porosities $P$. Therefore ion track filters are generated preferentially with an ion beam of large divergence (Figure 6-23). In this way the probability of an overlap of individual pores throughout the thickness of the filter membrane can be drastically reduced. For estimating the suppression-factor for double pores in an ion track filter we observe Figure 6-24.

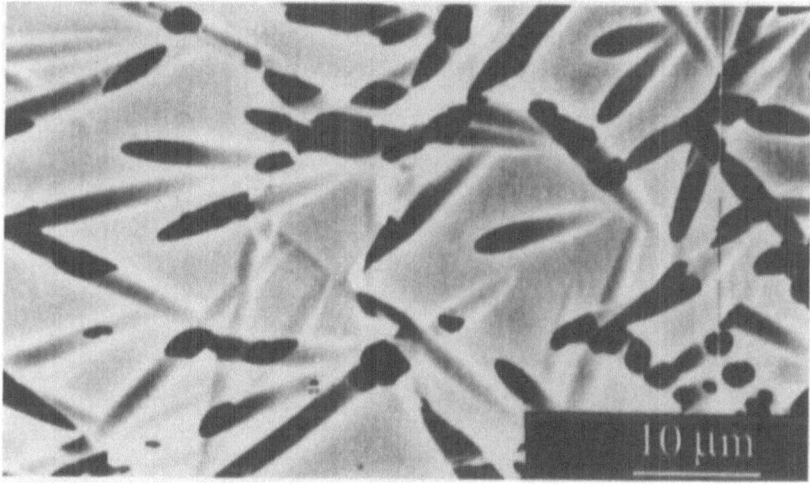

**Figure 6-23** *Multi-angular track filter*, irradiated while turning the track recorder about a fixed axis [2]. In this way highly porous filter matrices with well defined pore sizes can be obtained.

[1]    Riedel, C., R. Spohr: "Statistical Properties of Etched Nuclear Tracks. I. Analytical Theory and Computer Simulation." Radiation Effects, **42**, 69, (1979).

[2]    Vetter, J., Gesellschaft für Schwerionenforschung (GSI), Planckstr. 1, D-6100 Darmstadt 11.

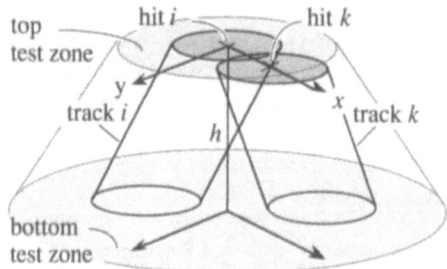

**Figure 6-24**  Estimating the probability of fully overlapping double-pores for an ion track filter.

Assuming an irradiation which fills homogeneously a solid angle $\Omega$, the fraction of fully overlapping double-pores is roughly reduced by a factor corresponding to the ratio of the top test zone of area $S = \pi\, d^2$ and the bottom test zone of area $S' = \pi\, h^2\, \Omega$ over which the ion tracks are distributed on the bottom surface:

$$R = \frac{\pi\, d^2}{\pi\, h^2\, \Omega} = \frac{d^2}{h^2\, \Omega} \quad \text{for} \quad R < 1 , \tag{6.22}$$

where $h$ is the thickness of the filter membrane, $d$ is the pore diameter, and $\Omega$ is the solid angle of the ion beam. A more intricate solution for $R \approx 1$ is provided by reference [1]. For estimating the fraction of remaining double-pore configurations at the bottom surface of a track filter, the fraction $P_2$, according to eq. (6.10) has to be multiplied by the reduction factor $R$. This rough approximation is shown in Figure 6-25.

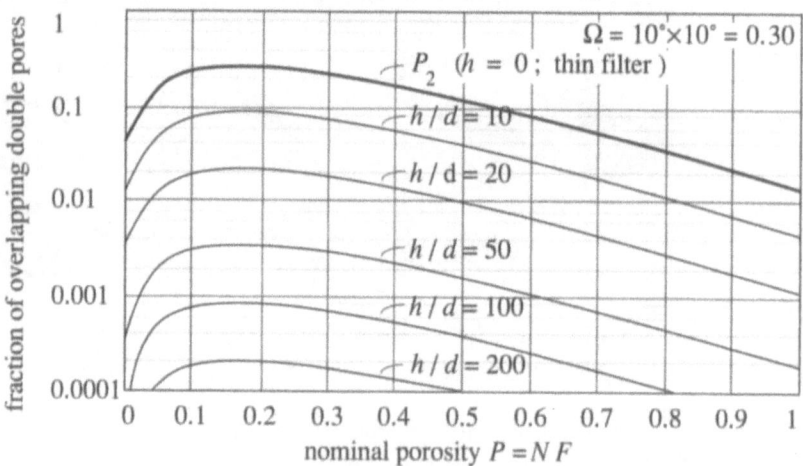

**Figure 6-25**  *Fraction of "fully overlapping" double pore configurations* — pores which overlap throughout the thickness $h$ of the track filter of thickness $d$ — for an irradiation within the solid angle $\Omega = 0.3$ sterad, equivalent to a beam of 10° by 10° opening angle, for different ratios $h\,/\,d$, as function of the parameter "nominal porosity" $P$.

[1]     Jaech, J.L.: "Interference in the Manufacture of Nuclepore Filters". Technometrics, **9**, no. 2, pp. 319-324, (1967).

# II  Track applications

# 7  Single-ion tracks

This chapter describes two examples where single-ion tracks are used as critical apertures ruling the flow between two adjacent reservoirs of a fluid. First, the classical flow of an aqueous suspension of microparticles. Second, the quantum mechanical flow of a superfluid.

## 7.1  Number, size, and deformability of particles

The characterization of suspended particles in fluids is of primary interest in several fields, ranging from water pollution studies to medical diagnostics. Among the most vitally required data are the number-density and the volumes of the suspended particles, for example in the counting and sizing of bacteria in drinking water.

The number and size of particles can be determined using the resistive pulse technique [1], a central tool in the field of flow cytometry [2]. In contrast to optical observations or light scattering in particle ensembles, this technique is based on the sequential observation of individual particles passing through an aperture. The ion track technique enlarges the scope of the technique to microparticles [3]. More recently [4], also the deformability of individual cells was observed, using apertures smaller in diameter than the cell size.

[1]    Coulter, W.H.: "High Speed Automatic Blood Cell Counter and Cell Size Analyzer." Nat. Electron. Conf. Proc. **12**, 1034 - 1042, (1956).
[2]    Melamed, M.R., P.F. Mullaney, M.L. Mendelsohn (Editors): **"Flow Cytometry and Sorting,"** Wiley, New York, 716 pp., (1979).
[3]    DeBlois, R.W., C.P. Bean: "Counting and Sizing of Submicron Particles by the Resistive Pulse Technique." Rev. Sci. Instrum. **41**, 909 - 916, (1970).
[4]    Roggenkamp, H.G., H. Kiesewetter, R. Spohr, U. Dauer, L.C. Busch: "Production of Single-Pore Membranes for the Measurement of Red Blood Cell Deformability." Biomedizinische Technik, **26**, 167 - 169, (1981).

## 7.1.1    Basic relations

The passage of suspended particles through capillaries is essentially governed by the relations outlined in the following.

### Ideal hydrodynamic flow through capillaries

The principle of energy conservation is the starting point for deriving the Bernoulli equation for an incompressible ideal fluid. The external work, $(p_1 - p_2)\, \delta V$, performed on the system corresponds to the kinetic energy gain, $\rho\, \delta V\, (v_2^2 - v_1^2)\,/\,2$, of the volume element $\delta V$ on which the work is performed, where $\rho$ is the mass density, $p_1$ and $p_2$ are the initial and final pressure, and $v_1$ and $v_2$ are the initial and final velocity:

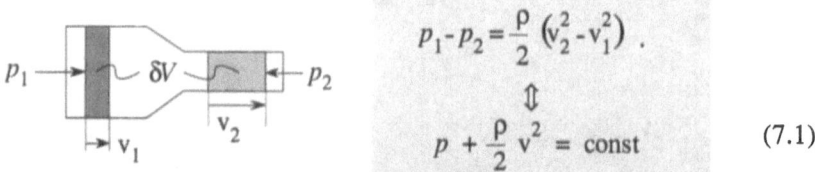

$$p_1 - p_2 = \frac{\rho}{2}\left(v_2^2 - v_1^2\right).$$

$$\updownarrow$$

$$p + \frac{\rho}{2}\,v^2 = \text{const} \qquad (7.1)$$

In eq. (7.1) the pressure can be interpreted as the potential ruling the acceleration of one unit of volume of the fluid.

### Viscosity

The force $F$ on a test plate of area $\sigma$ sliding with velocity $v$ on a film of thickness $dy$ is proportional to the viscosity $\eta$ of the medium, the area $\sigma$ of the test plate, and the velocity gradient at right angle with respect to the direction of the velocity vector:

$$F = \eta\,\sigma\frac{\partial v}{\partial y}. \qquad (7.2)$$

The viscosity $\eta$ of the medium is defined as force per unit area and unit velocity gradient.

### Viscous flow through capillaries

The mass current of a fluid through a capillary of inner diameter $r_c$, length $L_c$ depends on the viscosity $\eta$ of the fluid and the pressure difference $p_1 - p_2$ between entrance and exit of the capillary. The external force $F = (p_1 - p_2)\,\pi\,r^2$ applied on a rigid, axially aligned cylinder of radius $r$ and length $L_c$, moving at the velocity $v$ along the capillary is equal to the frictional force $F' = -\eta\,(2\pi\,r\,L_c)\,\partial v\,/\,\partial r$. The integration of the resulting differential equation — with the boundary condition $v(r_c) = 0$ — yields the velocity profile inside the capillary, corresponding to an inverted parabola:

$$v = \frac{p_1 - p_2}{4\,\eta\,L_c}\left(r_c^2 - r^2\right). \qquad (7.3)$$

For obtaining the total volume flow $J = dV / dt$ through the capillary, the differential current $2\pi r\, dr\, v(r)$, due to a cylinder of radius $r$, thickness $dr$, and velocity $v(r)$ is integrated between $r = 0$ and $r = r_c$, leading to the law of Hagen-Poiseuille:

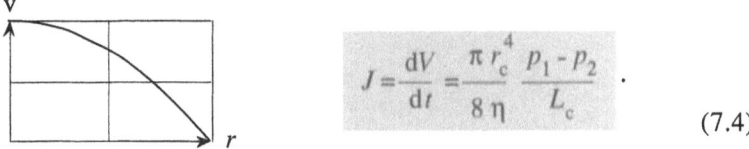

$$J = \frac{dV}{dt} = \frac{\pi\, r_c^4}{8\,\eta}\, \frac{P_1 - P_2}{L_c}.$$

(7.4)

The average velocity $\langle v \rangle$ of the fluid within the capillary is obtained by dividing eq. (7.4) through the cross sectional area $\pi\, r_c^2$ of the capillary.

### Sedimentation of particles

The sedimentation of particles — such as red blood cells in saline aqueous suspensions without convection currents— is governed by Stoke's equation:

frictional force $F$

$$F = 6\pi\eta R_0 v \ , \ F^* = \frac{4\pi}{3} R_0^3 \rho\, g$$

dynamic equilibrium $\Rightarrow$

gravitational force $F^*$

$$v = \frac{2}{9}\frac{\rho\, g}{\eta}\, R_0^2$$

(7.5)

The observation of suspended particles through a critical aperture can be severely influenced by sedimentation, according to eq. (7.5), which at the same time is the basis for particle separation in the acceleration of a centrifuge.

### Hydrodynamic focussing effects

Hydrodynamic focussing effects are employed in flow cytometry for localizing and orienting suspended cells in capillaries [1].

**Local confinement.** Under laminar flow conditions, the entrance aperture of the cylindrical capillary employed in the resistive pulse technique represents a current sink, acting as a central force on suspended particles contained in the half-space before the entrance aperture (Figure 7-1). As a result of the central force the fluid contained within the corresponding hemisphere of solid angle $2\pi$ — after being transported to the sink — appears mapped onto the cross-section of the capillary. This phenomenon enables a very precise localization of a stream of suspended particles within the cross-section of the cylindrical capillary. A volume element $\delta V = \pi\, r^2\, v$ originating from a distance $R$ within the hemisphere before the aperture of radius $r_c$ will be transformed into an elongated volume element $\delta V = \pi\, r^2\, \langle v \rangle$ of the same magnitude. Since an analogous relation connects the complete hemispherical volume element $2\pi\, R^2\, v$ and the complete cylindrical volume element $\pi\, r_c^2\, \langle v \rangle$, one obtains a relation for the final radius $r'$ of the elongated volume element:

$$\pi r^2 v = \pi r'^2 \langle v \rangle \ \text{and} \ 2\pi R^2 v = \pi r_c^2 \langle v \rangle \Rightarrow r' = \frac{r}{R}\frac{r_c}{\sqrt{2}}.$$

(7.6)

The final radius $r'$ is proportional to the ratio of the radius $r$ of the original volume ele-

[1]    Kachel, V., E. Menke: "Hydrodynamic Properties of Flow Cytometric Instruments." pp. 41 - 59, in **Flow Cytometry and Sorting**, Wiley, New York, 716 pp., (1979).

ment and its distance $R$ from the entrance aperture. In this way suspended particles —
under laminar flow conditions — can be precisely confined to a specific location within
the capillary. This corresponds to a self-acting micromanipulation mechanism which pre-
pares the suspended particles for observation along a precisely defined straight line.

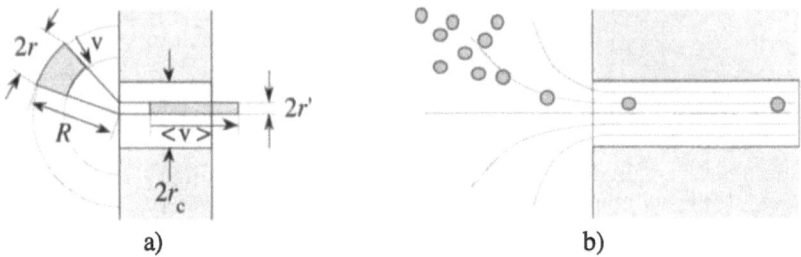

a)                                                              b)

**Figure 7-1** *Local confinement of suspended particles.* (a) Deformation of fluid volume
element $\delta V$ during its passage through a critical aperture. (b) Corresponding situation for
fluid-suspended particles.

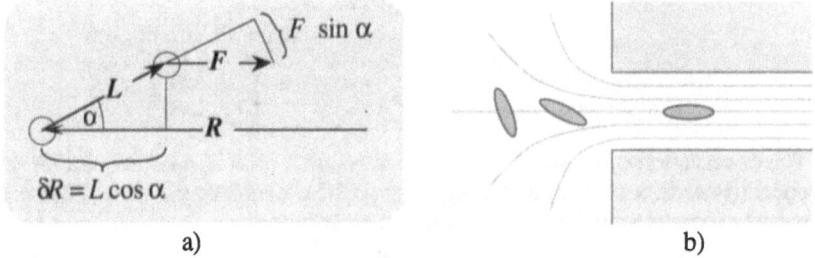

a)                                                              b)

**Figure 7-2** *Angular confinement of elongated suspended particles* (a) Determination of the
torque acting on two particles rigidly connected at the fixed distance $L$ in a radial flow field.
(b) Corresponding axial alignment of an elongated particles.

**Angular confinement.** In the radial flow field in front of the critical aperture,
elongated particles are subject to a torque which aligns them axially parallel to the flow direction. A simple model of an elongated particle is an assembly of two particles bound together at a fixed distance $L$. The difference between the flow velocities experienced by the
particles in the gradient of the flow field (Figure 7-2 ) can be calculated from eq. (7.6):

$$\delta v = -\frac{r_c^2 <v>}{R^3} \, \delta R = -\frac{r_c^2 <v>}{R^3} \, L \, \cos \alpha \quad . \tag{7.7}$$

According to eq. (7.5) this corresponds to the force $\delta F = 6\pi \, \eta \, R_0 \, \delta v$ and to the torque
or moment of force:

$$|\delta F \times L| = 3\pi \, \eta \, R_0 \, r_c^2 <v> \frac{L^2}{R^3} \, \sin \left( 2\alpha \right) \quad . \tag{7.8}$$

The main feature of this equation is a quadratic increase of the torque with the elongation $L$ of the particle, and the inverse cubic dependence of the distance $R$ from the critical
aperture. Furthermore, the torque is positive for $0 \le \alpha \le \pi / 2$ and attains its maximum
for $\alpha = 45°$. Elongated particles therefore are unstable for $|\alpha| > 0$ and attain their equi-

librium orientation for $\alpha = 0$. In other words, the axis of an elongated particle orients itself in the direction of the gradient of flow.

### Limits of diffusion transport

While the driving force of diffusion is a concentration gradient, the driving force of convection is a pressure gradient. Diffusion is governed by Fick's first law. According to eq. (4.27), $j = - D \nabla n$, where $j$ is the flux density, $D = \text{const } m^{-1/2}$ is the diffusion constant, $m$ is the mass and $n$ is the number density of the diffusing particles. Convection transport is ruled by a similar equation involving, however, quite different constants. In the case of suspended particles transported through a capillary one obtains the equation $j = -\{r_c^2 / (8 \eta)\} \nabla p$, derived by dividing eq. (7.4) through $\pi r_c^2$ and replacing the ratio $(p_1 - p_2) / L_c$ by the negatiye gradient of the pressure, $- \nabla p$. For large particle masses $m$ or small concentration gradients $\nabla n$ — or alternatively large transport distances $L_c$ at finite concentration differences $n_2 - n_1$ — the transport by diffusion becomes negligibly small and transport by convection becomes much more efficient. Therefore in macroscopic organisms convective transport is required, while on the cellular and sub-cellular level transport by diffusion suffices. In other words, the inefficiency of diffusion transport over long distances is the physical reason for the evolution of a circulatory system in animals using blood as carrier medium.

## 7.1.2    Resistive pulse technique

### Determination of particle volume

The resistive pulse technique [1], [2], [3], [4] is based on the observation of the change of the resistance of a capillary filled with an electrolyte during the passage of a non-conductive dielectric particle. Thereby one assumes that the particle radius $R_0$ is small in comparison with the capillary radius $r_c$ and that the capillary radius is small in comparison with the capillary length $L_c$.

According to a crude approximation, the resistance change due to a spherical particle of radius $R_0$ is attributed to that of a circular disk of diameter $2 R_0$ and thickness $2 R_0$, axially aligned within the critical capillary, blocking a fraction $\pi R_0^2$ of the cross section from the flow of electrical current (Figure 7-3 a). This approximation leads to a resistance change $\delta W = (2 / \pi) \rho_w (R_0^3 / r_c^4) = \{3 / (2\pi^2 r_c^4)\} \rho_w V_0$, proportional to the particle volume $V_0 = (4\pi / 3) R_0^3$. A more elaborate model [5] leads to a resistance change:

$$\delta W = \frac{2}{\pi} \rho_w \frac{R_0^3}{r_c^4} \left(1 + 6\frac{r_c^4}{L_c^4}\right) = \frac{3}{2\pi^2} \rho_w \frac{V_0}{r_c^4} \left(1 + 6\frac{r_c^4}{L_c^4}\right). \tag{7.9}$$

[1]    DeBlois, R.W., R.K.A. Wesley: "Sizes and Concentrations of Several Type C Oncornaviruses and Bacteriophage T2 by the Resistive-Pulse Technique." J. Virology, **23**, No. 2, pp. 227-233, (1977).

[2]    DeBlois, R.W., C.P. Bean: "Electrokinetic Measurements with Submicron Particles and Pores by the Resistive Pulse Technique." J. Colloid and Interface Sci., **61**, No. 2, 323, (1977).

[3]    DeBlois, R.W., E.E. Uzgiris, D.H. Cluxton, H.M. Mazzone: "Comparative Measurements of Size and Polydispersity of several Insect Viruses." General Electric Technical Information Series Report 77CRD250, (1977).

[4]    Kachel, V.: "Electrical Resistance Pulse Sizing (Coulter Sizing)" pp. 61 - 104, in **Flow Cytometry and Sorting**, Wiley, New York, 716 pp., (1979).

[5]    DeBlois, R.W., C.P. Bean: "Counting and Sizing of Submicron Particles by the Resistive Pulse Technique." Rev. Sci. Instrum. **41**, 909 - 916, (1970).

where $\rho_w$ is the specific electric resistance of the electrolyte filling the pore, $V_0 = (4\pi / 3) R_0^3$ is the particle volume, $L_c$ is the length of the capillary, and $R_0 \ll r_c \ll L_c$. The eq. (7.9) applies roughly also to non-spherical particles.

Figure 7-3 b shows the electric circuit used for determining the resistance change $\delta W$. The capillary resistance $W$ with or without the dielectric particle is large in comparison to the resistance $W'$ on which the corresponding voltage change $\delta U$ is observed. In this case the current $I$ through the circuit is dominated by the resistance $W$ and one obtains

$$\delta U' = -U\,W'\,\frac{\delta W}{W^2} \; , \tag{7.10}$$

which according to eq. (7.9) is roughly proportional to the particle volume $V_0$. For suppressing the influence of polarization of the electrodes and suppressing the pick-up of external electromagnetic noise the d.c. voltage source $U$ is often replaced by an a.c. voltage $U_\sim$ and the change of the measured voltage $\delta U_\sim'$ is observed phase-synchronously with respect to $U_\sim$, using the lock-in principle.

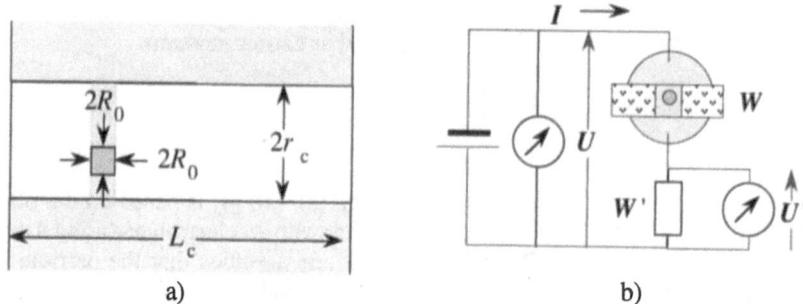

a)                                                                    b)

**Figure 7-3** *Principle of resistive pulse technique.* **(a)** Crude model replacing a spherical particle of radius $R_0$ by a circular disk of radius $R_0$ and thickness $2R_0$. **(b)** Electric circuit used for observing the change $\delta W$ of the resistance $W$ during the passage of the particle.

### 7.1.3    Deformability and interface energy

There exists a close relation between the interface energy — or the surface tension — and deformability of particles. This becomes obvious during their forced passage through an aperture which is smaller in size than the particles themselves.

If a capillary is dipped into a non-wetting liquid, a level difference $h$ is observed between the level inside the capillary, measured from the top of the spherical meniscus $A$, and the normal liquid level $B$ at sufficient distance from the capillary wall (Figure 7-4 a). Static equilibrium exists if the pressure at depth $h$ of the liquid is equal to the pressure inside a droplet of radius $r_c$, where $r_c$ is the capillary radius.

$$\rho\,g\,h = \frac{2\,\gamma}{r_c} \quad \Leftrightarrow \quad h = \frac{2\,\gamma}{\rho\,g}\,\frac{1}{r_c} \; . \tag{7.11}$$

This equation is equivalent to the statement that any increase of the level difference $h$ by $dh$ increases the gravitational energy by an amount $\rho\,(\pi\,r_c^2\,dh)\,g\,h$, which at equilibrium is equal to the decrease of the interface energy, $2\pi\,r_c\,dh\,\gamma$, where $\gamma$ is the interface energy per unit interface area.

A non-wetting liquid droplet entering a capillary which is smaller in diameter (Figure 7-4 b) is subject to a deformation increasing its surface — more precisely, the interface area between the droplet and the suspension medium. This requires the input of interface energy proportional to the growth of its area.

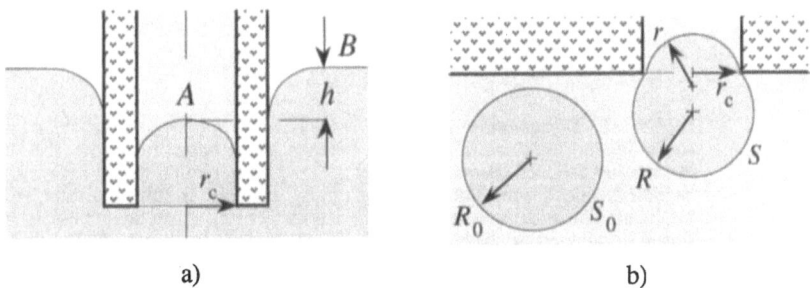

a)                                                          b)

**Figure 7-4** *Capillary depression and deformability of a droplet.* Similarity between the phenomenon of capillary depression and the deformation of a non-wetting liquid droplet during its passage through a cylindrical pore. (**a**) The deformation energy of the spherical meniscus inside the capillary, with respect to a flat surface, corresponds to the potential energy difference between level $A$ and level $B$ of the liquid. (**b**) The deformation of the droplet increases its surface and thus requires the input of external energy.

### Deformability of a spherical droplet

A hydrophobic oil droplet suspended in water, penetrating a cylindrical pore, is a useful model for describing the passage of a red blood cell through a cylindrical pore. While the water is required to wet the capillary, the oil droplet must not wet the capillary. For this purpose the pore surface has to be rendered hydrophilic, for example by chemical etching. The deformation of the droplet can be treated as a pure surface tension problem as long as the deformation takes place slowly. In this case the deformation is a reversible process in which the energy-loss due to internal friction — usually proportional to the velocity of the displacement — is neglected. The calculation is based on the spherical or cylindrical shape of the minimum-energy interfaces of a droplet confined by a circle or a cylinder, respectively.

The deformation of the droplet during its passage through the pore corresponds to an increase of its surface which is directly related to the energy input during the deformation. The displacement $x$ is defined as the ratio of the displaced volume $\delta V$ and the cross sectional area $\pi r_c^2$ of the water column replaced by the intruding droplet:

$$\text{displacement } x \equiv \frac{\delta V}{\pi r_c^2} \ . \tag{7.12}$$

For a sufficiently slow deformation the change of the surface energy of the droplet as function of the displacement $x$ represents a potential energy $U(x) = \gamma \{S(x) - S_0\}$, where $\gamma$ is the interface energy per unit area, $S(x)$ is the interface area of the deformed droplet as function of $x$, and $S_0$ is the original surface of the spherical droplet before entering into the capillary. The negative gradient of this potential energy, $-\partial U / \partial x$, thus represents the force counteracting the deformation. For determining the surface change $S(x) - S_0$ of the droplet as function of the position $x$ we observe Figure 7-5.

**Figure 7-5** *Deformation stages during the passage of a droplet through a cylindrical pore.*
(**a**) Initial configuration. The droplet just touches the edge of the pore. (**b**) The right meniscus has just reached its maximum volume and its minimum radius $r = r_c$. (**c**) A cylindrical
section develops while the radius of the right meniscus stays constant. (**d**) The stage of
maximum deformation is reached, where both meniscus-radii are identical and the deformation is completed. (**e**) The left meniscus-radius starts to shrink. (**f**) The droplet has recovered its original shape.

**Relative change of surface energy.** The relative change of the surface between
the original stage (Figure 7-5 a) without deformation and the stage of maximum deformation (Figure 7-5 d) corresponds to the maximum change of surface energy and is
given by:

$$\frac{S_{max} - S_0}{S_0} = \frac{1}{3}X^2 + \frac{2}{3}\frac{1}{X} - 1 \rightarrow \frac{2}{3}\frac{1}{X} \text{ for } X \rightarrow 0, \text{ where } X = \frac{r_c}{R_0} \qquad (7.13)$$

is the ratio between the capillary radius $r_c$ and the droplet radius $R_0$. It is equal to the relative change of surface energy between the initial state and the stage of maximum deformation in a slow deformation process (Figure 7-6).

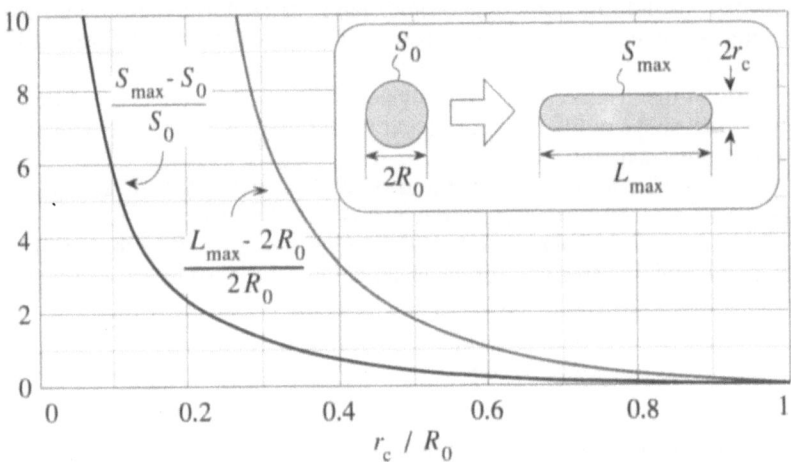

**Figure 7-6** *Deformation and surface energy.* Maximum change of surface area $S$ and length
$L$ of the deformed droplet as function of the capillary radius $r_c$, given in units of the undeformed-droplet radius $R_0$, between the initial stage and the stage of maximum deformation.
Both, $S_{max} - S_0$ and $L_{max}, - 2 R_0$, are given in units of the respective values, $S_0$ and
$2 R_0$, of the original droplet. The maximum change of the surface is directly proportional
to the maximum change of the interface energy.

**Surface areas.** The minimum-energy interfaces of the deformed droplet are composed of spherical sections of radius $r_c$ and height $h$ a nd — during later deformation stages — also of cylinders of radius $r_c$ and length $l$ with the following volumes $V$ and interface areas $S$:

$$V = \frac{\pi h}{6}\left(3\,r_c^2 + h^2\right) \qquad\qquad V = \pi\,r_c^2\,l$$

$$S = \pi\left(r_c^2 + h^2\right) \qquad\qquad\qquad S = 2\pi\,r_c\,l \qquad\qquad (7.14)$$

**Surface energy as function of deformation.** The result of a numerical calculation of the relative surface increase and its derivative with respect to $x$ — corresponding to the potential energy $U(x)$ and to the resulting force - $F(x)$ — is shown in Figure 7-7.

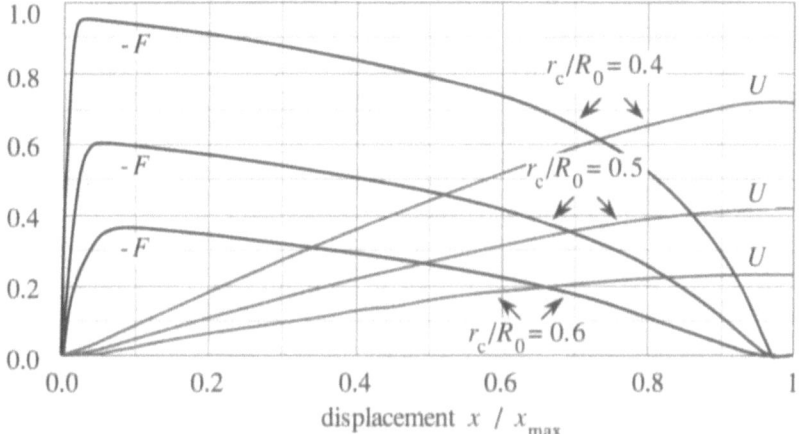

**Figure 7-7** *Surface energy and counteracting force during deformation. Potential energy $U$ and resulting force - $F$, counteracting the deformation of a droplet, in arbitrary units, for three different pore sizes $r_c$, measured in units of the undeformed-droplet radius $R_0$, as function of the displacement $x$, measured in units of the displacement $x_{max}$ at maximum deformation.*

**Fundamental equation of motion.** According to eq. (7.1) the kinetic energy of a laminar flow system is concentrated in the constrictions where the flow velocity is high. An extreme case is a system in which a small capillary separates two large volumes, where the kinetic energy is concentrated in the capillary and very close to its entrance and exit aperture, where the flow velocity decreases roughly quadratically with the distance from the capillary. The kinetic energy of the system corresponds to a moving mass $M = \rho\,(\pi\,r_c^2\,L_{eff})$, where $L_{eff}$ is the hydrodynamic length of the capillary of actual length $L_c$. An estimate of $L_{eff}$ is obtained by integrating the differential kinetic energy of a coaxial cylindrical volume element, obtained from eq. (7.3), over the whole volume of the capillary from $r = 0$ to $r = r_c$, and requesting that the resulting total kinetic energy of the liquid in the capillary is equal to $(M / 2) <v>^2$, where $<v>$ is the average velocity inside the capillary obtained by dividing eq. (7.4) by the cross section $\pi\,r_c^2$ of the capillary. The result is $L_{eff} = 10/3\, L_c \approx 3\, L_c$. In the following, the hydrodynamic length of the capillary, $L_{eff}$, is considered as a fitting parameter.

During the deformation of the droplet, the kinetic energy $T = (M / 2) \, v^2$ of the system, concentrated in the mass $M$, is converted into the potential energy $U(x)$ of the deformed surface. By differentiation of $T + U =$ const with respect to the time, considering that $dU / dt = (\partial U / \partial x) (dx / dt)$, one obtains the fundamental equation of motion of the system:

$$M \frac{d^2 x}{dt^2} = F(x) \; . \tag{7.15}$$

In order to treat also frictional energy losses for an inelastic droplet movement, this equation can be extended by a term which is proportional to the displacement velocity $v = dx / dt$. For the elastic droplet movement — considered here exclusively — the multiplication of eq. (7.15) by $dx / dt$ and integration with respect to time, leads back to the energy conservation equation:

$$\frac{dT}{dt} = F(x) \frac{dx}{dt} \quad \Rightarrow \quad T(x) = \int_0^x F(\xi) \, d\xi = T(0) - U(x) \; , \tag{7.16}$$

where $T(x) \equiv (M / 2) (dx / dt)^2$ is the remaining kinetic energy of the the equivalent hydrodynamic column of mass $M$ after a displacement by $x$, $T(0)$ is the initial kinetic energy of the mass $M$ at $t = 0$, and $U(x)$ is the potential energy of the droplet in contact with the capillary corresponding to a displacement $x$. Rearrangement of eq. (7.16) yields:

$$dt = \frac{dx}{\sqrt{\dfrac{2}{M} \left( T(0) - U(x) \right)}} \quad \Rightarrow \quad t = \int_0^x \frac{dx}{\sqrt{\dfrac{2}{M} \left( T(0) - U(x) \right)}} \; , \tag{7.17}$$

For $T(0) < U_{max}$ the kinetic energy $T(0)$ is insufficient to surmount the barrier $U_{max}$. In this case, close to the point-of-return where $T(0) - U(x) \to 0$, special care has to be taken during the integration. In the numeric calculation, the final integration step $dt$ before the point-of-return has been calculated from the movement of the mass $M$ in a homogeneous field. In this case the final time step before the point-of-return is proportional to the square root of the step length $dx$ and the singularity is overcome.

It is obvious from eq. (7.17) that the total time required to complete the deformation — in other words, the "time-of-deformation" — is proportional the square-root of the mass $M$, corresponding to the hydrodynamic length of the capillary. Furthermore, the time-of-deformation will decrease with increasing mass-current $J$ through the capillary, corresponding to an increasing kinetic energy $T(0)$ of the hydrodynamic volume of mass $M$. Finally, the time-of-deformation will increase with increasing interface energy $\gamma$, since $U(x) = \gamma \, S(x)$. Therefore an observation of the time required to complete the deformation process — characterizes the specific interface energy of the oil droplet with respect to the suspension medium.

The numeric integration of eq. (7.17) with a numerically obtained tabulation of $U(x)$ for a capillary which has half the radius of the original droplet radius ($r_c = R_0 / 2$), assuming $\gamma = 1$ and $M = 1$, is shown in Figure 7-8, using as initial condition different kinetic energies $T(0)$ with respect to the potential energy $U_{max}$ at maximum deformation.

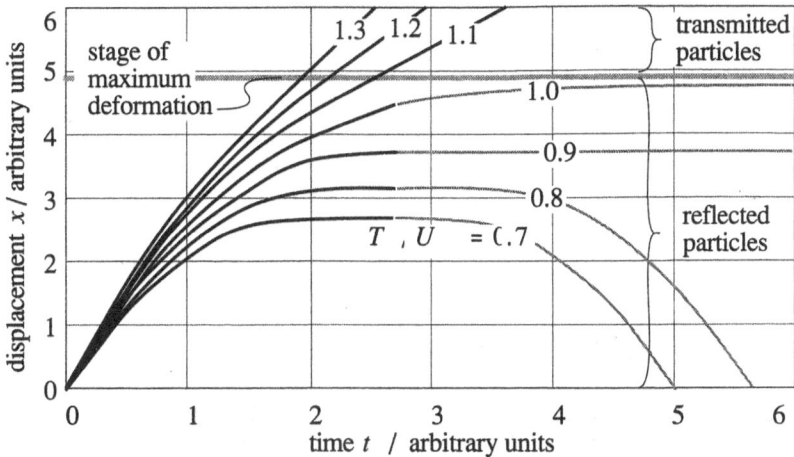

**Figure 7-8** *Deformation dynamics of suspended droplets*, approaching a given capillary with different kinetic energies $T_0$ , given in units of the maximum energy of deformation $U_{max}$. For kinetic energies $T_0 < U_{max}$ the oil droplet is elastically reflected in the backward direction. For $T_0 = U_{max}$ the time of passage will become infinite and the oil droplet will get stuck. For $T_0 > U_{max}$ the oil droplet will be retarded during the deformation process at the entrance of the capillary and accelerated back to its original velocity at the exit of the capillary. Due to the singularity of eq. (7.17) at the point-of-return, where $T_0 - U_{max} \rightarrow 0$, the corresponding trajectories past the point-of-return have a larger uncertainty, indicated by gray lines.

## 7.1.4    Red blood cell deformability

In the following an outline of the relevance and observation technique of red blood cell deformability is given [1].

During the evolution of life — besides diffusion — the active convection of liquids became necessary already on the microscopic scale of a cell. Much later the multi-cellular animals developed the active convection of liquids through the body on a macro-scopic scale, using the circulation of blood. Blood is a suspension of different cells — such as erythrocytes, thrombocytes, or leucocytes — in a carrier medium, the blood plasma. The primary task of the blood is to transport respiratory gases, nutritive sub-stances, end products of the metabolism, salts, hormones, macromolecules of diverse de-fense systems and heat from the supplier to the consumer. Mass and heat transport thereby are adapted to the specific needs of the different tissues by varying the flow rate of blood through them. The regulation of the blood flow therefore plays a crucial role for maintaining the vitality of the various tissues.

According to eq. (7.4) the flow of blood depends on the pressure difference between the arterial and the venous section of the specific vessel, of the length and diame-ter of the vessel, and of the dynamic viscosity of the blood itself. The adaptation of the flow rate thereby occurs by varying the beat frequency and the throughput of the heart and by varying the diameter of the vessels. The upper limit of the flow rate is character-ized by the term "vasomotoric limit".

[1]    Roggenkamp, H.G.: "Selektierendes Erythrozyten-Rigidometer." Dissertation, Universität Aachen, D-5100 Aachen, ca. 250 pp, (1985).

If the vasomotoric limit is reached, the normally uncritical viscosity of blood becomes a decisive factor, especially in the capillary system with pore diameters between 3 and 8 µm and pore lengths up to about 1 mm. Besides other factors, blood viscosity depends on the deformability of the red blood cells — the erythrocytes.

Up until now, red cell deformability is still more phenomenon than a well defined physical parameter. For example, the red cells adapt their shape to different flow conditions of the blood and are able to pass through very small capillaries with diameters smaller than their own diameter (Figure 7-9). Although this phenomenon of shape adaptation has been observed already during the time of the first microscopes, it became a research topic not until recent years. The reason for this belated appraisal is the recognition of diseases correlated with a change of red cell deformability.

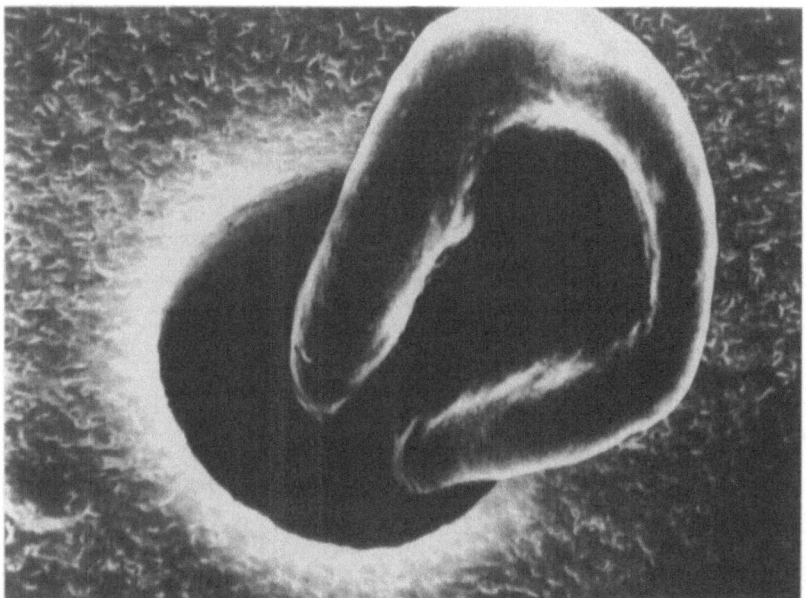

**Figure 7-9** *Passage of a red blood cell* through a single-pore [1]. Photo-mounting of two independent scanning electron micrographs.

One example is the sickle cell anemia, in which a direct causal relation exists between the reduced deformability of the sickle cells and the observed symptoms. The disease, frequent in African nations and thought to be associated with a simultaneous increase of the resistance to malaria, is characterized by a modified hemoglobin associated with a dramatic decrease of the red cell deformability.

### Blood — a non-Newtonian fluid

Blood behaves as a highly thixotropic liquid with several viscosity anomalies, depending on the dynamic structure of the suspension under specific flow conditions. With increasing shear stress the viscosity of blood decreases. Therefore one can only define a dynamic viscosity of blood which depends on plasma viscosity, red cell concentration, and red cell aggregation.

[1]     Roggenkamp, H.G.: "Selektierendes Erythrozyten-Rigidometer." Dissertation, Universität Aachen, D-5100 Aachen, ca. 250 pp, (1985).

Red cells in large vessels — due to the parabolic velocity profile according to eq. (7.3) and the finite size of the cells — are subject to low shear stresses and form large aggregates, a phenomenon corresponding to a high initial viscosity at $v \approx 0$. At intermediate shear stress the large cell aggregates segregate into individual cells which still maintain their quiescent shape. Thereby the viscosity of the blood decreases. At high shear stress the individual cells adapt their shape to the shear stress of the surrounding liquid decreasing the viscosity further more. All aggregates are dispersed and the dynamic viscosity depends mainly on the concentration and deformability of the red cells. The blood then behaves similar as an emulsion of fine droplets.

A still more drastic change of the dynamic viscosity occurs in vessels with diameters small compared to the diameter of the red cells. The red cell thereby behaves similar as an oil droplet suspended in water, pressed through a fine capillary. According to eq.(7.13), the energy input required for adapting a red cell of effective radius $R_0$ to a capillary of radius $r_c$ should be roughly inversely proportional to the capillary radius $r_c$. Therefore the driving pressure must exceed a certain value which is inversely proportional to the radius $r_c$ of the capillary. This enables a switching mechanism comparable to a safety valve which opens the capillary beyond a certain pressure or by adapting the capillary radius — depending on the actual needs of the various tissues.

### Definition of red blood cell deformability

Red cell deformability depends first of all on the excess surface of the cell as compared to a sphere of identical volume. The standard volume of red cells is about 91 $\mu m^3$. A sphere of this volume would have a surface of about 98 $\mu m^2$. But the actual surface of the red cell is bout 135 $\mu m^2$, or about 38 % more than the surface of the corresponding sphere of identical volume.

The increased surface is due to the expulsion of the cell nucleus during the maturing of the cell. The quiescent form of the red cell is an oblate disk and has a diameter of about 7.5 $\mu m$ and a thickness of about 2 $\mu m$ (Figure 7-10). The new shape provides much more freedom with respect to shape changes than a sphere, since the surface of the cell membrane may remain constant. The red cell adapts its shape passively to the shearing forces of an external flow field in which it is immersed. In this way it minimizes its hydrodynamic resistance and adapts itself to the geometrical boundary conditions of the microcirculation in the capillaries.

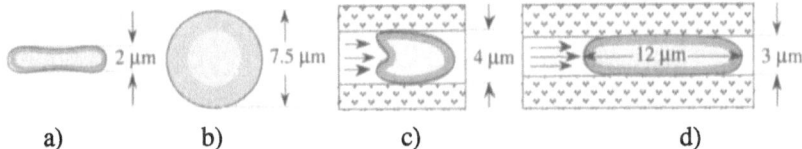

a)                    b)                    c)                            d)

**Figure 7-10** *Shapes of red blood cells.* (**a**) Cross section of quiescent shape. (**b**) Top view of quiescent shape (**c**) Shape under moderate constraint. (**d**) Shape under strong constraint, physiological limit [1].

Under the restriction that the volume of the red blood cell is constant, there exists a broad range of possible shapes. The strongest deformation occurs in the spleen which can be considered as a filter, where the red cells squeeze through elongated slits with sizes down to 1 $\mu m$ width and about 3 $\mu m$ length.

[1]    Gregersen, M.I., C.A. Bryant, W.E. Hammerle, S. Usami, S. Chien: "Flow Characteristics of Human Erythrocytes through Polycarbonate Sieves." Science, **157**, No. 3786, 825-827, (1967).

If the capillaries are smaller than the lower accessible limit which can be passed by the red cell, or if the red cell deformability is pathologically decreased, the red cell can get stuck in the capillary, blocking the blood flow. Therefore already a small fraction of stiff cells can actually decrease and in extreme cases lethally obstruct the perfusion of tissue.

The deformability of the red cells in capillaries with diameters smaller than 7 μm depends mainly on three factors. The ratio between surface and volume of the cell, the stiffness of the cell membrane, and the viscosity of the red cell plasma. Beyond that, the friction between the cell plasma and the cell membrane may in small vessels influence the viscosity. In some diseases the adhesion between membrane and cell plasma can drastically deteriorate the cell deformability. The viscosity of the red cell plasma — which is free of organelles and mainly consists of a highly concentrated hemoglobin solution — is usually low.

### Methods related to red blood cell deformability

Since deformability up to now is more a phenomenon than a defined physical parameter accessible to observation, the problem is how to perform meaningful observations and how to define a relevant parameter within reasonable physiological boundary conditions corresponding to the macrocirculation or to the microcirculation. According to this division, there exist two main groups of techniques, those related to the macrocirculation and those related to the microcirculation.

The first group of rheological techniques mimic the flow conditions of the macrocirculation and quantify the deformability of the red cells indirectly or directly. Viscosimetry determines the integral viscosity of the red blood and therefore enables only rather indirect conclusions on the deformability of the red cells. Laser light scattering from suspended red cells enables to determine the average deformability of the red cells by observing their elongation within the shear field of a macroscopic flow system.

The second group of rheological techniques tests the ability of the red cells to adapt to geometrical constraints within microscopic boundaries, for example within a pore. Filtration determines the flow rate of red blood in an integral way. It depends strongly on minor concentrations of leucocytes or other cells of decreased deformability. Micropipetting of individual red cells is capable to measure accurately the minimum pressure difference required to aspirate a red cell into a pipette. Finally, single-pore filtration is capable to repeat this aspiration procedure rapidly for different cells, using the sequential passage of red cells through a precise microcapillary.

### Red blood cell counter

In Figure 7-11 a conductivity cell for observing the deformability of red cells is shown [1]. It consists of two chambers filled by an electrolyte physiologically compatible with the red cells, separated by a single-pore membrane with a pore diameter smaller than the red cell. The pore diameter should be precisely defined between 3 and 5 μm and, in order to decrease the adhesion between the cell membrane and the pore, its surface should be as smooth as possible. The red cells enter through the upper chamber, pass through the single-pore, and leave the lower chamber. Additional openings in the two chambers enable their filling and cleaning. The passage of the red blood cell is observed using the decrease of the electrical current through the conductivity cell during the passage of the red cell through the critical aperture according to the resistive pulse technique.

[1]     Roggenkamp, H.G.: "Selektierendes Erythrozyten-Rigidometer." Dissertation, Universität Aachen, D-5100 Aachen, ca. 250 pp, (1985).

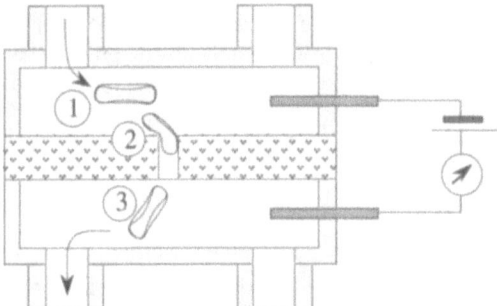

**Figure 7-11** *Cell counter* used for characterizing the deformability of individual red blood cells.

### Passage time observations

For determination of the passage times fresh venous blood has been antico-agulated with heparine (50 I.E./cm3) and suspended in SER buffer (193) (pH = 7.4, os-molarity = 300 milli-osmol/liter) at a relative cell volume of about 3%. If the passage time exceeds 100 ms, automatically the single-pore was flushed by applying a pressure pulse. Figure 7-12 compares a passage time spectrum of normal red cells with that due to artificial rigidification and that due to partial recovering of the red cell deformability.

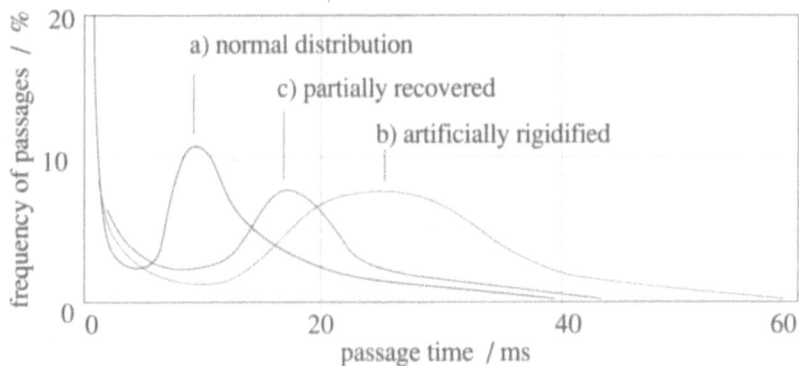

**Figure 7-12** *Passage time spectra* of red blood cells under the influence of artificial rigidification and mobilization of the red cell membrane [1]. (**a**) The distribution corresponding to "normal" blood peaks around 10 ms. (**b**) The peaks is shifted to around 28 ms by increasing the Ca ion concentration. (**c**) The stiffening effect is partially reversed by a suitable pharmaceutical increasing the flexibility of the cell membrane.

More information can be obtained by observing the pulse shape in the resistive pulse technique. During the passage of a red cell the resistance changes in a characteristic way. Three distinct phases of the passage can be discerned. During the entrance phase of the red cell the resistance rises steeply to its maximum value, during the passage of the red cell through the long channel of the pore the resistance drops to a smaller value, during the exit the resistance rises again without reaching the maximum of the entrance

[1]     Roggenkamp, H.G.: "Selektierendes Erythrozyten-Rigidometer." Dissertation, Universität Aachen, D-5100 Aachen, ca. 250 pp, (1985).

phase and drops ultimately to the original low value of the unobstructed capillary. The three phases are also related to different shapes of the red cell during the passage.

The passage time for a given red cell depends on the pressure difference between the upper and the lower chamber of the conductivity cell, on pore diameter, pore length, temperature of the suspension medium and on the composition of the suspension medium.

## 7.1.5   Suggested pore shapes

Cylindrical pores described above [1] provide a simple possibility to define a deformability parameter for red blood cells. A disadvantage of this pore geometry is the variation of the observed signal due to different cell orientations during entering the single-pore channel.

It should be possible to improve the technique by more precisely defined experimental conditions. Of special concern is the alignment of the red cells during their impact onto the critical aperture. Two possible pore shapes for improving the technique are shown in Figure 7-13. In addition to the determination of a deformability parameter, they provide the possibility to determine the cell volume during the transit, using the resistive-pulse signal during the cell passage through the wide part of the channel.

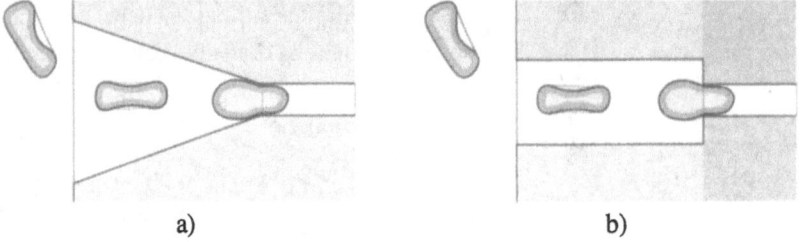

a)                                                 b)

**Figure 7-13** *Suggested pore shapes for the observation of cell deformability.* (a) A funnel-shaped entrance aperture enables the axial orientation of the cell before its impact onto the critical aperture. (b) A stepped two-section capillary enables to correlate the volume of the cells with their deformability cell by cell. The first capillary section has a wide diameter and is used for determining the cell volume. At the same time the cells are aligned axially. The second capillary section has a smaller diameter than the cell and is used for determining the deformability of the cell.

[1]   Roggenkamp, H.G., H. Kiesewetter, R. Spohr, U. Dauer, L.C. Busch: "Production of Single-Pore Membranes for the Measurement of Red Blood Cell Deformability." Biomedizinische Technik, **26**, 167 - 169, (1981).

## 7.2    Single-pores and superfluidity

Ion tracks provide obstacles sufficiently small to be interfere with superfluids and superconductors — the macroscopic coherent-phase objects of low temperature physics. According to the Josephson relation for electrical superconductors, potential differences can be measured as frequencies. This basic finding has resulted in a variety of different applications, for example in high-precision gauges for voltage and magnetic flux. While the exploitation of the superconductive Josephson effect is already quite advanced, the realization of its superfluid analog has just begun [1]. The resulting interference phenomena bear the promise to be applied as sensors for measuring pressure and absolute rotation in units of time. One obstacle in this development, however, is the reliable manufacture of sufficiently small apertures, needed as "weak links" — potential barriers through which tunneling occurs between two neighboring superfluids. Due to the smaller coherence lengths of superfluids in comparison with superconductors, these structures have to be smaller. The ion track technique is an alternative [2], [3] to other techniques — like conventional lithography — to produce these fine structures.

### 7.2.1    Basic phenomena

**Superfluids.** For the observation of clean interference effects in superfluids, the size $d$ of the obstacles interacting with the flow field has to approach the size of the coherence length $\xi_0$ and be at least of the order of $d \approx 10 \cdot \xi_0$. The existing superfluids — comprising superconductors — are characterized by quite different coherence lengths $\xi_0$ (Figure 7-14). Accordingly, quite different obstacle sizes are required.

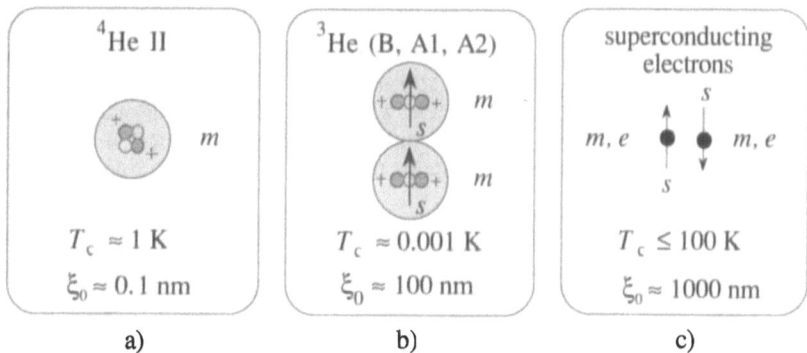

**Figure 7-14** *Three types of superfluids.* (a) Superfluid helium-4. (b) Superfluid helium-3, phases $B$, $A1$, and $A2$. (c) Superconducting electrons.

[1]    Varoquaux, E., O. Avenel: "Quantum Phase Slippage in Superfluid $^4$He." Physica Scripta, vol. **T19**, pp. 445 - 452, (1987).

[2]    Packard, R.E., J.P. Pekola, P.B. Price, R.N.R. Spohr, K.H. Westmacott, Zhu Yu-Qun: "Manufacture, Observation, and Test of Membranes with Locatable Single Pores." Rev.Sci.Instrum., **57** (8), 1654-1660, (1986).

[3]    Pekola, J.P., J.C. Davis, Zhu Yu-Qun, R.N.R. Spohr, P.B. Price, R.E. Packard: "Suppression of the Critical Current and the Superfluid Transition Temperature of $^3$He in a Single Submicron Cylindrical Channel." Journal of Low Temperature Physics, **67**, 47, (1987).

**Superfluid flow.** Figure 7-15 a shows the flow of a superfluid through a small aperture connecting two reservoirs $A$ and $B$ [1]. As soon as the mass current between the two reservoirs exceeds a critical value given by the size of the aperture, energy-loss processes occur by vortex generation (Figure 7-15 b). Simultaneously a pressure difference between the two reservoirs is established. The elementary processes of vortex generation are observed at the rate of the "Josephson frequency".

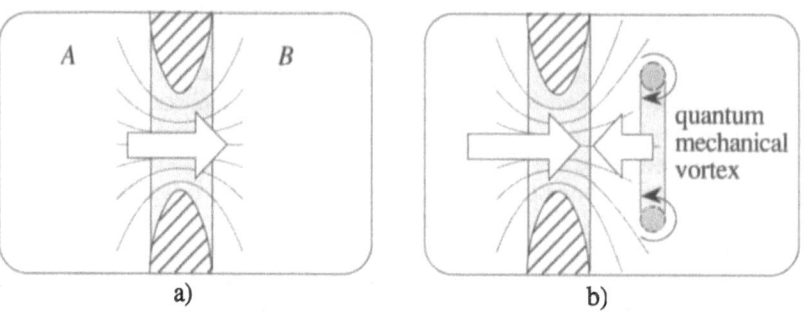

a)                                                                b)

**Figure 7-15** *Superfluid flow through an aperture* separating two reservoirs $A$ and $B$. (a) At subcritical velocities, $v_s < v_c$, the flow can be derived from a potential field and occurs without any observable friction. (b) At supercritical velocities, $v_s \geq v_c$, the flow field is characterized by the creation of vortex rings and a rotational field has to be superposed on the potential field. The rate of vortex production determines the frictional energy-loss rate of the system.

**Macroscopic wave function.** In contrast to normal many-particle systems, superfluid systems can be described as a condensate of many strongly correlated particles occupying one and the same preferred quantum state. The condensation becomes effective below a critical temperature $T_c$ at which the aggregation of the particles due to an attractive potential starts to overcome their thermally excited disaggregation. According to the two-fluid model, below $T_c$ the system becomes a mixture of a superfluid of number density $\rho_s$ and a normal liquid of number density $\rho_n$, whereby the total density $\rho = \rho_n + \rho_s$. The wave function of an arbitrary particle within the superfluid is identical with the wave function of all the other particles. By multiplying this wave function with the total number of particles in the superfluid one obtains the macroscopic wave function of the superfluid condensate:

$$\psi(r,\, t) = \sqrt{\rho_s(r,t)}\; e^{\,i\, \varphi(r,\, t)}\,, \qquad\qquad (7.18)$$

where $\rho_s = \psi\, \psi^*$ is the probability or number density of the particles in the superfluid state, $\psi^* = \rho_s^{1/2} \exp(-\,i\, \varphi)$ is the conjugate complex of $\psi$, and $\varphi$ is the quantum mechanical phase of the particles. For a sufficiently large number of particles in the superfluid state the wave function $\psi$ — alternatively known as the order parameter — is macroscopic in two respects. It is slowly varying in space and it is a directly observable macroscopic quantity. Both, the probability density $\rho_s$ and the phase $\varphi$ of the superfluid condensate, except for a constant phase factor, can be determined with high accuracy. While the probability density $\rho_s$ — by multiplication with the particle mass $m$, — can be converted into the density $m{\cdot}\rho_s$, the phase $\varphi$ of the macroscopic wave function, except

---

[1]     Varoquaux, E.: Direct communication. Laboratoire de Physique des Solides, Bat. 510, Université Paris-sud, F-91405, France, July (1988).

for the factor $[- h / (2\pi m)]$, represents the potential of the velocity field. As shown below, the negative gradient of this potential is equal to the velocity $v_s$ of the superfluid.

## 7.2.2    Basic relations

**Coherence length and superfluid density.** The description of superfluidity by a smooth macroscopic wave function $\psi = \rho_s^{1/2} \exp (i\, \varphi)$, according to eq. (7.18), breaks down at short distances of the order of the coherence length $\xi_0$ — for example within the core of a vortex ring. The coherence length is the distance over which two particles within the superfluid are coupled together — forming a "Cooper pair" in superconductors and $^3$He — and corresponds to the ideal size of a microscopic obstacle required to induce interference effects.

The coherence length increases to infinity close to the critical temperature $T_c$ of the superfluid. This implies that close to the critical temperature even large apertures can lead to interference phenomena. Unfortunately, at the same time the superfluid density $\rho_s$ decreases to zero while normal fluid density $\rho_n$ increases to the total density $\rho$. Therefore the interference effects are washed out by the superposed flow of the normal liquid.

The coherence length $\xi$ and the superfluid density $\rho_s$ are roughly inversely proportional according to:

$$\xi \approx \frac{\xi_0}{\rho_s} \,, \quad \text{where } \rho_s \approx \sqrt{1 - \frac{T}{T_c}} \quad \text{and} \tag{7.19}$$

$\xi_0$ is the coherence length at $T = 0$ which is characteristic of the specific type of superfluid.

**Josephson effect**

The Josephson equations and the associated concept of phase slippage are the most fundamental and exact consequences of the present theoretical understanding of superfluidity [1]. They can be derived using a simple phenomenological model [2] based on a two-level system consisting of two superfluid zones $A$ and $B$, separated by a potential barrier (Figure 7-16).

This potential barrier has to be sufficiently transparent for enabling the tunneling of particles — due to the overlapping wave functions — between the system $A$ and $B$. At the same time it must still be sufficiently opaque to suppress the establishment of a complete phase-coherence between the two reservoirs which would be an uninteresting situation. Such a potential barrier is called a "weak link". It couples the wave function $\psi_A$ on the left side of the barrier moderately with the macroscopic wave function $\psi_B$ on the right side of the barrier.

[1]    Anderson, P.W.: "Considerations on the Flow of Superfluid Helium." Rev. Mod. Phys. **38**, No. 2, pp. 298 - 310, (1966).
[2]    Feynman, R.P., R.B. Leighton, M. Sands: "The Schrödinger Equation in a Classical Context: A Seminar on Superconductivity." in "**The Feynman Lectures on Physics**" Vol. **3**, Quantum Mechanics, Addison-Weseley Publishing Company, Reading, Massachusetts, USA, (1965).

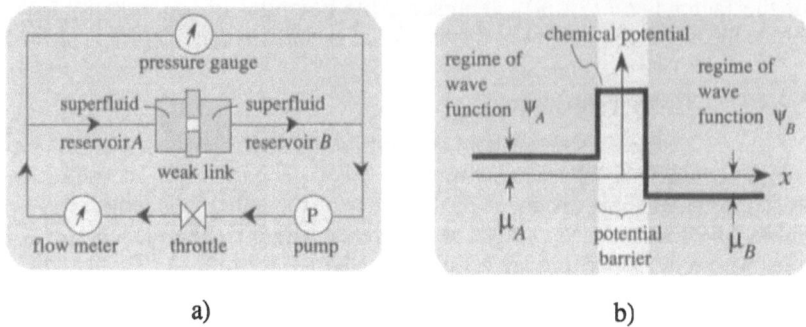

a)                                                                          b)

**Figure 7-16** *Principle of Josephson experiment* for superfluids (**a**) and equivalent two-level model (**b**).

For an infinitely high potential barrier, the particle will be either somewhere left or somewhere right of the potential barrier, corresponding to the standing waves $\psi_A$ and $\psi_B$, which are divergence-free eigenstates of the Schrödinger equations:

$$i\frac{h}{2\pi}\frac{\partial \psi_A}{\partial t} = \mu_A \, \psi_A \, , \quad \text{where } \psi_A = \sqrt{\rho_A} \, e^{i\,\varphi_A} \, , \qquad (7.20)$$

$$i\frac{h}{2\pi}\frac{\partial \psi_B}{\partial t} = \mu_B \, \psi_B \, , \quad \text{where } \psi_B = \sqrt{\rho_B} \, e^{i\,\varphi_B} \qquad (7.21)$$

Here $\mu_A$ and $\mu_B$ are the chemical potentials of the reservoirs $A$ and $B$, respectively.

**Chemical potential.** The chemical potential $\mu$ is defined as the energy gain of the system per added particle, $\mu \equiv \partial E / \partial N$. The simplest form of a chemical potential $\mu$ is the potential energy gain $\mu = m \, g \, h$ of a particle of mass $m$ in the gravitational field of acceleration $g$, lifted from zero height to the liquid level $h$.

In situations where the particle possesses a locally dependent kinetic energy, the term $m \, v^2 / 2$ has to be added to the potential energy term $\rho \, g \, h$ corresponding to the energy input to accelerate the particle to the velocity of the flow field, similar to the Bernoulli equation describing the hydrodynamic flow in ideal liquids.

If the height of the potential barrier is reduced to a level where tunneling starts to become important, the rate of change of the wave function the left side, $\partial \psi_A / \partial t$, is influenced by the amplitude $\psi_B$ on the other side of the barrier, and vice versa:

$$i\frac{h}{2\pi}\frac{\partial \psi_A}{\partial t} = \mu_A \, \psi_A \; + \; K \, \psi_B \quad , \qquad (7.22)$$

$$i\frac{h}{2\pi}\frac{\partial \psi_B}{\partial t} = \mu_B \, \psi_B \; + \; K \, \psi_A \quad , \qquad (7.23)$$

where the coupling strength $K$ depends on the height and width of the specific barrier. If we insert wave functions of the type used in eqs. (7.20) and (7.21), separate the real and imaginary parts, assume that $\rho_A \approx \rho_B \approx \rho$, and recognize that the time derivation of the superfluid density, $\partial \rho_A / \partial t$, in reservoir $A$ corresponds to the probability current $J$ flowing from $B$ into $A$, we obtain two basic equations ruling the Josephson effect:

$$\frac{\partial \delta}{\partial t} = -\frac{2\pi}{h}\mu \;\Rightarrow\; \delta(t) = \delta_0 - \frac{2\pi}{h}\int_0^t \mu(\tau)d\tau, \text{ where } \mu \equiv \mu_B - \mu_A \;, \qquad (7.24)$$

$$J = J_c \sin\delta, \quad \text{where } J_c = \frac{4\pi}{h}K\rho \text{ and } \delta \equiv \varphi_B - \varphi_A \;. \qquad (7.25)$$

Multiplication of the probability current $J$ by the particle mass $m$ yields the mass current. The phase $\delta_0$ is arbitrary and depends on the specific choice of $t = 0$. The eqs. (7.24) and (7.25) reflect the most important features of the Josephson effect and are known as Josephson's a.c. and d.c. relations, respectively. According to eq. (7.24) the phase $\delta$ — except for an arbitrary phase factor $\delta_0$— can be determined as function of the time history of the chemical potential difference $\mu(t)$, and in turn, according to eq. (7.25), the probability current $J$ can be determined as function of the phase $\delta$.

**DC Josephson effect.** Roughly speaking, it enables the generation of a high frequency current by applying a d.c. potential difference over a weak link.

The evaluation of eq. (7.24) depends on the chemical potential difference $\mu$. Assuming a constant gravitational potential $\mu = m\,g\,h$, the height $h$ corresponds to a pressure difference $p = \rho\,g\,h.$, where $\rho$ is the mass density. This leads to $\mu = p/N$, where $N$ is the number density of the superfluid. Inserting $N \approx 2 \cdot 10^{22}/cm^3$ for superfluid $^4$He II and $p = 1$ bar $= 10^6$ dyne $/cm^2 = 10^5$ Pa, one obtains $(2\pi/h)\mu \approx 5 \cdot 10^{10}$ s$^{-1}$ which is a large angular frequency. Therefore the phase difference $\delta$ changes rapidly with time. Inserting this phase difference into eq. (7.25) thus yields a probability current $J$, which oscillates with the so-called Josephson frequency:

$$\omega_J = \frac{2\pi}{h}\mu = \frac{p}{N}, \qquad (7.26)$$

where $p$ is the pressure difference and $N$ is the number-density of the particles within the superfluid. In the example of superfluid $^4$He II the Josephson frequency corresponds to roughly 8 kHz per $\mu$bar. Due to the rapid oscillation no net mass current would be observed on the average in this idealized tunnel-effect situation, as long as the pressure difference $p$ is not zero. Only for $p = 0$ a mass current can be observed, the size of which depends on the arbitrary phase $\delta_0$ (Figure 7-17).

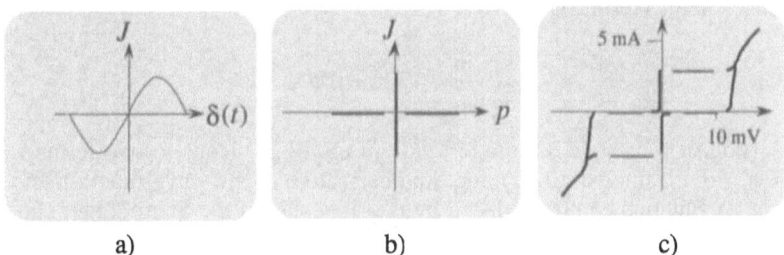

a)                                    b)                                    c)

**Figure 7-17** *d.c. Josephson effect.* (a) Ideal current-phase relation. (b) Ideal current-pressure relation. (c) Observed effect for a superconductor after reference [1].

[1]    Barone, A., G. Paterno: **"Physics and Applications of the Josephson Effect."** Wiley, New York, 529 pp.,(1982).

**AC Josephson effect.** Roughly speaking, it is the inverse of the d.c. Josephson effect. If a high frequency current is applied to a weak link, a d.c. potential difference develops across it. The effect is well established in superconductors, however, it has not yet been observed in superfluids.

If — in addition to a constant pressure difference $p$ — a small oscillating pressure difference $p_{osc} = a \cos(\omega t)$ is applied, a current $J$ should be observed only if the frequency $\omega = \omega_J$ is equal to the Josephson frequency according to eq. (7.26). This relation should enable the conversion of a pressure difference $p$ into a frequency, similar to the conversion of an electric potential difference into a frequency in the case of super-conducting Josephson junctions.

### Velocity field and phase of wave function

According to the d.c. Josephson equation, eq. (7.25) the current through the "weak link" is determined by the phase difference $\delta = \varphi_B - \varphi_A$ between the right and left side of the potential barrier. A more general relation exists in the bulk of an arbitrary su-perfluid system between the local flux density $j_s = \rho_s \, v_s$ — or the local velocity $v_s$ — and the gradient of the phase $\varphi$, which is derived in the following.

The number density $\rho_s$ in a one-dimensional volume element $dx$ increases during the time $dt$ by the difference of the flux densities passing through the left and right boundaries of the volume element, $j_s(x) - j_s(x + dx)$, according to :

$$d\rho_s = \frac{\left(j_s(x) - j_s(x + dx)\right) dt}{dx} \Rightarrow \frac{\partial \rho_s}{\partial t} = -\frac{\partial j_s}{\partial x} . \qquad (7.27)$$

In three dimensions we obtain the equation of continuity:

$$\frac{\partial \rho_s}{\partial t} = -\nabla j_s = -\nabla \rho_s v_s , \qquad (7.28)$$

considering that the flux density $j_s$ is equal to the number density $\rho_s$ times the velocity of the superfluid $v_s$.

While $\rho_s$ is already defined by the relation $\rho_s = \psi \psi^*$, the locally dependent quantities flux density $j_s$ and flux velocity $v_s$ have to be defined through the complete Schrödinger equation — which is analogous to eq. (7.20) but includes the divergence (or kinetic energy) term $\Delta \psi \equiv \nabla^2 \psi$:

$$i \frac{h}{2\pi} \frac{\partial \psi}{\partial t} = -\frac{1}{2m} \frac{h^2}{4\pi^2} \nabla^2 \psi + \mu \psi , \qquad (7.29)$$

Insertion of $\rho_s = \psi \psi^*$ on the left side of eq. (7.28) leads to two terms $\psi^* \, \partial \psi / \partial t$ and $\psi \, \partial \psi^* / \partial t$. Using eq. (7.29), the time derivative of the wave function and its conjugate complex function can be replaced by the right side of the Schrödinger equation [1]. This yields:

[1]    Feynman, R.P., R.B. Leighton, M. Sands: "The Schrödinger Equation in a Classical Context: A Seminar on Superconductivity." in "**The Feynman Lectures on Physics**" Vol. 3, Quantum Mechanics, Addison-Weseley Publishing Company, Reading, Massachusetts, USA, (1965).

$$\frac{\partial \rho_s}{\partial t} = \psi^* \frac{\partial \psi}{\partial t} + \psi \frac{\partial \psi^*}{\partial t} = -\nabla \left( \frac{h}{4\pi i m} \left( \psi^* \nabla \psi - \psi \nabla \psi^* \right) \right). \tag{7.30}$$

Comparing eq. (7.28) with eq. (7.30), the term within the large brackets following the $\nabla$ operator is identified as the flux density $j_s$. Inserting, according to eq. (7.18), the explicit form of the wave function, $\psi = \rho_s^{1/2} \exp(i\,\varphi)$ one obtains:

$$\frac{\partial \rho_s}{\partial t} = -\nabla \left( \rho_s \left( \frac{h}{2\pi} \frac{1}{m} \nabla \varphi \right) \right) \;\Rightarrow\; v_s = \frac{h}{2\pi} \frac{1}{m} \nabla \varphi . \tag{7.31}$$

This final result is the important relation between the phase $\varphi$ of the wave function and the local velocity $v_s$ of the superfluid. It is inferred by comparing eq. (7.28) with eq. (7.31). According to eq. (7.31) the velocity vector $v_s$ of the superfluid flow is directed towards increasing phase $\varphi$.

**Fundamental equation of motion.** Following reference [1], the macroscopic fundamental equation of motion for superfluids is derived in the following way. Inserting the wave function $\psi$ of eq. (7.18) into the Schrödinger equation, eq. (7.29), yields two equations, the real and to the imaginary part of the Schrödinger equation. While the imaginary part leads to the equation of continuity, eq. (7.28), the real part yields:

$$\frac{h}{2\pi} \frac{\partial \varphi}{\partial t} = \frac{h^2}{8\pi^2 m} \frac{\nabla^2 \sqrt{\rho_s}}{\sqrt{\rho_s}} - \frac{h^2}{8\pi^2 m} \left( \nabla \varphi \right)^2 - \mu . \tag{7.32}$$

Within the bulk of the superfluid the superfluid density $\rho_s$ is spatially slowly varying and its gradient — the first term on the right side of eq. (7.32) therefore can be neglected. Inserting $\nabla \varphi$ from eq. (7.31) into the second term on the right side of eq. (7.32) yields the first representation of the fundamental equation of motion for superfluids:

$$\frac{h}{2\pi} \frac{\partial \varphi}{\partial t} = -\left( \mu + \frac{1}{2} m v_s^2 \right) . \tag{7.33}$$

The gradient of this equation leads to the second representation of the fundamental equation of motion — the Euler Landau equation — for superfluids:

$$m \frac{\partial v}{\partial t} = -\nabla \mu - m v_s \nabla v_s \quad \text{or} \quad \frac{\partial v_s}{\partial t} + \nabla \left( \frac{\mu}{m} + \frac{v_s^2}{2} \right) = 0 . \tag{7.34}$$

**Quantization of circulation**

According to eq. (7.31) the velocity $v_s$ is derived from a potential which is directly related with the gradient of the phase $\varphi$ of the wave function $\psi$. In a topologically singly connected system therefore the velocity field is free of rotation:

$$\text{rot } v_s = \lim_{S \to 0} \frac{1}{S} \oint v_s \, dr = 0, \text{ since } \oint v_s \, dr = C \oint \frac{\partial \varphi}{\partial r} \, dr = C \oint \partial \varphi = 0 . \tag{7.35}$$

[1]   Tilley, D.R., J. Tilley: "**Superfluidity and Superconductivity.**" Adam Hilger Ltd, Bristol, 429 pp., (1986).

However, in a multiply connected system — such as a torus or a superfluid system surrounding a vortex line— this is not necessarily the case which allows for a quantized rotation of the superfluid around a torus or around a vortex line. Following the argumentation of reference [1], by going once around the torus, the macroscopically observable, single valued, wave function $\psi$ has to return to the same value. Thereby its phase $\varphi$ only needs to return to its original value modulo $n\,2\pi$, where $n$ is an integer. This degree of freedom enables a quantized hydrodynamic circulation around the torus:

$$\oint v_s\,dr = \frac{h}{2\pi m} \quad \oint \nabla\varphi\,dr = \frac{h}{2\pi m} \quad \oint \partial\varphi = \frac{h}{2\pi m}\,n\,2\pi = n\frac{h}{m} \ , \ \text{or} \qquad (7.36)$$

$$\oint v_s\,dr = n\,\kappa_0 \ , \quad \text{where} \quad \kappa_0 \equiv \frac{h}{m} = \text{quantum of circulation} . \qquad (7.37)$$

Since the phase $\varphi$ increases in the direction of the velocity vector $v_s$, the quantum number $n$ is positive if the turning sense of the velocity is the same as the turning sense of the integration path surrounding the vortex and negative if the turning senses are opposite.

Figure 7-18 a shows lines of equal phase around a right-turning vortex of $n = 1$. Integrating $\partial\varphi$ counter-clock-wise once around this vortex changes the phase by - $2\pi$. Whenever such a right-turning vortex drifts into a left-turning closed path of integration, the closed integral $\int d\varphi$ will change by - $2\pi$ (Figure 7-18 b):

$$\oint d\varphi = \int d\varphi + \int d\varphi = (\varphi_B - \varphi_A) - (\varphi_A - \varphi_B) = -2\pi .$$
$$\quad\ \ \ _{A\rightarrow B}\quad _{B\rightarrow A}\quad \text{lower path}\quad \text{upper path} \qquad\qquad\qquad (7.38)$$

If the upper path is far from the entering vortex, the upper part of the integral will remain unchanged while the lower path integral will fully reflect the phase slip by - $2\pi$.

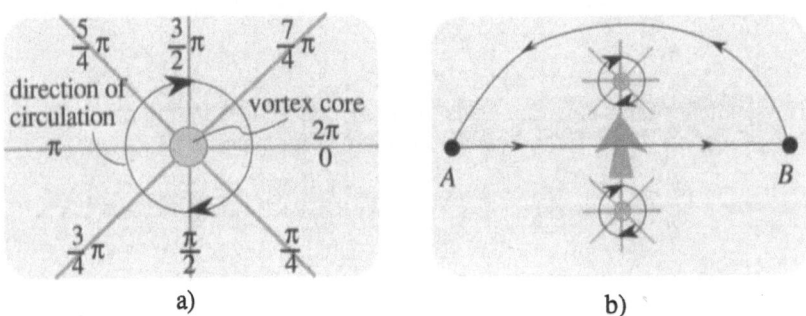

a)                                                    b)

**Figure 7-18** *Quantization of circulation in superfluids.* (a) Lines of equal phase around a right-turning vortex line of quantum number $n = 1$. (b ) Whenever this vortex crosses the chosen path of integration between the two points $A$ and $B$, the line integral $\int d\varphi$ between these two points changes by - $2\pi$.

**Nucleation of a vortex ring.** Figure 7-19 shows the nucleation of a vortex ring in a cylindrical channel. The vortex ring expands at right angle from the axis of symmetry, has a turning sense such that the flow velocity is decreased along the axis and drifts

[1]    Anderson, P.W.: "Considerations on the Flow of Superfluid Helium." Rev. Mod. Phys. **38**, No. 2, pp. 298 - 310, (1966).

in the direction of the flow velocity — the direction of increasing phase.

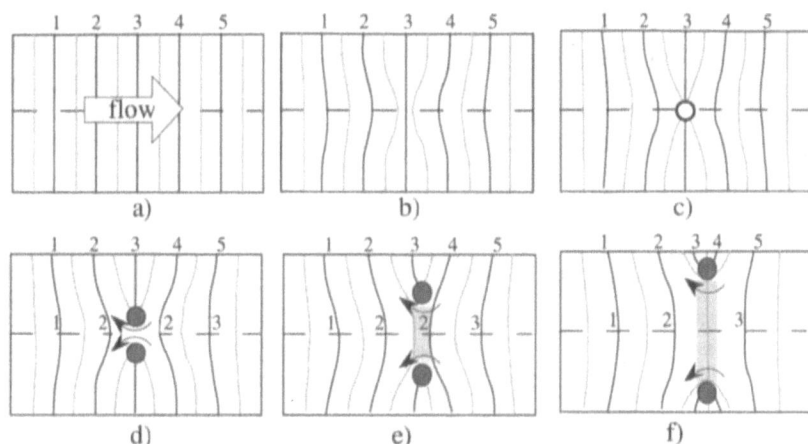

**Figure 7-19** *Nucleation and growth of a vortex ring in a cylindrical channel* [1]. Surfaces of constant phase φ in units of π for an integration path along the boundary of the system (top labels) and along the axis of symmetry (center labels). (**a**) Undisturbed flow. (**b**) Disturbed flow increasing toward center. (**c**) Nucleation of vortex ring. (**d, e, f**) Axial drift and expansion of vortex ring.

### Phase slip rate and energy loss

According to the generalization of eq. (7.24) the difference μ of the chemical potential difference — or pressure difference — between two points $B$ and $A$ in a superfluid is given by the rate of increase of the phase difference between the points $B$ and $A$. In turn, according to eq. (7.38), this corresponds to - $2\pi$ times the the rate at which vortex lines cross the line connecting $A$ and $B$ — in other words to the phase slip rate $dn \, / \, dt$  (Figure 7-18 b) [2]:

$$\frac{d}{dt}\left(\varphi_B - \varphi_A\right) = -\frac{2\pi}{h}\left(\mu_B - \mu_A\right) = -\frac{dn}{dt}\,2\pi \quad \text{or} \quad \mu_A - \mu_B = \frac{dn}{dt}\,h \;. \qquad (7.39)$$

**Energy-loss per phase slip.** Following the argumentation of reference [3], according to eq. (7.37), during a phase slip event the circulation $\int v_s \, dr = \int v_s \, dr$ across the weak link between point $A$ and point $B$ decreases by one quantum of $\kappa_0$. Before the phase slip event occurs, the flow has reached a critical velocity $v_c$. During the phase slip event the flow velocity is reduced from the critical velocity $v_c$ by an amount $\delta v$ to $v_s = v_c - \delta v$. Two simplifications are used for determining the change of the circulation and the kinetic energy. First, the contributions to the circulation outside the critical aperture of length $l$ can be neglected for sufficiently small apertures for which the velocity decreases rapidly outside the aperture. Thus the change in the circulation is:

[1]    Langer, J.S. Fisher, M.E.: "Intrinsic Critical Velocity of a Superfluid" Phys. Rew. Lett. **19**, 560 - 563, (1967).
[2]    Tilley, D.R., J. Tilley: "**Superfluidity and Superconductivity.**" Adam Hilger Ltd, Bristol, 429 pp., (1986).
[3]    Varoquaux, E., O. Avenel: "Quantum Phase Slippage in Superfluid ⁴He." Physica Scripta, vol. **T19**, pp. 445 - 452, (1987).

$$\int_A^B (v_c - \delta v)\, dr - \int_A^B v_c\, dr = -\delta v\, l = -\kappa_0 \quad \Rightarrow \quad \delta v = \frac{\kappa_0}{l} \; . \tag{7.40}$$

Second, for not too small apertures the change of the velocity is small in comparison with the critical velocity. Thus, in the limit $\delta v_s \ll v_c$, the corresponding kinetic energy-loss in the weak link is given by:

$$\delta E = \frac{M}{2\, l}\left(\int_A^B (v_c - \delta v)^2 dr - \int_A^B v_c^2\, dr\right) = -\rho\, \kappa_0\, v_c\, s \; , \tag{7.41}$$

where $M = \rho\, s\, l$ is the effective mass of the superfluid within the weak link and $s$ is the cross section of the weak link. According to eq. (7.41) the energy loss during a phase slip event is proportional to the cross section $s$ of the weak link.

### 7.2.3    Flow through a single pore

**Single-pore cell.** In the following a technique for observing the superfluid analog of the Josephson effect is described [1], [2] (Figure 7-20). In a preliminary experiment a weak link in the form of an etched ion track of about 0.7 µm diameter and about 7 µm length was used [3].

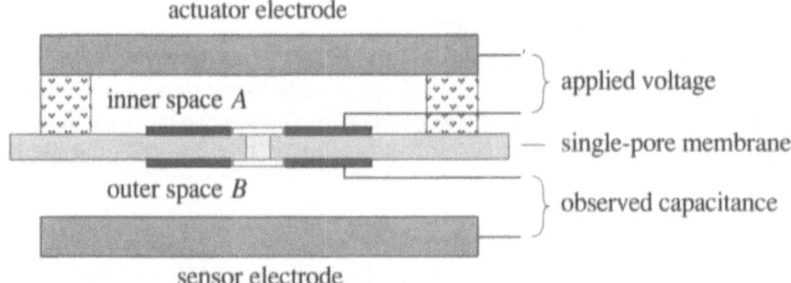

**Figure 7-20** *Single-pore capacitance-cell* for observing the superfluid analog of the Josephson effect.

The technique combines an electrically actuated elastic membrane with a distance sensor. The membrane separates an inner space $A$ from an outer space $B$, both filled with superfluid $^4$He or $^3$He at a temperature below their respective critical temperatures. In conjunction with the critical aperture the device represents a so-called "Helmholtz" resonator, the frequency of which is determined by the compliance $\kappa$ of the membrane and the accelerated mass $M = \rho\, s\, l$ in the vicinity of the critical aperture (Figure 7-21).

[1]    Varoquaux, E., O. Avenel: "Quantum Phase Slippage in Superfluid $^4$He." Physica Scripta, vol. **T19**, pp. 445 - 452, (1987).

[2]    Pekola, J.P., J.C. Davis, Zhu Yu-Qun, R.N.R. Spohr, P.B. Price, R.E. Packard: "Suppression of the Critical Current and the Superfluid Transition Temperature of $^3$He in a Single Submicron Cylindrical Channel." Journal of Low Temperature Physics, **67**, 47, (1987).

[3]    Packard, R.E., J.P. Pekola, P.B. Price, R.N.R. Spohr, K.H. Westmacott, Zhu Yu-Qun: "Manufacture, Observation, and Test of Membranes with Locatable Single Pores." Rev.Sci.Instrum., **57** (8), 1654-1660, (1986).

The equation of motion for the effective mass $M$ is obtained in the following way. (1) The displacement $X = F \kappa$ of the membrane is equal to the product of the restoring force $F$ and the compliance $\kappa$ of the membrane. (2) At low frequencies the pressure $p = F / S = f / s$ can be assumed to be equal everywhere within the resonator, where $f$ is the force acting on the effective mass $M$. (3) Assuming an incompressible fluid, the volume $S X$ displaced by the membrane is equal to the volume $s x$ displaced at the critical aperture.

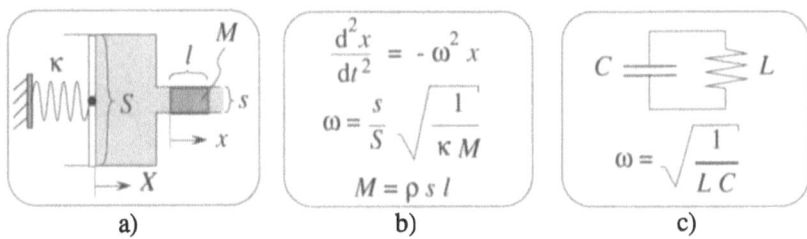

a)    b)    c)

**Figure 7-21** *Helmholtz resonator.* (**a**) Mechanical equivalent. (**b**) Equation of motion, frequency $\omega$, and mass $M$. (**c**) Electrical equivalent. The inductance $L$ is proportional to the accelerated mass $M$. The capacitance $C$ is proportional to the compliance $\kappa$ of the membrane.

In order to observe phase slip events according to reference [1], the Helmholtz resonator is excited by an external a.c. voltage. Thereby its amplitude increases linearly with time. Ultimately, when the maximum velocity of the superfluid pulsing back and forth through the aperture exceeds the critical velocity $v_c$, a phase slip event occurs, removing, according to eq. (7.41), the finite energy $\delta E = \rho \, \kappa_0 \, v_c \, s$ from the resonating system and thus reducing the maximum amplitude by a finite amount.

## 7.2.4    Suggested shapes of weak links

Future experiments are very likely to use simple planar geometries such as thin film apertures as weak links. Up to now, classical fabrication techniques, such as conventional lithography or ion beam sputtering have been emploied for manufacturing thin-film apertures. The limiting size of these apertures is of the order of 0.1 μm or larger. Which possibilities are offered by the nuclear track technique in order to overcome the present size limit and possibly enable the observation of a Josephson effect even for superfluid $^4$He which has the disadvantage of a small coherence length but provides the advantage of much higher critical temperature?

Figure 7-22 gives some possible aperture geometries based on the ion track technique which could be used in the future for observing the superfluid analog of the Josephson effect. Figure 7-23 shows a conical funnel obtained by one-sided etching of an ion track in a thin glass plate as well as a thin pore in polycarbonate. Other interesting geometries would be provided by a regular array of track pores generated by means of a scanning ion microbeam.

[1]    Varoquaux, E., O. Avenel: "Quantum Phase Slippage in Superfluid $^4$He." Physica Scripta, vol. **T19**, pp. 445 - 452, (1987).

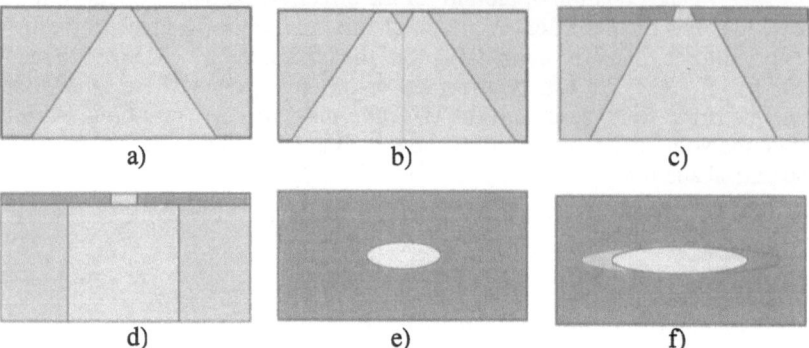

**Figure 7-22** *Potential weak-link aperture geometries.* (**a**) Cross-section through single cone obtained by one-sided etching. (**b**) Overlapping cones due one-sided etching with more than one ion track present. (**c**) Stepped cones obtained by one-sided etching, sputter deposition of the top layer and further enlargement of the bottom cone by a second etch step. (**d**) Double-layer material with different lateral etch-rates. (**e**) Top view of elliptical pore due to non-isotropic lateral etch rates. (**f**) Pore due to the oblique incidence of the ion onto a thin film.

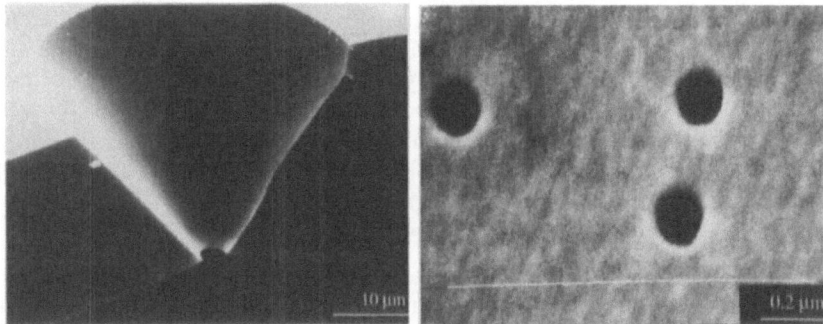

**Figure 7-23** (**left**) Conical funnel in glass, according to scheme of Figure 7-22 (a). (**right**) Metallic pores according to scheme of Figure 7-22 (d) [1]. Top layer 0.1 μm evaporated gold film, pore diameter 0.1 μm. Bottom layer: ca. 30 μm polycarbonate, pore diameter ca. 2 μm.

[1]     Trautmann, C., Gesellschaft für Schwerionenforschung (GSI) Darmstadt, Planckstr. 1, D-6100 Darmstadt.

# 8     Multiple ion tracks

Solid state properties depend on size and shape as soon as these are only becoming small enough:

*"When the dimensions of artificially created structures approach or become smaller than certain characteristic distances — such as grain size, domain size, wavelength, mean free path, coherence length — it becomes possible to access new phenomena or manipulate materials in new ways"* [1].

Ion tracks can be used to modify existing properties and create qualitatively new properties. Thereby contributions of individual tracks merge synergetically to new global properties. The transformation of the solid can be achieved step by step, starting with latent tracks, continuing with track etching and ultimately ending with the arbitrary modification of the etched tracks (Figure 8-1). For example, latent tracks provide increased diffusion, etched tracks can be used as highly selective filters, and the attachment of ionic groups or catalysts to the inner surface of etched tracks may lead to new technical membranes.

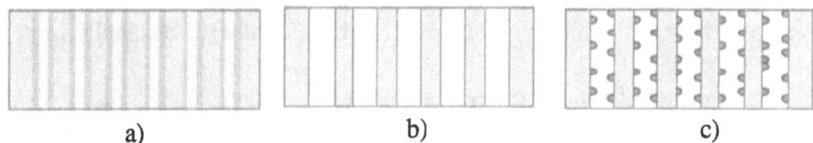

a)                              b)                              c)

**Figure 8-1** *Modification of global properties of solids* by stochastically distributed track arrays. (**a**) Latent tracks provide enhanced diffusion. (**b**) Etched tracks enable the manufacture of homoporous filters with very well defined por sizes. (**c**) Etched tracks with modified walls are a nicely defined study object in membrane science and may open up new applications.

## 8.1     Enhanced diffusion

Latent tracks provide enhanced diffusion, whereby the permeability depends strongly on the track recording material and on the type of the transmitted substance. The diffusion in polymers is much larger than the diffusion in inorganic crystals and glasses and can be easily modified by latent ion tracks of sufficient areal density. The increased throughput of track-irradiated membranes suggests their application as gas-selective membranes. Due to their compactness, latent tracks can provide the potential to develop microsized chromatographic separators.

[1]     Smith, H.T.: "The Impact of Submicrometer Structures in Research and Applications." Proceedings of the International Conference on Microlithography, Amsterdam, 30 Sept. - 2 Oct., 1980, edited by R.P. Kramer, Delft University, Delft, (1981).

## 8.1.1    Diffusion equations

Diffusion is a transport process based on the random walk of particles within a given medium, such as observed in the Brownian motion of suspended particles in a liquid. The diffusion of particles is very similar to  the diffusion of heat. Both processes are governed by the same mathematics. The central differential equations ruling both processes are Fick's first and second law of diffusion, derived by observing Figure 8-2.

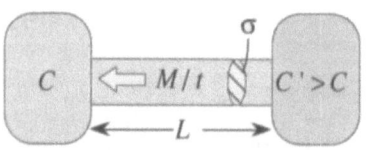
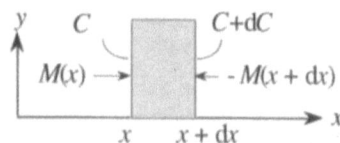

a) Fick's first law — diffusion in            b) Fick's second law — diffusion
a constant concentration gradient           in a varying concentration gradient

**Figure 8-2** Deriving Fick's first (**a**) and second (**b**) law of diffusion.

### Stationary diffusion equation

The starting point for deriving the stationary diffusion equation is the one-dimensional diffusion of a mobile species — for example a gas — along a cylindrical channel with cross section $\sigma$ and length $L$ filled with a stationary phase — for example a solid. The mass flow rate $I = M / t$ through this channel per unit time $t$  is proportional and antiparallel to the concentration gradient $(C' - C) / L$, where $C'$ and $C$ are the concentrations of source and drain reservoir, and $L$ is the length of the channel (Figure 8-2 a):

$$I = -D\frac{C' - C}{L}\sigma \;\rightarrow\; j \equiv \frac{I}{\sigma} = -D\frac{dC}{dx} \quad \text{or } j = -D\;\nabla C \;, \tag{8.1}$$

where the concentration, $C = dM / dx$,  is the mass concentration of the mobile — the diffusing — phase, given in mass units per unit volume. The total concentration is $C\,S$, where $S$ is the solubility of the mobile phase — the gas — within the stationary phase — the solid. In three dimensions, the mass flow density is a vector $j = I / \sigma$ which is proportional and antiparallel to the gradient $\nabla$ of the concentration $C$. The constant of proportionality, $D$ — called the diffusion constant or diffusion coefficient— has the dimension area per unit time and is usually given in units of $cm^2 / s$. The essence of Fick's first law, according to eq. (8.1), is that the diffusion current is proportional and antiparallel to the concentration gradient.

### Dynamic diffusion equation

Starting point for deriving the dynamic diffusion equation is the accumulation of the mass $M$ within the infinitesimally small, one-dimensional volume element $dx$ during the time $dt$ , due to the inflow of mass from the left and right boundaries of the volume element $dx$ (Figure 8-2 b):

$$dM \equiv dx\; S\; dC = M(x) - M(x + dx) \equiv \big(j(x) - j(x + dx)\big)\, dt \;, \quad \text{or} \tag{8.2}$$

$$dx\, S\, dC = -\left(\frac{dj}{dx}\, dx\right) dt \;\Rightarrow\; S\frac{\partial C}{\partial t} = D\frac{d^2 C}{dx^2} \quad \text{or}\; S\frac{\partial C}{\partial t} = D\,\Delta C \tag{8.3}$$

in three dimensions, where $S$ is the solubility of the mobile phase in the stationary phase, and the inflowing masses from the three spatial directions are assumed to occur independently of each other and thus can be linearly superposed. For a homogeneous medium the solubility constant $S$ can be incorporated within the diffusion constant $D$ by setting $S = 1$. In three dimensions, the change of the concentration per unit time, $(\partial C) / (\partial t)$, within the volume element $dx\, dy\, dz$ is proportional to the divergence $\Delta \equiv \nabla^2$ of the concentration $C$. In this derivation it is assumed that the diffusion constant $D$ is independent of the local concentration $C$. The essence of Fick's second law, according to eq. (8.3), is that the increase of the concentration per unit time is proportional to the second spatial derivation or curvature — more precisely, to the divergence — of the concentration.

## 8.1.2 Electric analogue of diffusion equations

The stationary diffusion eq. (8.1) is analogous to Ohm's law, eq. (8.4), if we replace the heat current by the electric current, the thermal energy by the electric charge, the thermal conductivity by the electric conductivity, and the temperature by the potential. The dynamic diffusion eq. (8.3) is analogous to a chain of infinitesimally small electric four-poles, eq. (8.6) in the form of RC circuits (Figure 8-3):

$$I_e = -\frac{1}{r}\frac{U' - U}{L} \quad \sigma \to j_e \equiv \frac{I_e}{\sigma} = -\frac{1}{r}\frac{dU}{dx} \quad \text{or} \quad j_e = -\frac{1}{r}\nabla U \text{, and} \tag{8.4}$$

$$dQ \equiv dx\, c\, dU = Q(x) - Q(x+dx) \equiv \left(j_e(x) - j_e(x + dx)\right) dt \text{, leads to} \tag{8.5}$$

$$dx\, c\, dU = -\left(\frac{dj_e}{dx}dx\right)dt \Rightarrow c\frac{\partial U}{\partial t} = -\frac{1}{r}\frac{d^2U}{dx^2} \quad \text{or} \quad c\frac{\partial U}{\partial t} = \frac{1}{r}\Delta U. \tag{8.6}$$

Here $j_e$ is the electric current density, $r$ is the resistance per unit path length, $U$ is the voltage, $Q$ is the electric charge, and $c$ is the capacitance per unit path length. The electric analogue can thus be used as a simple comprehensive alternative for solving the diffusion equations by an analogous electric circuit. This analogy can be especially helpful for solving three-dimensional diffusion problems.

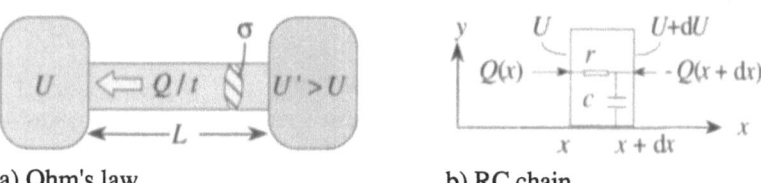

a) Ohm's law            b) RC chain

**Figure 8-3** *Ohm's law and the time-space evolution of a signal in an RC chain.* In the stationary case (**a**) the electric current $dQ / dt$ is equal to the gradient of the electric potential $U$ times the resistance $r$ per unit path length. In the dynamic case (**b**) the voltage increase $dU$ is proportional to the stored charge $dQ$ divided by the capacitance $c$ per unit path length.

### 8.1.3    Solution of dynamic diffusion equation

**Standard form, separation of variables, and partial solution**

Incorporating the solubility constant $S$ within the diffusion constant $D$ — or setting $S = 1$ — the dynamic diffusion eq. (8.3) can be simplified to its standard form by scaling the x-axis according to $x = D^{1/2} \xi$, yielding in the one-dimensional case the standard differential equation $\partial C / \partial t = \partial^2 C / \partial \xi^2$ [1]. A partial solution of this differential equation is found by factorization according to the method of the separation of variables, setting $C(\xi, t) = \varphi(\xi) \cdot \psi(t)$. For nontrivial solutions — those which are not identical zero — insertion into the standard form yields the condition $\varphi''(\xi) / \varphi(\xi)$ $= \psi'(t) / \psi(t) = \text{const} = \pm v^2$, where the dashes indicate differentiation with respect to the function variable given in brackets and $v$ is an arbitrary constant. The standard equation thus can be separated into two independent differential equations, $\varphi''(\xi) = \pm v^2$ $\varphi(\xi)$ and $\psi'(t) = \pm v^2 \psi(t)$, with the standard solutions $\varphi(\xi) = \sinh [v (\xi - \alpha)]$ for const $= + v^2$, $\varphi(\xi) = \sin [v (\xi - \alpha)]$ for const $= - v^2$ and $\psi(t) = \exp (\pm v^2 t)$ for const $= \pm v^2$. Since $\sinh [v (\xi - \alpha)] \exp (+ v^2 t)$ increases exponentially with $t$, the case const $= + v^2$ is physically excluded. Thus we obtain for one-dimensional problems as partial solution of eq. (8.3) the exponentially decaying sinus functions

$$C (\xi, t) = A \, \sin \left[ v (\xi - \alpha) \right] e^{- v^2 t} \text{ with } \xi = \frac{1}{\sqrt{D}} \, x \, , \tag{8.7}$$

where $A$, $v$ and $\alpha$ are arbitrary constants.

**Superposition principle**

If $C_1$ and $C_2$, according to eq. (8.7), are partial solutions of eq. (8.3) than the sum function $C_1 + C_2$ is also a solution of eq. (8.3) — the superposition principle — and further solutions can be obtained by summation or integration of linearly independent functions. A "complete" set of linearly independent functions — by which any arbitrary solution can be represented — is given by the functions $C(\xi, t) = \sin (v \, \xi) \exp (- v^2 t)$ and $C(\xi, t) = \cos (v \, \xi) \exp (- v^2 t)$. The general solution therefore can be represented by superposing solutions of the type of eq. (8.7):

$$C (x, t) = \sum_{m=1}^{\infty} \left( A_m \sin \left( \lambda_m x \right) + B_m \cos \left( \lambda_m x \right) \right) e^{- \lambda_m^2 D t} \, , \tag{8.8}$$

where the constants $A_m$, $B_m$, and $\lambda_m$ have to be adjusted to fulfill the preset boundary conditions.

**Concentration spikes or delta functions**

Of special interest are the one-dimensional concentration spikes — delta functions for $t \to 0$ — corresponding to a system into which a finite number of diffusing particles is injected at a precisely given time $t = 0$ and within a precisely localized, one-dimensional volume element $dx$. A one-dimensional spike is obtained by a superposing solutions according to eq. (8.7) [2]:

[1]    Boyce, W.E., R.C. DiPrima: "**Elementary Differential Equations and Boundary Value Problems.**" John Wiley & Sons, New York, 3rd edition, pp. 452-486 (1977).
[2]    Courant, R, D. Hilbert: "**Methoden der Mathematischen Physik.**" Band II, 2. Auflage, Springer-Verlag, Heidelberg, (1968).

$$C(\xi, t) = \int_{-\infty}^{+\infty} \cos(v\,\xi)\, e^{-v^2 t}\, dv \equiv \sqrt{\frac{\pi}{t}}\, e^{-\frac{\xi^2}{4t}}, \quad \xi = \frac{1}{\sqrt{D}}\, x \ . \tag{8.9}$$

It corresponds to the injection of $2\pi\,D^{1/2}$ mass units of diffusing particles at the time $t = 0$ and at the location $x = 0$. Applying a coordinate translation $x \to x - x_0$ and multiplying eq.(8.9) by a factor $M / (2\pi\,D^{1/2})$ yields a spike of strength $M$ originating at the site $x = x_0$:

$$C(x, t) = \frac{M}{2\sqrt{\pi D t}}\, e^{-\frac{(x - x_0)^2}{4 D t}} \ , \tag{8.10}$$

corresponding to the injection of a mass $M$ of diffusing particles at the location $x = x_0$, which is shown in Figure 8-4 for $M = 1$ and $x_0 = 0$.

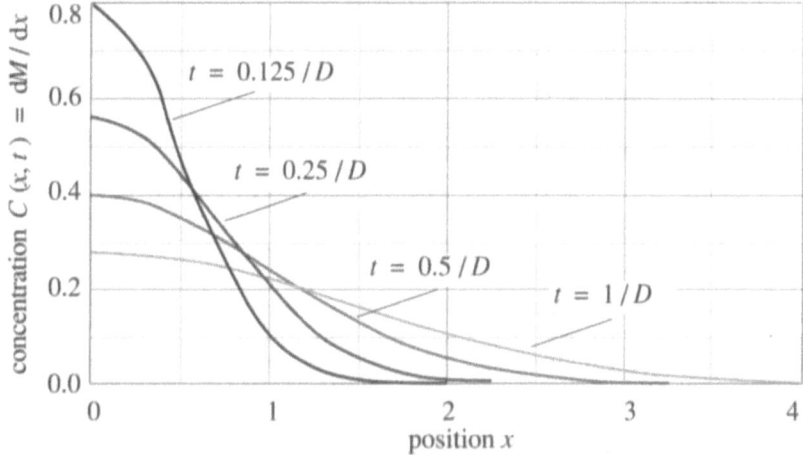

**Figure 8-4** *Time-space evolution of a one-dimensional concentration spike*, starting from the plane $x = 0$ and spreading symmetrically in both directions of the $x$ axis, for an injected mass $M = 1$. The time $t$ for which a specific distribution is obtained is inversely proportional to the diffusion constant $D$.

### Superposition of concentration spikes

Any arbitrary initial conditions $S(x) \equiv C(x, 0)$ can be met by a linear superposition of one-dimensional concentration spikes $C(x_0, t)$ according to eq. (8.10), where $S(x_0)\, dx$ is the individual strength of a spike at position $x_0$ and time $t = 0$:

$$C(x, t) = \frac{1}{2\sqrt{\pi D t}} \int_{-\infty}^{+\infty} S(\xi)\, e^{-\frac{(x - \xi)^2}{4 D t}}\, d\xi \ . \tag{8.11}$$

This equation will be applied in the following sections to the field of chromatography and to the selective transmission of substances through membranes.

### Chromatographic analysis of chemical substances

Small tracks exhibit a large interface area in comparison to their volume. In such a geometry adsorptive and diffusive phenomena dominate over convective phenomena. Therefore ion tracks may be applied as miniature columns performing chromatographic separations [1] — for example in microsensors. Their advantage would be the high cycling rate of these very compact separation columns. The principle of a chromatographic separation process is sketched in Figure 8-5.

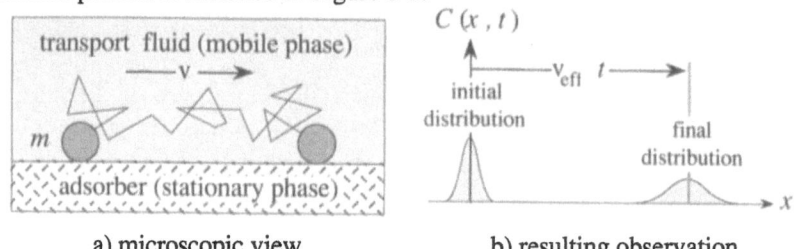

a) microscopic view                              b) resulting observation

**Figure 8-5** *Principle of chromatography.* (**a**) The temporarily immobilized molecule *m* desorbs from the stationary phase into the transport fluid where it is subject to diffusion and simultaneously to a drift with the velocity v of the transport fluid until it is again immobilized by readsorption. (**b**) The combined action of adsorption, diffusion and drift leads a reduced effective drift velocity $v_{eff}$ of the molecule which is characteristic for its adsorption to the stationary phase.

The main features of a chromatographic separation column can be described on the basis of a concentration spike according to eq. (8.10). We assume that a chemically pure substance is injected into the transport fluid of the separation column at the time $t = 0$, and at the position $x = 0$. We assume that the transport medium flows at the velocity v and that the diffusion constant of the injected substance within the transport medium is $D$. Furthermore we assume, that the particles composing the substance are adsorbed — and thus immobilized — during a fraction $\tau$ of one time unit (second) and that they are drifting at the velocity v of the transport fluid for the remaining time fraction $1 - \tau$. Due to the subsequent absorption and desorption processes the substance travels at a reduced effective velocity $v_{eff} = v (1 - \tau) < v$. During the absorbed time fraction, the diffusion process is effectively halted — due to the coupling of the particle to the large mass of the stationary phase of the column — resulting in a reduced effective diffusion constant $D_{eff} = D (1 - \tau) < D$.

For describing the concentration of the substance — the eluant — at the end of the column of length $L$ as function of the time, we assume that, according to eq. (8.10), the centerpoint $x_0$ of the concentration spike is drifting at the effective velocity $v_{eff} = v (1 - \tau) = x_0 / t$, that the diffusion constant is reduced to $D_{eff} = D (1 - \tau)$ and that the concentration is observed at the position $x = L$, yielding:

$$C (L, t) = \frac{M}{2\sqrt{\pi D (1 - \tau) t}} \, e^{- \frac{(L - v (1 - \tau)t)^2}{4 D (1 - \tau) t}} , \tag{8.12}$$

shown in Figure 8-6 for $D = 1$, $M = 1$, $L = 10$, v $= 10$, for three different values of $\tau$.

[1]    Ovchinnikov, V.V., V.D. Seleznev, V.V. Surguchev, V.I. Tokmantsev: "Investigation of Separation Efficiency for Gases on Nuclear Membrane with Hyperfine Pores." submitted to Journal of Membrane Science, (1989).

The forefactor $M / \{2 [\pi D (1 - \tau)]^{1/2}\}$ is slowly varying with time and reflects the decreasing peak height of the resulting chromatogram due to the diffusional broadening of the initial peak. The resolution of a chromatograph is given by its ability to separate neighboring substances. The ability of the chromatograph to separate neighboring peaks is defined by the terms within the exponential factor, that is, by the length $L$ of the column, by the velocity v of the transport fluid, by the time fraction during which the substance is absorbed to the stationary phase of the column, and by the diffusion constant $D$ within the transport medium. For example, one recognizes from the exponential factor of eq. (8.12) that the separation of neighboring peaks can be increased either by simultaneously increasing the length $L$ as well as the drift velocity of the transport fluid or alternatively by increasing the interaction between the substance and the stationary phase, leading to an increased absorbed time fraction $\tau$. This last factor $\tau$ depends strongly on the size of the absorbing cavities — in other words on the structure of the column which can be made very fine using ion tracks.

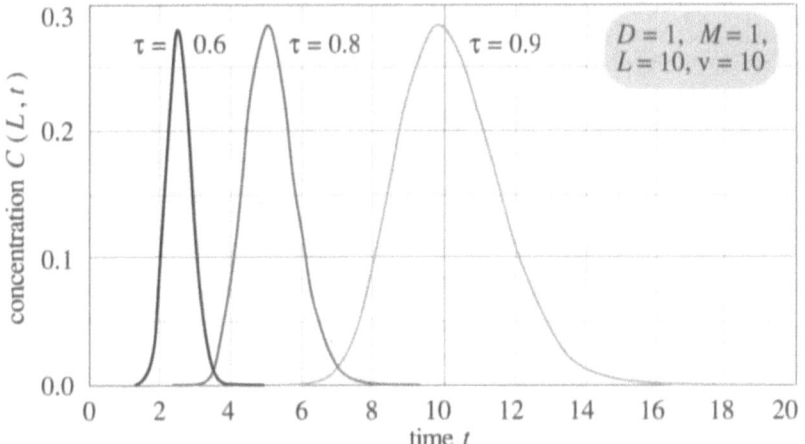

**Figure 8-6** *Separation of three substances* with different adsorption times $\tau$, according to eq. (8.12), for a diffusion constant $D = 1$ in the transport fluid of the column, an injected mass $M = 1$, a column length $L = 10$, and a transport fluid velocity v $= 10$.

### Selective transmission through membranes

The diffusion along ion tracks can be roughly approximated by long cylinders with impermeable walls. This problem is equivalent with a one-dimensional diffusion through a planar membrane and leads to the possibility of a selective transmission of substances through the membrane.

In the following, the time-space evolution of a concentration step is described corresponding to the sudden access to a large reservoir of the substance at time $t = 0$ on the left side of the membrane while the substance is continuously removed from the right side of the membrane (Figure 8-7). Thereby for all times $t > 0$ the left side is kept at $C = C_1$ and the right side is kept at $C = 0$.

In the following, first, the time-space development of the diffusion front within the membrane, second, the time-dependent flux density $j$ of the diffusing substance at the exit of the membrane, and third, the steady state flux density through the membrane is given.

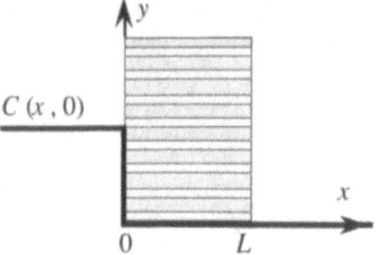

**Figure 8-7** *Initial conditions for time-space evolution of a diffusion step*, spreading along the ion tracks in a thin membrane. For $t > 0$ the left side of the membrane is kept at the constant concentration $C = C_1$ and the right side at the constant concentration $C = 0$.

### Time-space evolution of a concentration step

According to eq. (8.8) the solution has the form of a trigonometrical series, given in reference [1] — which provides an extensive introduction into the mathematics of diffusion — and shown in Figure 8-8:

$$C = C_1 \left\{ 1 - \frac{x}{L} - \frac{2}{\pi} \sum_{n=1}^{\infty} \left( \frac{1}{n} \sin \frac{n \pi x}{L} e^{-\frac{D n^2 \pi^2}{L^2} t} \right) \right\}, \qquad (8.13)$$

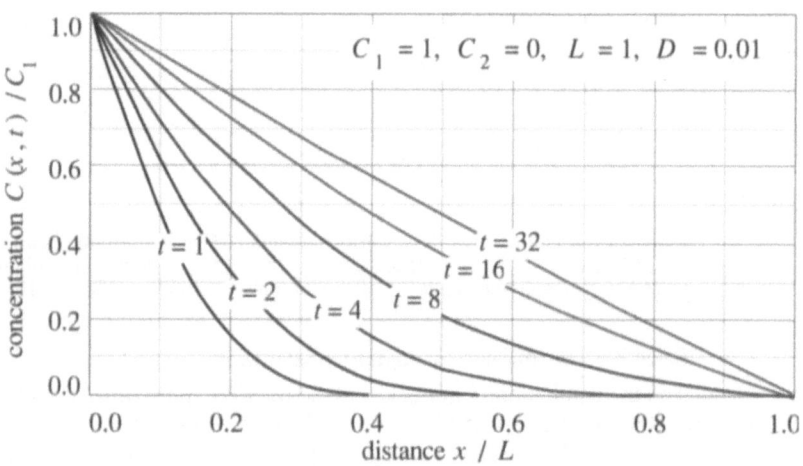

**Figure 8-8** *Time-space evolution of a concentration step* within a membrane according to eq. (8.13). The diffusion front spreads from left to right and reaches asymptotically a straight line corresponding to a constant gradient throughout the membrane.

### Flux density at the exit of the membrane

The flux density at the exit of the membrane depends on the gradually increasing gradient of the concentration at the position $x = L$. Inserting the concentration from eq. (8.13) into eq. (8.1) we obtain for $x = L$ (shown in Figure 8-9 for $C_1 = 1$, $C_2 = 0$, $L = 1$, $D = 0.01$):

---

[1]    J. Crank: **"The Mathematics of Diffusion."** Clarendon Press, Oxford, 414 pp., (1986).

$$j \equiv \frac{1}{\sigma}\frac{\partial M}{\partial t} = -D\frac{\partial C}{\partial x} = D\frac{C_1}{L}\left\{1+2\sum_{n=1}^{\infty}\left((-1)^n\ e^{-\frac{D n^2 \pi^2}{L^2}t}\right)\right\}\quad \text{at } x = L,\quad (8.14)$$

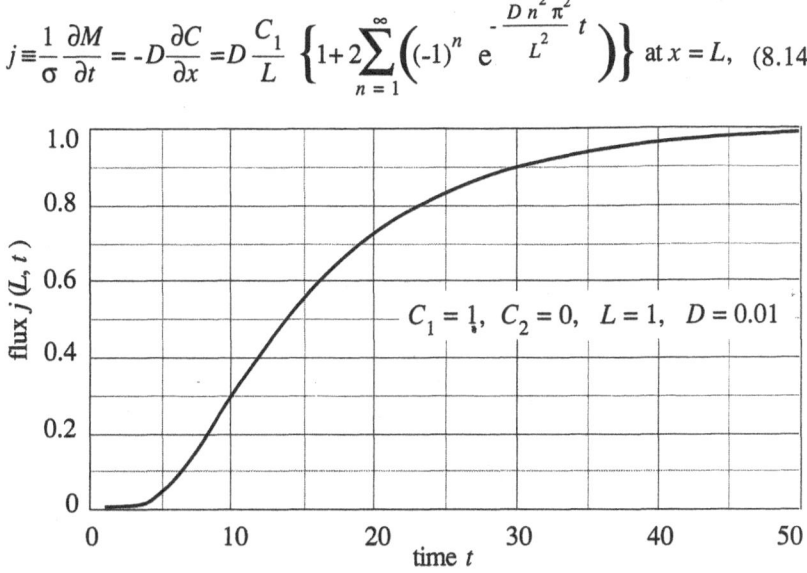

**Figure 8-9**  *Flux density* at exit plane of membrane (for $x = L$) according to eq. (8.14).

**Steady state flux density**

According to eq.(8.14) the steady state flux density is given by:

$$j \equiv D\frac{C_1}{L}\quad \text{for } t \to \infty \text{ at } x = L \ . \tag{8.15}$$

**Permeability and sorption constant**

In many cases the surface concentration $C_1$ at the left boundary of the membrane — which cannot be observed directly — can be expressed as the product of the external vapor pressure $p$ and the sorption or solubility constant $S$ of the specific material:

$$C_1 = S\ p \ , \quad \text{where } S = \text{sorption constant, } p = \text{vapor pressure.} \tag{8.16}$$

In these cases the permeability constant of the membrane is defined by:

$$j = D\frac{C_1}{L} = D\frac{S\ p}{L} \equiv P\frac{p}{L} \ , \quad \text{where } P = \text{permeability constant.} \tag{8.17}$$

## 8.1.4   Connected-cavity model

Assuming a track structure in the form of a cylindrical track with a Gaussian radial density dependence [1], [2], [3] one obtains the statistical distribution of defects —

[1]   Albrecht, D., P. Armbruster, M. Roth, R. Spohr: "Small Angle Neutron Scattering Observations from Oriented Latent Nuclear Tracks." Radiation Effects, 65, 145-148 (1982).
[2]   Albrecht, D., P. Armbruster, R. Spohr, M. Roth, K. Schaupert, H. Stuhrmann: "Investigation of Heavy Ion Produced Defect Structures in Insulators by Small Angle Scattering." Applied Physics, A37, 37-46 (1985).
[3]   Albrecht, D.: "Untersuchung der von schweren Ionen in Dielektrika erzeugten Defektstrukturen mittels Kleinwinkelstreuung." Dissertation, Technische Hochschule Darmstadt, D-6100 Darmstadt, GSI-Report 83-13, (1983).

here the missing atoms in a regular matrix of atoms for given track radius and track density — in a cross-section vertical to the ion track (Figure 8-10).

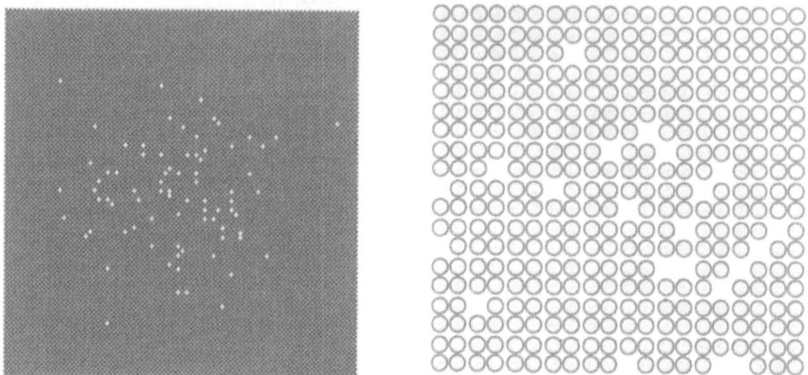

**Figure 8-10** *Cross-section through equivalent track.* (**left**) Gaussian defect distribution across an ion track calculated for an ion track of 4 nm radius and 10% density deficit at the track axis. The used matrix has a size of 20 nm by 20 nm corresponding to 80 by 80 = 6400 atoms with a mutual lattice spacing of 0.25 nm. (**right**) Central blow-up of an equivalent defect distribution with 5 nm side length corresponding to an unrelaxed lattice of 20 by 20 = 400 atoms of which 22 atoms or roughly 5% are missing.

In accordance to Figure 8-10 (right) an ion track can be described as a network of defects (synonymously: voids, cavities, traps) interconnected by paths of increased mobility (Figure 8-11).

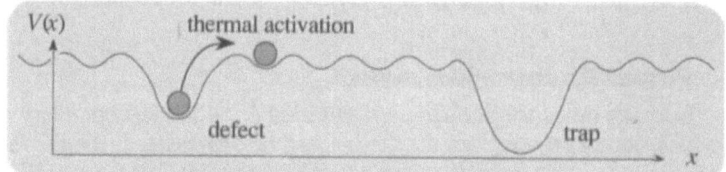

**Figure 8-11** Concept of diffusion between defects or traps of different trough depth.

In the case of organic polymers — such as polyethylene terephthalate — the voids can be filled almost completely by small molecules such as $H_2O$ which can be proven by the method of contrast variation in neutron small-angle scattering experiments (Figure 8-12 ) [1], [2]. According to this technique the different cross sections (scattering lengths) of hydrogen ($^1H$ or H) and deuterium ($^2H$ or D) with respect to neutron scattering is used to generate a mixture of ligth water ($H_2O$) and heavy water ($D_2O$) matching the "refractive index" of the virgin polymer and suppressing the small angle scattering from the impregnated latent tracks.

[1]    Schaupert, K.: "Untersuchungen der Bewegung von Atomen und Molekülen in latenten Kernspuren in Polymeren mit Hilfe von Permeations- und Neutronenstreuexperimenten." Disseration, Universität Gießen, 1-144 (1986).
[2]    Schaupert, K., D. Albrecht, P. Armbruster, R. Spohr: "Permeation Through Latent Nuclear Tracks in Polymer Films." Appl. Phys. **A 44**, 347-352, (1987).

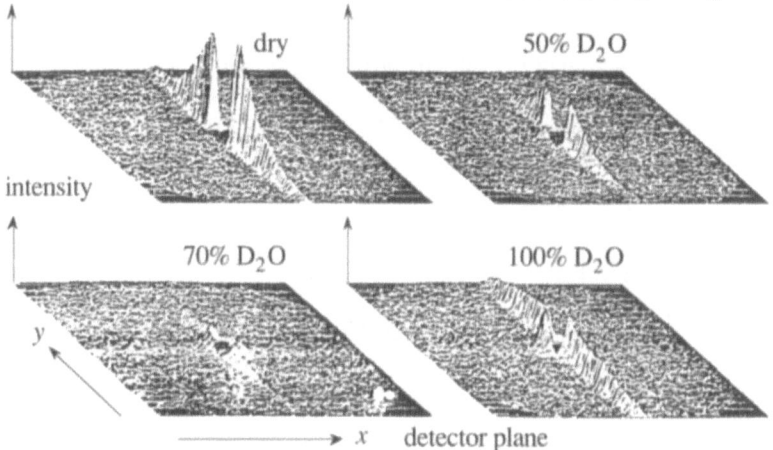

**Figure 8-12** *Filling of latent tracks in a polymer* (polyethylene terephthalate) as observed by contrast variation in neutron small-angle scattering, before and after filling with three different mixtures of $H_2O$ and $D_2O$. For a mixture of 30% $H_2O$ and 70% $D_2O$ the diffraction contrast vanishes almost completely which is an indication that the defect zone associated with the ion track is nearly replenished by this medium.

Replacing the defect distribution of Figure 8-10 (right) by a network of interconnected voids, we arrive at the network model and its electric analogue (Figure 8-13).

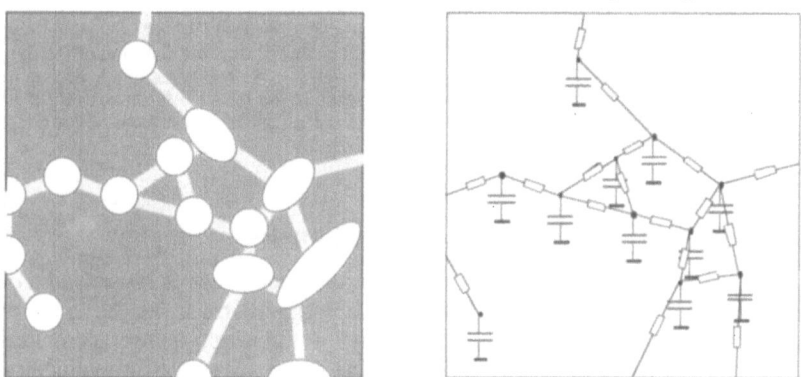

**Figure 8-13** Track defect pattern replaced by a network of interconnected voids (**left**) and by an electric analogue (**right**).

The simplest assumptions in the sense of the electric analogue of Fick's equations are a linear chain of voids connected by diffusion paths (Figure 8-14 a) and a branched chain of voids connected by diffusion paths (Figure 8-14 b).

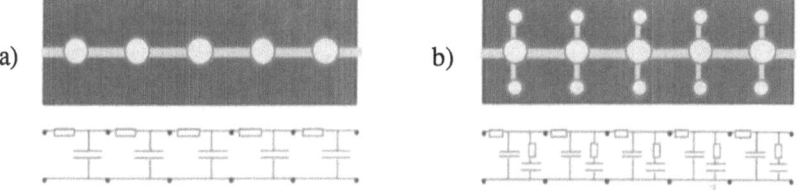

**Figure 8-14** *Defect chains and electric analogues.* (**a**) Linear chain. (**b**) Branched chain.

The results of both models are practically equivalent, at least if the time constants of the corresponding RC-circuits differ by not more than a factor three between the main path and the side branches (Figure 8-15) and correspond to the solution of the dynamic diffusion equation (8.14) for a pressure step at $t = 0$ from 0 to $p$ on one side of a membrane of thickness $d$ and observing the flux $j(t)$ at the other side at zero pressure.

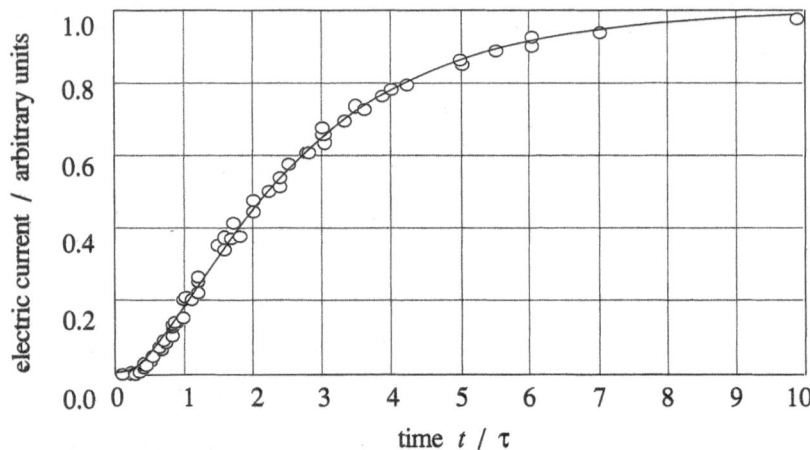

**Figure 8-15** *Comparison of diffusion response with electric analogue.* Response of different RC chains to a step function. Electric current — analogous to a diffusion flow — versus time $t$ measured in units of a characteristic time $\tau$ of the system. Circles: electric analogue of linear chain and of a branched chain. Black line: Solution of the dynamic diffusion equation — Fick's second law of diffusion — adapted by a suitable time shift. At time $t = 0$ a voltage of constant amplitude is applied at the left-side terminals of the circuit while observing the output voltage (or current) at the right-side terminals of the circuit for times $t \geq 0$.

## 8.1.5    Gas permeation through latent tracks

Latent ion tracks have a typical diameter of about 10 nm and a density that is by about 10 % smaller than the density of the virgin material [1], [2]. In ion tracks the diffusion and sorption constants are larger than in the virgin membrane material. This has been observed for some atomic and molecular gases and simple inorganic liquids.

### Experimental setup

Figure 8-16 shows the experimental setup for determining the permeability of track-irradiated foils. It measures essentially the gas flow through an ion-irradiated polymer membrane after admission of a gas at constant pressure on the left side of the membrane at time $t = 0$.

[1]    Albrecht, D. P. Armbruster, R. Spohr, M. Roth, K. Schaupert, H.B. Stuhrmann: "Small Angle Scattering from Oriented Latent Nuclear Tracks" Nucl. Instr. Meth. in Physics Research, B2 **230**, pp. 702 - 705, (1984).
[2]    Albrecht, D. P. Armbruster, R. Spohr, M. Roth, K. Schaupert, H.B. Stuhrmann: "Investigation of Heavy Ion Produced Defect Structures in Insulators by Small Angle Scattering." Appl. Phys. **A 37**, pp. 37 - 46, (1985).

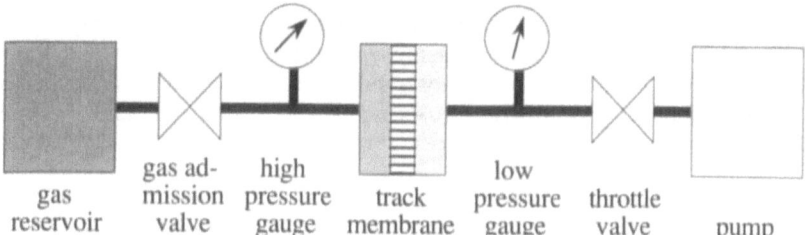

gas | gas ad-mission | high pressure | track | low pressure | throttle | pump
reservoir | valve | gauge | membrane | gauge | valve |

**Figure 8-16** *Observation of gas permeation through track-irradiated membranes.* At time $t = 0$, gas is admitted to the left side of the ion irradiated membrane at the pressure $p$. The steady increase of the gas flux through the membrane is observed using a low pressure gauge — a quadrupole mass filter — in connection with a throttle valve.

### Experimental results

By uranium ion tracks in polyethylene terephthalate membranes, the permeability of oxygen, water, carbon dioxide, neon, and argon could be enhanced by about one order of magnitude [1] (Table 8-1).

**Table 8-1** permeability ratio of 19 μm thick Hostaphan with respect to virgin material at areal dose of $5 \cdot 10^{11}$ 8.6 MeV/u $^{238}$U ions per cm$^2$. The permeability is defined as the flux $j$ per unit pressure difference between entrance and exit of a membrane of unit thickness.

| gas | permeability-ratio |
|-----|--------------------|
| $O_2$ | 22 |
| $H_2O$ | 13 |
| $CO_2$ | 37 |
| Ne | 26 |
| Ar | 35 |

Figure 8-17 shows the gas permeation through membranes with four different track densities. The permeation curves can be individually described by eq. (8.14) with the diffusion constant $D$ and the sorption constant $S$ as fitting parameters.

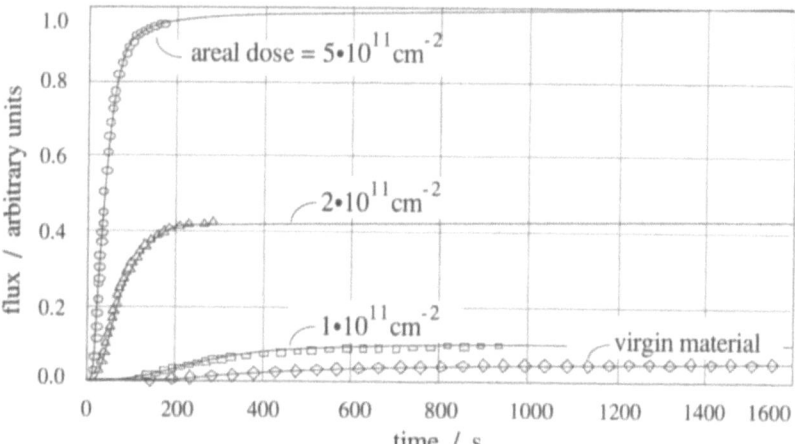

**Figure 8-17** *Track enhanced diffusion* for argon gas through polyethylene terephthalate membranes of roughly 10 μm thickness, irradiated with three different fluences of uranium ions of roughly 10 MeV/u specific energy.

[1]    Schaupert, K., D. Albrecht, P. Armbruster, R. Spohr: "Permeation Through Latent Nuclear Tracks in Polymer Foils." Appl. Phys. A, **44**, pp. 347 - 352, (1987).

The diffusion of Ne, $O_2$, $H_2O$, and $CO_2$ through polyethylene terephthalate follows eq. (8.14), linearly increasing with the ion dose. However, the diffusion of methanol shows a time-lag of the observed flux which cannot be described by eq. (8.14) and is attributed to a chemical reaction between methanol and the damaged zone (Figure 8-18).

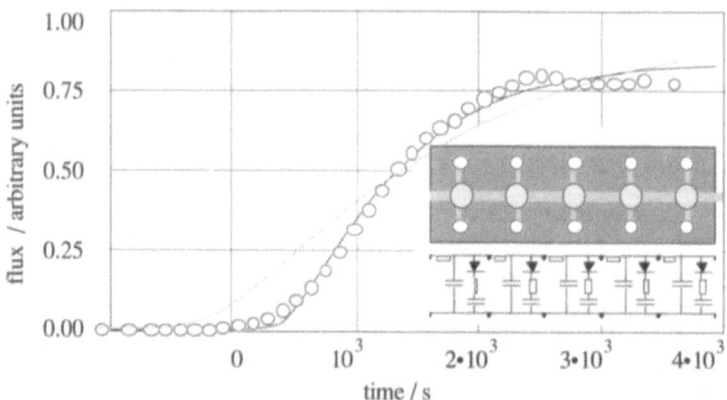

**Figure 8-18** *Diffusion of methanol through latent tracks in polyethylene terephthalate.* Circles: experimental points. Black line: Fit according to the dynamic diffusion equation, however, with a time-delay of 1100 s indicating the occurrence of a chemical reaction between methanol and the chemically activated damage zone associated with the ion track. Gray line: Attempted fit without time delay. Inset: Suggested diffusion model and electric analogue describing the permanent trapping of the diffusing species by chemical reactions.

# 8.2    Membrane technology

## 8.2.1    Ion track filters

Almost immediately after the first observation of etched ion tracks in mica single crystals, the potential of ion track filters for the mechanical separation of small particles suspended in a liquid or gaseous medium was recognized [1]. Ion track filters offer distinct advantages over conventional filters, which usually consist of bonded fibers or porous foams and are ruled by several mutually interdependent processing parameters (Figure 8-19). In contrast, ion track filters are defined by very few, almost independent parameters, the track length, the track diameter, and the areal density of the ion tracks. These three main parameters can be varied in an easily controllable manner over several orders of magnitude.

[1]    Price, P.B., R.M. Walker: "Molecular Sieves and Methods for Producing Same." United States Patent Office, No. 3,303,085, Feb. 7, (1967)

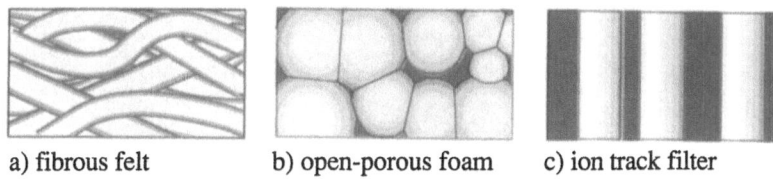

a) fibrous felt          b) open-porous foam          c) ion track filter

**Figure 8-19** *Comparison of filter structures.* Fiber materials (**a**) and open-porous foams (**b**) provide high porosities but relatively low selectivities. Ion track filters (**c**) provide relatively low porosities but very high selectivities.

Since about 1972, fission fragments from nuclear reactors have been used for the commercial manufacture of ion track filters up to a thickness of about 15 μm [1]. This is historically the first example of a large-scale commercial use of individual atomic particles. Very recently, ion accelerators are starting to replace the fission fragment sources. The concept of an on-line facility for filter production is shown in Figure 8-20. It is principally possible to irradiate several square meters of virgin material per second. However, off-line treatment is more appropriate when the sensitization by ultraviolet light and the chemical etching are the throughput-limiting steps.

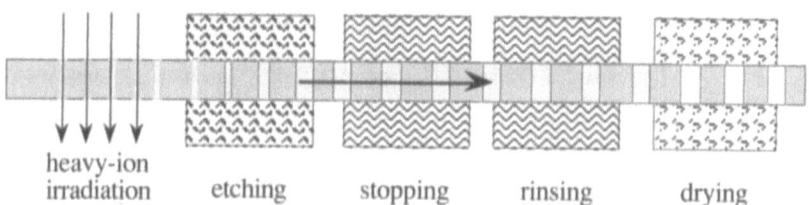

heavy-ion
irradiation       etching       stopping       rinsing       drying

**Figure 8-20** *Principal steps for on-line filter production.*

The main advantage of ion track filters over conventional filters is their well-defined pore size. This feature will probably dominate their future applications. The retained particles can be observed directly on the filter surface and can be removed from it easily. Well-defined pore size is a necessary condition for selecting biological cells, which are characterized by virtually identical sizes. Oscillating piezo-electric foils enable the on-line gravimetry of the deposited particles [2].

One apparent disadvantage of ion track filters is the occurrence of multiple pores with larger sizes than the individual single pores. This problem can be circumvented by irradiating the virgin material with an ion beam of wide angular spread. In this way many independent track arrays can be generated which do not interfere mutually (Figure 8-21).

[1]     Nuclepore Corporation, 7035 Commerce Circle, Pleasanton, California 94566, USA.
[2]     Vater, P.: "Production and Applications of Nuclear Track Microfilters." Nuclear Tracks and Radiation Measurements, **15**, pp. 743-749, (1988).

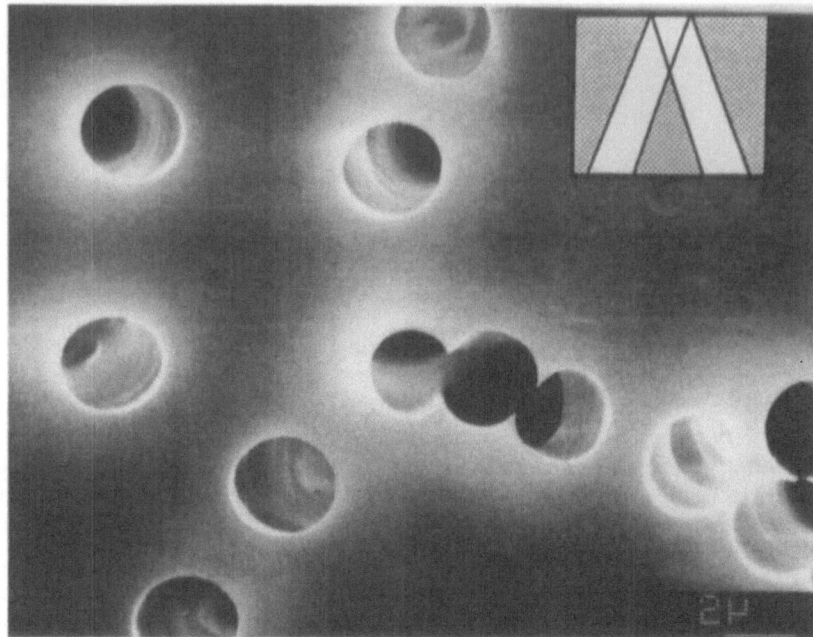

**Figure 8-21** *Circumventing pore overlap at high permeabilities.* Multi-angular track filter in polycarbonate (Makrofol N, Bayer AG, D-5090 Leverkusen). **Inset:** schematic cross section.

### Filter materials

Muscovite mica is a mechanically brittle mineral with extremely high track etch rate  (Figure 8-22). Allyl diglycol carbonate, due to its excellent homogeneity — indicated by its clear transparency — is a highly polymerized track recording material with extremely planar surfaces and smoothness of the pore walls (Figure 8-23) [1]. Polycarbonate is another quite homogenoeus polymer used for commercial track filters [2]. Polyethylene terephthalate provides a higher mechanical strength but has a somewhat more uneven chemical attack due to its higher crystallinity. Polyimide is a chemically inert, high strength, radiation resistant material (Figure 8-24 a and Figure 8-25). Polyvinylidene fluoride [3] is a very promising material due to its high chemical inertness and due to its piezo- and pyro-electric properties (Figure 8-24 b).

[1]     Berndt, M. G. Siegmon, R. Beaujean, W. Enge: "A New Nuclear Track Filter of CR-39." Nuclear Tracks and Radiation Measurements **8**, No. 1-4, 589, (1983).
[2]     Nuclepore Corporation, 7035 Commerce Circle, Pleasanton, California 94566, USA.
[3]     Komaki, Y.: "Growth of Fine Holes by the Chemical Etching of Fission Tracks in Polyvinylidene Fluoride." Nucl. Tracks **3**, 33, (1979).

**Figure 8-22** *High temperature, high precision track filter.* Parallel channels of rhombic cross section in muscovite mica $(KAl_2Si_3AlO_{10}(OH,F)_2)$, etched in hydrofluoric acid [1].

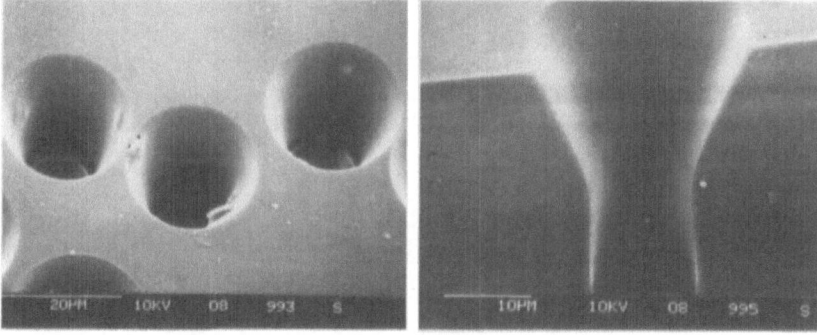

**Figure 8-23** *Track filter with high surface smoothness* [2]. (**left**) Allyl diglycol carbonate (CR39) microfilter with funnel-shaped 10 μm diameter pores of 20 μm maximum funnel diameter. Pore length 90 μm. (**right**) Cross section through one pore.

[1]    Vater, P., G. Tress, R. Brandt, B. Genswürger, R. Spohr: "Nuclear Track Microfilters Made of Mica." Nuclear Instruments and Methods, **173**, pp. 205-210, (1980).
[2]    Berndt, M., J. Krause, G. Siegmon, W. Enge: "Investigation on a Modified CR-39 Microfilter." Nuclear Tracks and Radiation Measurements, **12**, No. 1-6, pp. 985-988, (1986).

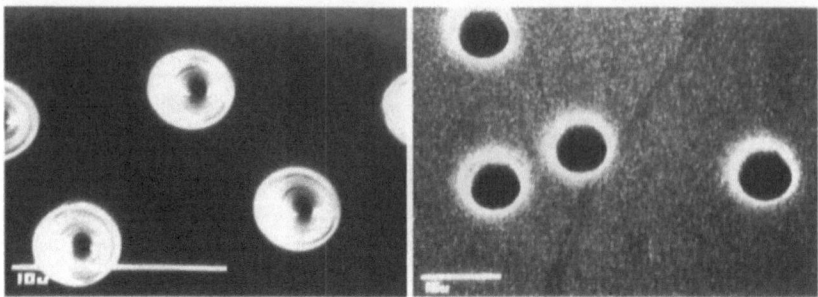

**Figure 8-24** *Track filters of high mechanical and chemical strength.* (**left**) Conical channels in polyimide (Kapton, DuPont, CH-1200 Geneva), etched in NaClO solution [1]. (**right**) Nearly cylindrical channels in polyvinylidene fluoride (PVDF, Solef, Solvay, B-1000 Bruxelles), etched in KOH solution with added $KMnO_4$ [2].

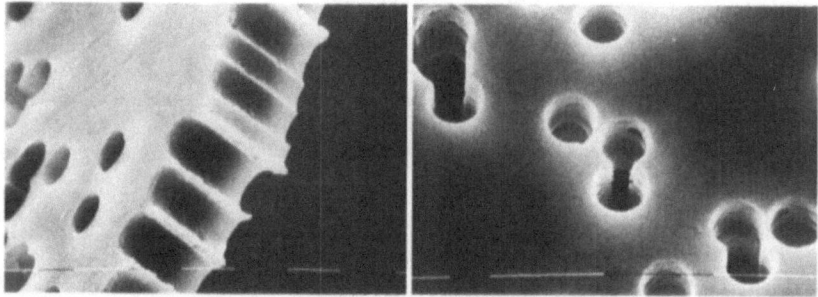

**Figure 8-25** *Cylindrical pores in 15 μm thick polyimide* (Kapton, DuPont, CH-1200 Geneva), etched in NaClO solution [3]. Bar unit = 10 μm.

## Suggested pore shapes

Ion track technology provides — besides cylinders, single and double cones — several possibilities for shaping the individual pores (Figure 8-26). In this way, it is possible to adapt the filter to the specific properties of the suspended particles.

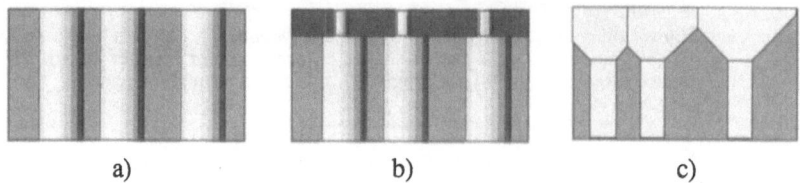

a)                              b)                              c)

**Figure 8-26** *Suggested pore shapes for ion track filters.* (**a**) Cylindrical pores. (**b**) Two-layer material with stepped pores for improved throughput and mechanical strength. (**c**) Funnelled cylinders increasing the hydrodynamic flow conditions.

[1]     Zhu, T.-C., P. Vater, R. Brandt, J. Vetter: "Microfilters with Tiny Holes in Kapton, GSI Scientific Report 1987, GSI-88-1, p. 253, (1988).
[2]     Holländer, W., W. Dunkhorst, R. Brandt, P. Vater: "Measurement of Absorbing Aerosols with Piezo-Electric Filters from Polyvinylidene Fluoride (PVDF): First Results." J. Aerosol Sci. **18**, 907-910, (1987).
[3]     Vetter, J., Gesellschaft für Schwerionenforschung (GSI), Planckstr. 1, D-6100 Darmstadt 11.

## 8.2.2 Prospects of track membranes

Membranes are structures acting as a barrier to the flow of matter, electrical charge, or heat between two compartments. They exist in a wide range of complexity, from very simple, homogeneous phases to highly asymmetric composites.

Passive membranes can selectively transmit various components of a solution by applying an external force in the form of a gradient in pressure, voltage, temperature, or concentration. Active membranes are still the exclusive domain of biology, working against concentration gradients, whereby biochemical energy is consumed.

Artificial membranes [1] are finding large-scale application in a variety of industrial fields where liquid or gas mixtures have to be separated. The main advantage of membrane separation processes is their low energy consumption in comparison to alternative methods. Examples are the separation of $^{238}U$ and $^{235}U$, the separation and purification of helium, the desalination of sea water by reverse osmosis, the filtration of waste water in galvanotechnic and photographic processes, the separation of liquid/liquid and liquid/gas phases in batteries and fuel cells.

The task of membrane science is to explain observable properties on the basis of observable membrane structures. Their structure, however, is often quite complex. Conventional membranes can be only partially characterized by their pore size and by the tortuosity of their network. Solvent and solute molecules diffuse through this network, jumping in succession from adsorption site to adsorption site. In order to describe the observed phenomena, model parameters must be fitted to the experiments.

### Potential of ion track membranes

In contrast to the complex structure of conventional membranes, ion tracks represent structures of well defined shape and thus provide study objects well suited to theoretical treatment. Important are the uniformity of the pore diameter and length, the large surface to volume ratio, and the possibility to create defined individual channels.

Extremely fine, uniform channels of almost molecular dimensions can be created, enabling the systematic study of transport phenomena as function of chemical or ionic groups attached to the channel walls. Thereby the ion tracks can serve as inert supports either for chemically active groups or for ionic groups. The channels can also be coated by monolayers of polar molecules to decrease the pore diameters, using Langmuir's film coating technique. Another possibility is to use the track structure as a mechanical support for very thin conventional membranes. Two possible geometries are shown in Figure 8-27.

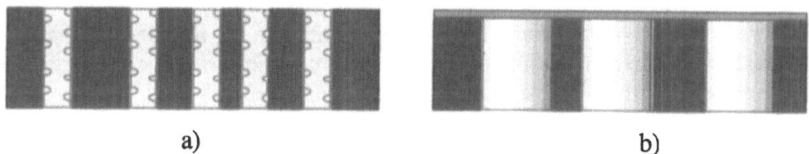

a)                                                    b)

**Figure 8-27** *Possibilities in membrane technology.* (a) Chemical groups — ionic molecules, encymes, even monolayer films — can be attached to the inner walls of etched tracks. (b) Track structures can serve as mechanical supports for very thin conventional membranes. In this way the chemical separation efficiency and the mechanical properties of the supporting structure can be separately optimized.

[1]   Pusch, W., A. Walch:: "Synthetic Membranes — Preparation, Structure, and Application." Angew. Chem. Int. Ed. Engl. **21**, p. 660 - 685, (1982).

# 9    Bulk properties

.While single tracks lead to property changes that are extremely localized, ensembles of many tracks, distributed over a given volume, induce globally distributed property changes which gradually determine the gross behavior of the solid with increasing density. Thus the integrated effect of many distributed ion tracks is capable to change existing properties and to induce new properties on a global scale. Simultaneously the direction of the track array defines the axis of a new, track-induced anisotropy in the solid.

The influence of ion tracks on the magneto-optic properties of iron garnets represents a study example how bulk properties can be influenced by ion tracks [1], [2], [3] and is described in this chapter exclusively.

## 9.1    Adjusting magnetic properties

### 9.1.1    Magnetic properties of matter

#### Magnetic moment of the electron

The magnetic moment of the electron — Bohr's magneton — is the origin of all magnetic phenomena in matter. It is a vector parallel to the electron spin and its magnitude is given by:

$$\mu = \frac{e\,h}{4\pi\,m\,c}$$

(9.1)

#### Magnetization

The magnetization $M$ is defined as the magnetic moment per unit volume (Figure 9-1). In magnetostatic terms it corresponds to the number of magnetic monopoles on the top surface of the unit cube. In electrodynamic terms it corresponds to an equivalent number of current loops of unit area $\sigma$ and unit current $I$.

[1]  Krumme, J.P. I. Bartels, B. Strocka, K. Witter, Ch. Schmelzer, R. Spohr: "Pinning of 180° Bloch Walls at Etched Nuclear Tracks in LPE-Grown Iron Garnet Films." Applied Physics, **48**, 5191-5196 (1977).
[2]  Hansen, P. H. Heitmann: "Influence of Nuclear Tracks on the Magnetic Properties of a (Gd,Bi)3(Fe,Ga)5O12 Garnet Film." Phys.Rev.Lett. **43**, 1444-1447 (1979).
[3]  Hansen, P., H. Heitmann, B. Strocka, R. Spohr: "Der Einfluß von Ionenstrahlen auf die magnetischen Eigenschaften von ferrimagnetischen Schichten." BMFT Bericht FB T83-048 Fachinformationszentrum Karlsruhe, D-7514 Eggenstein-Leopoldshafen 2, 1-180 (1983).

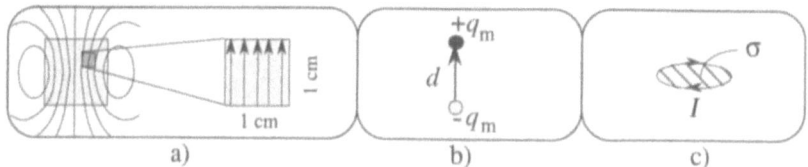

**Figure 9-1** *Basic concept of magnetization.* (**a**) Definition of magnetization *M*, corresponding to the sum of the magnetic monopoles on the top surface of the unit cube. (**b**) Magnetostatic model of a magnetic dipole in the form of two magnetic monopoles of opposite sign and magnitude $q_m$, connected at the fixed distance *d*. (**c**) Electrodynamic model of a magnetic dipole in the form of a current loop of area σ and current *I*.

### Susceptibility

The magnetic susceptibility χ corresponds to the magnetic momentum *M* created per unit external magnetic field strength *H*. The susceptibility measures the degree of alignment of the elementary magnetic moments achieved per unit field strength:

$$M = \chi\, H \tag{9.2}$$

For diamagnetic and paramagnetic substances, the susceptibility χ is a constant and independent of the external magnetic field strength *H*.

### Theory of paramagnetism

Before describing a ferromagnetic system of mutually coupled dipoles a short glimpse on a system of independent dipoles coupling exclusively with an external magnetic field seems valuable.

While in diamagnetic materials the elementary magnetic dipoles compensate each other mutually already within the atom, paramagnetic materials possess elementary magnetic dipoles capable to interact with the external magnetic field *H* (but not with each other). Such a paramagnetic system exhibits already the phenomenon of global ordering by the interaction of the dipoles with the external magnetic field *H*. In contrast to ferromagnetic systems, however, the self-generated field of the magnetic dipoles has a negligible influence on the dipoles themselves.

If the elementary magnetic dipoles interact exclusively with the external magnetic field, the magnetic susceptibility χ reflects the thermodynamic equilibrium between the alignment of the magnetic dipoles by the external magnetic field *H* and their disalignment by thermal agitation. The potential energy *E* of a classical dipole of moment μ in a homogeneous magnetic field of strength *H* is:

$$E = -\,\mu\, H\, \cos \vartheta \tag{9.3}$$

The Boltzmann probability for an arbitrary magnetic dipole at the angle ϑ with respect to an external magnetic field *H* is given by:

$$w\,(\vartheta)\ d\vartheta\ =\ \text{const}\ e^{-\frac{E}{k\,T}}. \tag{9.4}$$

According to this classical picture, an elementary dipole of orientation ϑ contributes the dipole moment μ cos ϑ to the magnetization. The average of cos ϑ is determined from eq. (9.4) by integrating the probability $w(\vartheta)$ times $\cos \vartheta$ over the full solid angle, whereby $d\Omega = 2\pi \sin \vartheta \, d\vartheta$:

$$<\cos \vartheta> = \frac{\int w(\vartheta) \cos \vartheta \, d\Omega}{\int w(\vartheta) \, d\Omega} = \frac{e^{x} + e^{-x}}{e^{x} - e^{-x}} - \frac{1}{x} = \coth(x) - \frac{1}{x} \, , \tag{9.5}$$

where $x = \mu H / (kT)$. This function (Figure 9-2) is known as the Langevin function $L(x)$ and has the asymptotic behavior:

$$L(x) \to 1 - \frac{1}{x} \quad \text{for } x \to \infty \quad \text{and} \quad L(x) \to \frac{x}{3} \quad \text{for } x \to 0 \, . \tag{9.6}$$

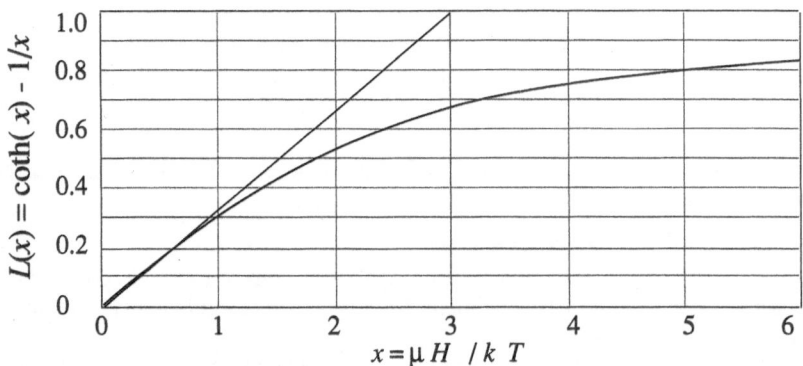

**Figure 9-2**  Langevin function describing paramagnetism.

If there are $n$ magnetic dipoles per unit volume, they add up to a total magnetization $M = n \mu <\cos \vartheta>$, which is the product of the number $n$ of magnetic dipoles per unit volume times the dipole moment $\mu$ of an elementary dipole times the Langevin function $L(x)$:

$$M = n \mu L(x) \, , \quad \text{where } x = \frac{\mu H}{kT} \, . \tag{9.7}$$

**Theory of ferromagnetism**

If the elementary magnetic dipoles interact not only with the external magnetic field but also mutually with each other, a wealth of cooperative or synergetic phenomena [1] occurs — such as ferromagnetism, ferrimagnetism, antiferromagnetism — depending on the type of the interaction. In ferromagnetism, the local magnetization $M$, given by eq. (9.7), depends on the local magnetic field $F$, which itself corresponds to the sum of the external field $H$ and the resulting magnetization $M$, multiplied by an empiric "amplification" factor $\lambda$ [2] — known as Weiss constant or molecular field constant — :

[1]  Haken, H.: "**Synergetics**." Springer, Berlin, 371 pp., (1983).
[2]  Flügge, S.: "**Lehrbuch der Theoretischen Physik**." Band III, klassische Physik II, das Maxwellsche Feld, Springer Verlag, Heidelberg, pp. 45-51, pp. 83-87, (1961).

$$M = n\,\mu\,L\left(\frac{\mu\,F}{k\,T}\right)\;,\qquad \text{where}\quad F = H + \lambda\,M \tag{9.8}$$

The two relations contained in eq. (9.8) represent a coupled pair of equations, equivalent to an electronic feedback loop, which — due to the amplification factor $\lambda$ — is only stable for certain, sufficiently small values of $M \le n\,\mu = M_s$, where $M_s$ is the saturation magnetization. For obtaining graphical solutions of the problem, the two eqs. (9.8) are transformed into one pure Langevin function and one straight line which cuts the Langevin function at certain locations:

$$y = L\,(x)\;\text{ and }\;y = \frac{T}{3\,T_c}\,x - y_0\;,\text{ where} \tag{9.9}$$

$$y = \frac{M}{M_s}\;,\quad y_0 = \frac{H}{\lambda\,M_s}\;,\quad x = \frac{\mu\,F}{k\,T}\;,\quad T_c = \frac{\lambda\,\mu\,M_s}{3\,k}\;,\quad M_s = n\,\mu\;. \tag{9.10}$$

Thereby $T_c$ represents the critical or Curie temperature beyond which the cooperative phenomena vanish, and $M_s$ represents the saturation magnetization. For $T > T_c$ the straight lines have a slope larger than the maximum slope $1/3$ of the Langevin function occurring at the origin of the coordinate system. Beyond the Curie temperature $T_c$ therefore only one, well defined, solution of the eqs. (9.9) exists, corresponding to a non-cooperative, paramagnetic system. For $T < T_c$ there exist always three solutions, corresponding to the unmagnetized virgin state of a ferromagnetic system and to two states of parallel and antiparallel magnetization (Figure 9-3).

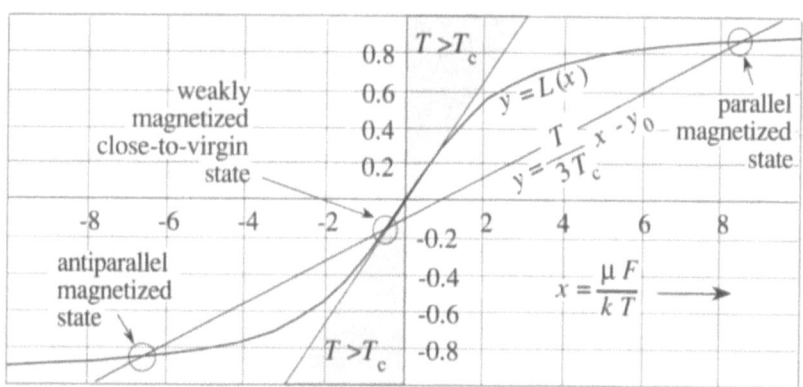

**Figure 9-3** *Classical theory of ferromagnetism.* Geometric solution of eqs. (9.9).

For $H = 0$ the straight line of eq. (9.9) crosses through the origin of the coordinate system, its slope being $T/(3\,T_c)$. The two other cross-over points between the straight line and the Langevin function $L(x)$ correspond to the spontaneous magnetization of the ferromagnetic system which — as function of temperature — has the typical shape of Figure 9-4.

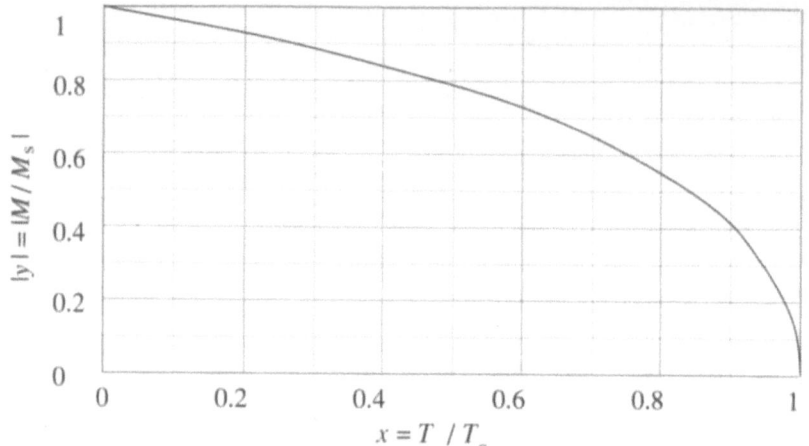

**Figure 9-4** *Spontaneous magnetization* |M|, *given in units of the saturation magnetization* $M_s$, *as function of temperature* $T$, *given in units of the critical or Curie temperature* $T_c$.

## Hysteresis

If one solves the two eqs. (9.9) as function of the magnetic field $H$ for different temperatures $T$ one obtains the typical hysteresis curves of a ferromagnet (Figure 9-5).

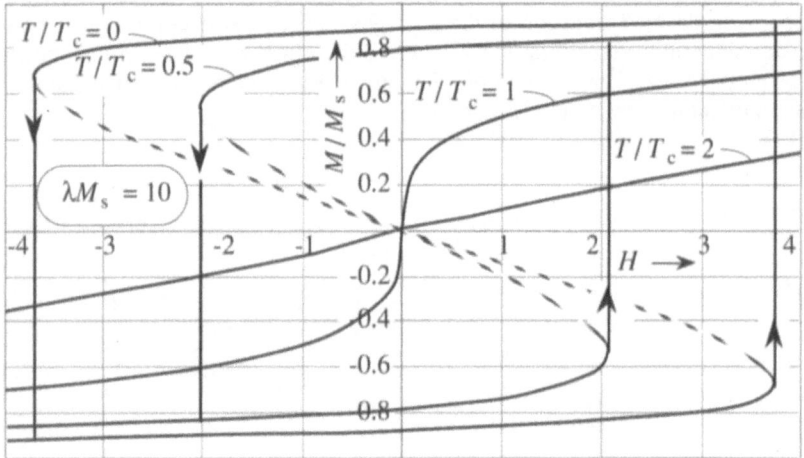

**Figure 9-5** *Hysteresis of a classical ferromagnet* for $\lambda M_s = 10$. For temperatures $T$ below the Curie temperature $T_c$ the magnetization $M$ depends on the magnetic history — the previous direction of the magnetization — of the sample. For $T < T_c$, the magnetization $M$ jumps abruptly to the lower (upper) branch of the curve if the external magnetic field $H$ is decreased (increased) beyond a certain critical value $H_c$. Below $T_c$ there exists therefore a spontaneous magnetization $M_r$ even at zero magnetic field $H$. Above $T_c$ the magnetization becomes an unambiguous, single-valued function of the external magnetic field $H$.

The two characteristic parameters of such hysteresis curves are the remaining magnetization $M_r$ or "remanence" for zero external magnetic field $H$ and the minimum magnetic field $H_c$ or "coercivity" required to force the magnetization $M$ into its antiparallel direction. At very high external magnetic fields $H$ the magnetization $M$ approaches the saturation magnetization $M_s$.

**Terminology of magnetic materials**

In the case of ferromagnetic iron garnets there exist mutually coupled sublattices, the magnetization of which compensate each other partially. Such system are termed ferrimagnetic. Other systems exist in which the sublattices compensate each other exactly and completely, forming a so-called antiferromagnet. Such systems are magnetically ordered although they exhibit no measurable external magnetization and are not susceptible to external magnetic fields.

## 9.1.2 Matrix model of magneto-optic films

In order to understand the principal behavior of thin ferromagnetic films used as storage medium, a simple model is used (Figure 9-6). It consists of a regular two-dimensional array of $N^2$ magnetic dipoles which are oriented either parallel or antiparallel with respect to the array normal. In this model only exchange and demagnetization interaction are considered. While the exchange interaction is a short-distance interaction which tends to align neighboring dipoles parallel, the dipole interaction has a longer range and tends to disaglign neighboring dipoles antiparallel. The influence of the crystal anisotropy on the elementary magnets is implicitly embodied in the model by allowing only up and down orientations of the elementary magnets. Therefore magnetic domain walls are infinitely thin within the model. This modest model enables to gain a feeling for the spontaneous creation of patterns in a cooperative system. At the same time it allows to give a very rough description of the principal effects observed in iron garnets. A more realistic treatment is given in reference [1].

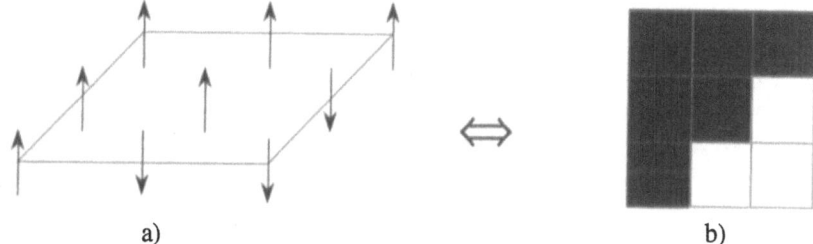

a)                                                                       b)

**Figure 9-6** *Two-dimensional dipole array — a simple model for magneto-optics.* (**a**) and equivalent matrix representation (**b**). The direction of an external magnetic field $H$ is chosen parallel to the dipoles pointing upward — the up domains.

The distance between neighboring dipoles is assumed to be one length unit. According to the model, the temperature $T$, the external magnetic field $H$, and the the alignment parameter $A$, given by the ratio of the exchange force and the dipole force, are the only three parameters defining the behavior of the system. While the exchange force is a quantum-mechanical phenomenon which tends to align the dipoles parallel to each other, the dipole interaction is a classical phenomenon and tends to align the dipoles antiparallel to each other. In the following, the parameters $T, H$, as well as $A$ are given always in arbitrary dimensionless units.

The matrix model described here assumes as disaligning force the magnetic dipole interaction which decays as $r^{-3}$. It has the tendency to minimize the total magnetic field generated by the dipoles by orienting them antiparallel. In addition, the matrix model

[1]  Tebble, R.S.: "**Magnetic Domains**." Methuen &Co. Ltd., 11 New Fetter Lane, London EC4, England, (1969).

assumes as aligning force the quantum-mechanical exchange force, which decays at a higher power, here $r^{-4}$. It has the tendency to maximize the total magnetic field generated by the dipoles by orienting them parallel. For any alignment parameter $A$ — the ratio between the exchange interaction and the dipole interaction — characteristic magnetic domain patterns are formed. A magnetic domain is a contingent region of uniformly aligned dipoles. According to the direction of the dipoles, one distinguishes up domains and down domains. Three typical degrees of alignment are shown in Figure 9-7.

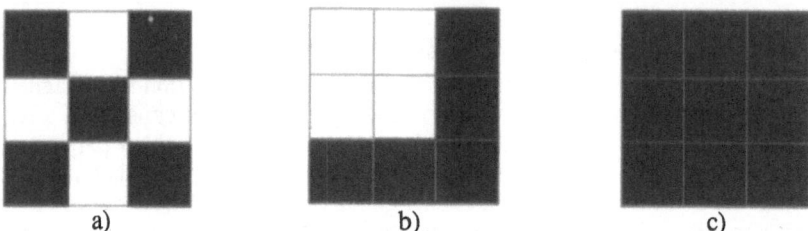

a)                                      b)                                      c)

**Figure 9-7** *Array of 3 by 3 magnetic dipoles.* With increasing exchange interaction — from left to right — the mutual alignment between the magnetic dipoles increases. (**a**) Antiparallel orientation of neighboring dipoles for $A = 0$. (**b**) Partial alignment of neighboring dipoles for $A = 1$. (**c**) Total alignment of neighboring dipoles for $A = 2$.

While the domain patterns for small arrays exhibit high symmetry, large arrays have a high degree of freedom and resemble surprisingly close the experimental patterns found in magneto-optic memories.

### Computer simulation

In order to calculate the total potential energy of a two-dimensional array of $N$ by $N$ dipoles, the dipole interaction as well as the exchange interaction has to be summed over all pairs of dipoles within the array. A dipole flip at an arbitrarily chosen site will occur exactly, if the total potential energy of the system decreases during the flip. A flow diagram of the used computer simulation is shown in Figure 9-8.

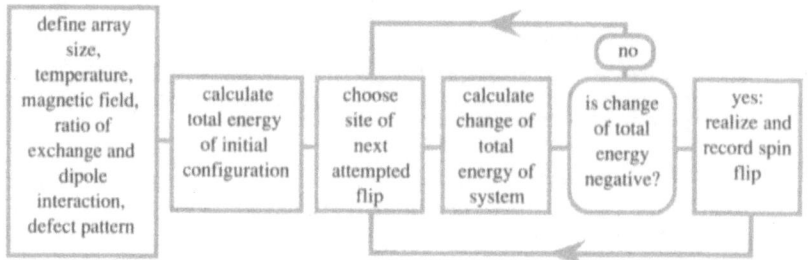

**Figure 9-8** Flow diagram for computer simulation of 2-dimensional magnetic dipole array.

**Dipole energy.** The potential energy for a pair of two dipoles $i$ and $k$ within the plane of the array depends on their relative alignment — parallel (+) or antiparallel (-) — and on the mutual distance of the dipoles, according to:

$$V_d(r_{ik}) = \pm \frac{q_m^2 \, d^2}{r_{ik}^3} = \pm \frac{\mu^2}{r_{ik}^3} = \pm \left(\frac{e \, h}{4\pi \, m \, c}\right)^2 \frac{1}{r_{ik}^3} \quad \text{for } r_{ik} \gg d \ . \tag{9.11}$$

In the computer simulation, all constants are set to one, so that the potential energy between the dipoles becomes $V_d(r_{ik}) = \pm 1 / r_{ik}^3$. Due to the positive sign for parallel dipoles, this potential energy corresponds to a disaligning force between the interacting dipoles $i$ and $k$.

According to eq. (9.11) the total potential energy $E_d$ due to dipole interaction becomes:

$$E_d = \frac{1}{2} \sum_{i \neq k} V_d(r_{ik}) , \qquad (9.12)$$

where the summation excludes the self-interacting term for $i = k$ and the factor $1/2$ circumvents the double-counting of identical pairs.

**Exchange interaction.** According to the matrix model, the exchange interaction — which is the aggregating force between the magnetic dipoles in ferromagnetism — is replaced by an exchange potential acting between the dipoles $i$ and $k$ and decaying as $r^{-4}$:

$$V_e(r_{ik}) = - A \frac{V_d(r_{ik})}{r_{ik}} \qquad \text{for } r_{ik} \gg d . \qquad (9.13)$$

In the computer simulation, all constants are set to one so that the exchange potential between the dipoles becomes $V_e(r_{ik}) = \pm A / r_{ik}^4$, whereby the plus sign corresponds to antiparallel alignment and the minus sign to parallel alignment. This exchange potential corresponds to an aligning force between the interacting dipoles which is typical for ferromagnetic materials. Due to the forth power in $r$ it decays more rapidly with distance than the disaligning dipole interaction which is an essential requirement for establishing domain patterns by local order. The size of the constants is chosen such that for $A = 1$ the the dipole interaction $V_d(r_{ik})$ and the exchange interaction $V_e(r_{ik})$ are equal in magnitude and opposite in sign for a rectilinear array of dipoles in which the neighbors are placed at unit distance from each other.

According to eq. (9.13), the total potential energy $E_e$ due to exchange interaction becomes:

$$E_e = \frac{1}{2} \sum_{i \neq k} V_e(r_{ik}) , \qquad (9.14)$$

where the summation excludes the self-interacting term for $i = k$ and the factor $1/2$ avoids double counting of identical pairs.

**Energy due to external magnetic field.** The potential energy $E_f$ of the magnetic dipoles in an external magnetic field $H$ is:

$$E_f = - \sum_i (\pm H) , \qquad (9.15)$$

where the plus (minus) sign corresponds to parallel (antiparallel) dipoles with respect to the external magnetic field $H$.

**Thermal energy.** Similar to the experimental procedure, the starting domain configuration corresponds to a completely disordered, paramagnetic system kept at $T \gg T_c$. The system is then settled down to thermodynamic equilibrium at a temperature $T \ll T_c$. Always a small but finite thermal excitation is applied in the computer simulation to overcome local potential energy barriers which requires the simultaneous rearrangement of several dipoles.

The probability that an arbitrarily chosen dipole will actually flip is influenced by its thermal energy which, according to the chosen model, reduces the total potential energy such that a flip can even occur if the total potential energy is slightly increasing. The probability $p(E)$ that a chosen dipole has the additional thermal energy $E_t$ is assumed to follow the Boltzmann distribution:

$$p(E_t) = e^{-\frac{E_t}{kT}} \quad . \tag{9.16}$$

Accordingly, for an arbitrary probability between 0 and 1, the energy $E$ is determined from the inverse function

$$E_t = -kT \, \log(p) \quad , \tag{9.17}$$

whereby, in the computer simulation, $p$ is replaced by a random value chosen arbitrarily within the interval between 0 and 1 and $k$ is arbitrarily set to one.

The described matrix model exhibits several important phenomena of magneto-optic films. In the following, the bistability of a small dipole array, the formation of magnetic domains, and the influence of disturbed paramagnetic zones — or ion tracks — on the resulting domain patterns is demonstrated.

### Bistability

For demonstrating the phenomenon of bistability, a not too large matrix of 5 by 5 elements is chosen with a moderate alignment parameter $A$ of 1.3, just sufficient to create two domains (Figure 9-9).

The total potential energy $E_{tot} \approx E_d + E_e + E_f$ for a given pattern and thus its stability depends on the external magnetic field $H$. With increasing $H$ the preferential pattern changes from a bistable pattern — a pattern which corresponds to relative magnetizations $M / M_s = \pm 0.2$ — to a completely aligned up-pattern. The smoothed total potential energy curve for different external magnetic fields $H$ and the corresponding most stable domain patterns are shown in Figure 9-9. Thereby the magnetic field tends to increase the potential energy of the antiparallel system and decrease the potential energy of the parallel system.

While the chosen system expresses bistability at $H = 0$, corresponding to two minima of the potential energy curve, more complex systems possess more than two potential energy minima depending on the number and shape of the magnetic domain patterns.

### Formation of magnetic domains at zero external magnetic field

According to the chosen model, the formation of magnetic domains corresponds to a minimum-energy configuration in which the sum of the dipole energy and the exchange energy is minimized. Depending on the alignment parameter $A$, domains of different size are formed, increasing very rapidly with increasing alignment parameter $A$ (Figure 9-10). At zero external magnetic field the domain pattern settles at zero magnetization, $M / M_s = 0$

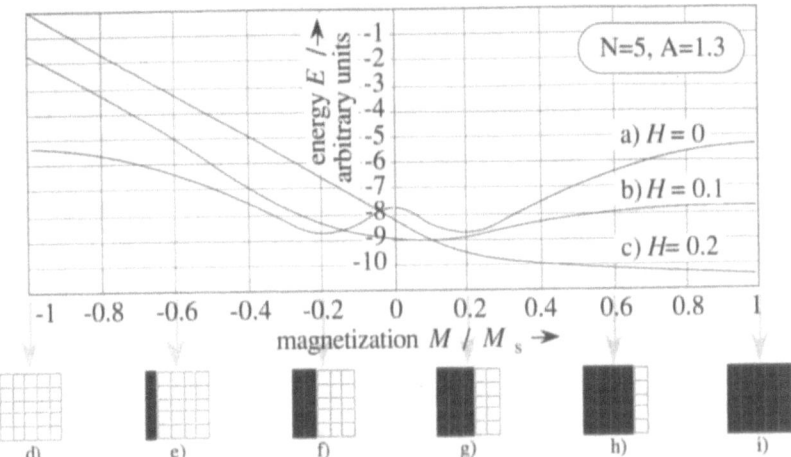

**Figure 9-9** *Bistability of a small dipole array.* Simulation of a system consisting of a 5 by 5 matrix. **Top figure: (a)** Without external magnetic field $H$ the system possesses two minima of the total potential energy $E$ and thus is a bistable system. Due to the moderate alignment parameter $A = 1.3$ the minima correspond to states of relatively small magnetization $M / M_s \approx \pm 0.2$. **(b)** For $H = 0.1$ the two energy minima merge into one broad minimum corresponding to a slightly preferred alignment in the direction of the external magnetic field $H$. **(c)** For $H = 0.2$ the system tends to complete alignment in the direction of the external magnetic field $H$. **Bottom figure: (d)** to **(i)** Corresponding matrix patterns.

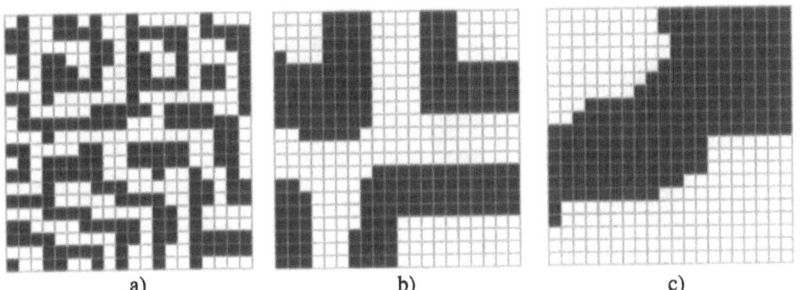

**Figure 9-10** *Influence of alignment parameter A.* Domain patterns as function of alignment parameter $A$ for zero external magnetic field $H$. Slightly different temperatures, T=0.1, 0.2, and 1.0, have been chosen in order to decrease the settling time of the corresponding systems during the simulation. **(a)** For small alignment parameter ($A = 1$) 22 domains are formed. **(b)** For medium alignment parameter ($A = 1.5$) 8 domains are formed. **(c)** For large alignment parameter ($A = 2$) 3 domains are formed.

### Domain wall energy

A dipole within the bulk of a domain is surrounded by parallel neighbors supporting its orientation and decreasing its potential energy. On the other hand, a dipole along the boundary between two domains has antiparallel neighbors which try to flip it over and increase its potential energy. This phenomenon is analogous to an existing interface tension between neighboring domains. The corresponding interface energy per unit length of the domain wall is termed domain wall energy. It can be decreased by ion tracks which remove a certain fraction of the interface. This effect can be used to stabilize domain patterns by the so-called domain wall pinning resembling a macroscopic frictional force (Figure 9-11).

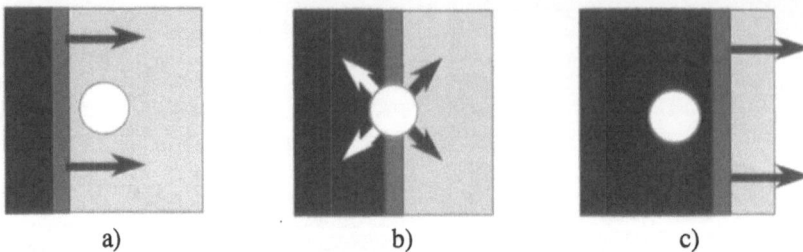

a)                          b)                          c)

**Figure 9-11** *Domain wall pinning* by a single-ion track (circle). (**a**) The up-domain (black) is enlarged by a parallel external magnetic field. The enlargement is associated with a shift of the domain wall. (**b**) The wall energy within the ion track zone is dissipated thermally. (**c**) While the domain boundary shifts further, the interrupted wall heals out again, requiring the input of external magnetic field energy.

### Influence of external magnetic field on domain pattern

If a virgin-state domain pattern is subject to an external magnetic field $H$, the potential energy of the up spins is decreased with respect to the down spins. Accordingly, the up domains are energetically favored with respect to the down domains. The resulting effect is to increase the size of the up domains and while decreasing the size of the down domains. Simultaneously the system is magnetized in the direction of the external magnetic field (Figure 9-12). With increasing magnetic field, ultimately discrete magnetic domains — so-called magnetic bubbles are formed.

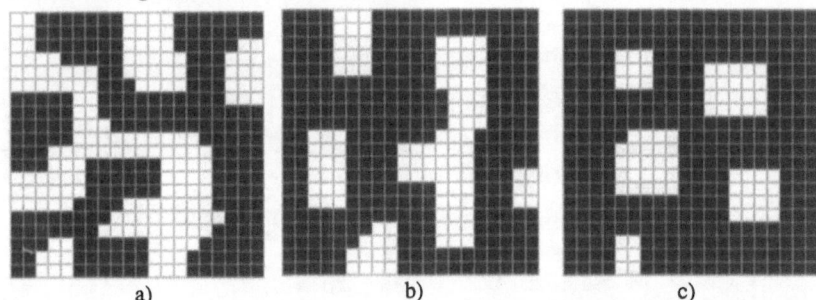

a)                          b)                          c)

**Figure 9-12** *Influence of external magnetic field $H$ on domain patterns and magnetization.* A medium alignment parameter, $A = 1.5$, and low settling temperature $T = 0.2$ was chosen in this example. With increasing field the down domains (white) shrink to discrete islands — so-called magnetic bubbles — while the up domains (black) merge to one contingent domain. (**a**) $H = 0.1$. (**b**) $H = 0.3$. (**c**) $H = 0.5$.

### Storage capacity of domain patterns

Magnetic domains (bubbles) can be related with the binary numbers "0" and "1", depending on their orientation. The storage capacity the resulting magneto-optic memory is proportional to the number of magnetic domains per unit area. Since the areal density of the domains decreases rapidly with increasing alignment parameter $A$, the exchange interaction — or alignment parameter $A$ — between neighboring dipoles should be sufficiently small for attaining high storage capacity. On the other hand there should still be sufficient "coherence" within the domains to maintain its integrity. Experimentally this can be achieved by irradiating the magneto-optic film with a certain areal dose of heavy ions. The main effect of the irradiation is to decrease the exchange interaction by creating defects in the lattice of the magneto-optic film.

### Influence of ion tracks on domain size

According to the matrix model, irradiated zones of the magneto-optic film correspond to zones of decreased alignment parameter $A$. This feature enables to mimic the paramagnetic behavior of track-damaged zones by decreasing the exchange interaction within certain regions of the matrix (Figure 9-13).

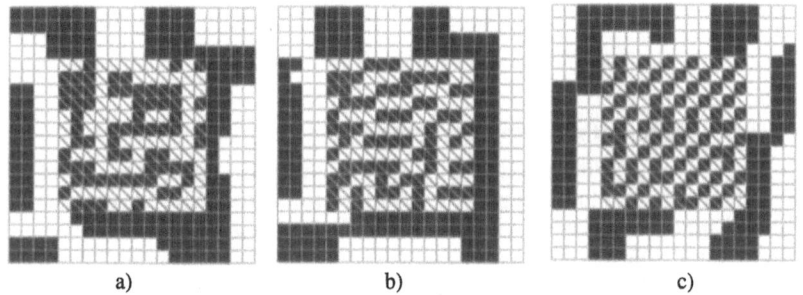

a)                          b)                          c)

**Figure 9-13** *Influence of ion irradiation* on the size of magnetic domains. The quadratic inner zone corresponds to the irradiated zone and is marked by 45° lines. While the nonirradiated outer zone has an alignment parameter $A = 1.3$, the alignment parameter $A'$ within the irradiated zone is assumed to be decreased to a smaller value by the ion tracks. (a) Inner-zone alignment parameter $A = 1.0$. (b) $A = 0.8$. (b) $A = 0.6$. A settling-down temperature $T = 0.2$ was chosen.

## 9.1.3    Experimental results

### Iron garnet films

In the following, the effect of ion irradiation on the magneto-optic properties of iron garnet films is outlined [1], [2], [3], [4], [5], [6], [7] . Such films are grown on nonmagnetic substrates using the technique of liquid phase epitaxy. They have a thickness of a few μm while the substrate is a rigid support of about 1 mm thickness (Figure 9-14).

[1] Krumme, J.P. I. Bartels, B. Strocka, K. Witter, Ch. Schmelzer, R. Spohr: "Pinning of 180° Bloch Walls at Etched Nuclear Tracks in LPE-Grown Iron Garnet Films." Applied Physics, **48**, 5191-5196 (1977).

[2] Hansen, P. H. Heitmann: "Influence of Nuclear Tracks on the Magnetic Properties of a $(Gd,Bi)_3(Fe,Ga)_5O_{12}$ Garnet Film." Phys.Rev.Lett. **43**, 1444-1447 (1979).

[3] Hansen, P., H. Heitmann, B. Strocka, R. Spohr: "Der Einfluß von Ionenstrahlen auf die magnetischen Eigenschaften von ferrimagnetischen Schichten." BMFT Bericht FB T83-048 Fachinformationszentrum Karlsruhe, D-7514 Eggenstein-Leopoldshafen 2, 1-180 (1983).

[4] Heitmann, H., P. Hansen, R. Spohr, K. Witter: "Properties of Magneto-Optic $(Gd,Bi)_3(Fe,Ga)_5O_{12}$ Films Irradiated with High-Energy Heavy Ions." J. Magnetism and Magnetic Materials, **15-18**, pp. 1543-1544, (1980).

[5] Hansen, P., H. Heitmann, P.H. Smit: "Nuclear Tracks in Iron Garnet Films." Phys. Rev. **B, 26**, No. 7, pp. 3539-3546, (1982).

[6] Heitmann, H., C. Fritzsche, P. Hansen, J.-P. Krumme, R. Spohr, K. Witter: " Influence of Irradiation with High Energetic Ions on Storage Properties of Magneto-Optic $(Gd,Bi)_3(Fe,Ga)_5O_{12}$ Epitaxial Films." J. Magnetism and Magnetic Materials, **7**, pp. 40-43, (1978).

[7] Hansen, P. J.-P. Krumme: "Magnetic and Magneto-Optical Properties of Garnet Films." Thin Solid Films, **114**, pp. 69-107, (1984).

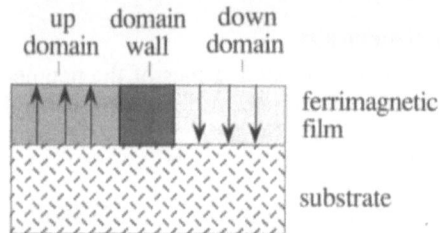

up        domain      down
domain     wall        domain

ferrimagnetic
film

substrate

**Figure 9-14**  Magneto-optic storage film supported by a nonmagnetic substrate.

### Composition of magnetic lattice

Garnets are cubic oxides of the general composition $\{A\}_3 [B]_2 (C)_3 O_{12}$. The symbols A, B, and C represent arbitrary metal ions and O represents oxygen. The different brackets indicate three different oxygen-coordination numbers. The $\{A\}$ ions possess eight, the $\{B\}$ ions six, and the (C) ions four oxygen atoms as their nearest neighbors, characterized by dodecaedric, octaedric, and tetraedric symmetry, respectively. The corresponding sublattices are characterized by the letters $c$, $a$, and $d$, respectively.

An classic example is the ferrimagnetic $Gd_3 Fe_5 O_{12}$ — gadolinium iron garnet — characterized by the formula $\{Gd^{3+}\}_3 [Fe^{3+}]_2 (Fe^{3+})_3 O_{12}^{2-}$, which has three, partially compensating, magnetic sublattices indicated in Figure 9-15.

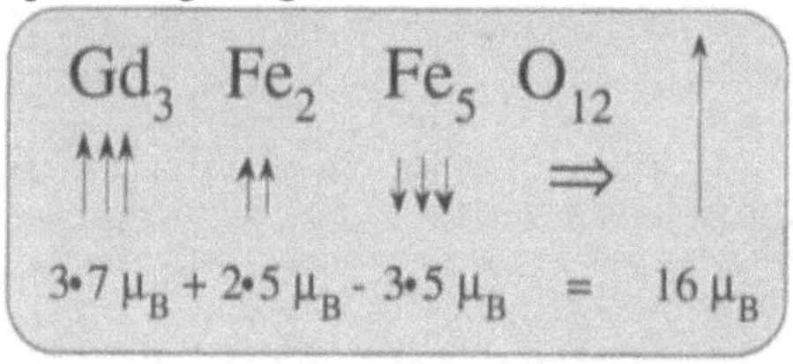

$$Gd_3 \quad Fe_2 \quad Fe_5 \quad O_{12}$$

$$\uparrow\uparrow\uparrow \qquad \uparrow\uparrow \qquad \downarrow\downarrow\downarrow \qquad \Rightarrow$$

$$3{\cdot}7\,\mu_B + 2{\cdot}5\,\mu_B - 3{\cdot}5\,\mu_B \;=\; 16\,\mu_B$$

**Figure 9-15**  Total magnetic moment of $Gd_3 Fe_5 O_{12}$ per formula unit in units of the Bohr magneton $\mu_B$ at $T = 0°$ K.

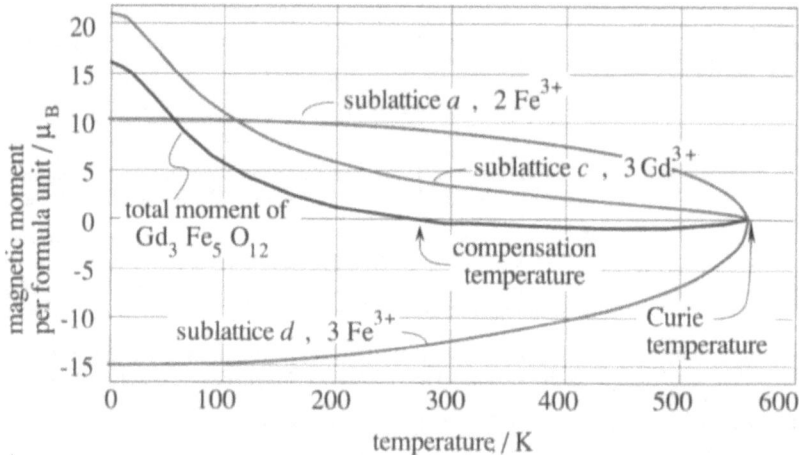

**Figure 9-16**  *Compensation temperature* of gadolinium iron garnet. Due to the different thermal dependence of the sublattice magnetizations, a compensation temperature $T_c$ exists, for which the magnetic order is maintained while the susceptibility of the domain patterns to an external magnetic field $H$ vanishes.

The magnetization of gadolinium iron garnet depends on temperature in a technologically useful way. A so-called compensation point exists at the temperature $T <$ $T_c$ at which the magnetization of the sublattices compensates mutually while maintaining the magnetic order of the lattice (Figure 9-16). This correspond to a temperature at which the iron garnet behaves as an antiferromagnet which is completely inert to the attack of external magnetic fields. This feature enables to store domain patterns permanently close to the compensation temperature, where the patterns are independent of the external magnetic field.

### Influence of ion tracks on domain size

Figure 9-17 left shows the influence of ion irradiation on domain size for an iron garnet film of composition $(Gd,Bi)_3(Fe,Ga)_5O_{12}$ [1].

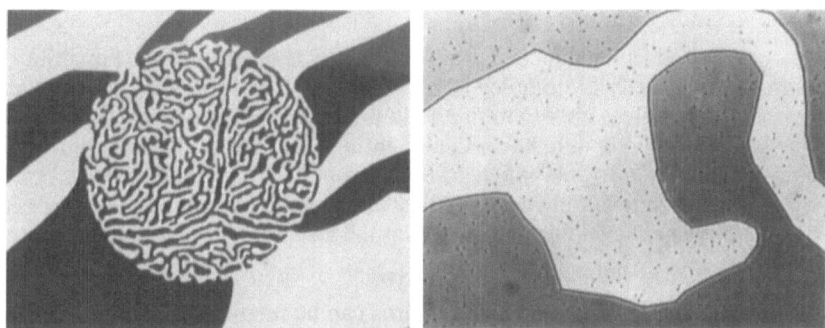

**Figure 9-17** *Influence of ion irradiation on domain patterns* in iron garnet films, track-etched to increase the effect of the ion irradiation. (**left**) Macroscopic observation of a film irradiated within a circular disk of 1 mm diameter. (**right**) Microobservation of a similar irradiated and etched zone. The domain walls (dark lines) are preferentially attached to the etched tracks (black spots).

The film was irradiated with $10^8$ / $cm^2$ uranium ions of 1.4 MeV/u within a circular disk of 2 mm diameter. The effect was magnified by etching the tracks for 30 min at 90° C in a medium consisting of 25% $HNO_3$, 25% $CH_3COOH$, and 50% $H_2O$, yielding channels of about 50 nm diameter which were about 10 times larger in diameter than the original damage zones of the latent ion tracks.

A microscopic observation of a similar iron garnet film irradiated with $10^6$ / $cm^2$ xenon ions of 1.4 MeV/u and etched to increase the effect of the ion irradiation is shown in Figure 9-17 right. The preferential attachment of the domain walls to etched ion tracks — known as domain wall pinning — corresponds globally to a frictional force which reduces the domain wall mobility.

### Compensation point writing

New information can be inscribed by rising the temperature locally above the compensation temperature while applying globally an external magnetic field. This technique is known as compensation point writing (Figure 9-18).

[1]  Hansen, P.: "Magnetic Anisotropy and Magnetostriction in Garnets." Italian Physical Society, Proc. of the Internat. School of Physics "Enrico Fermi", Course LXX, editor A. Paoletti, Varenna, June 27-July 9, 1977, **Physics of Magnetic Garnets**, North Holland Publishing Co., Amsterdam, (1978).

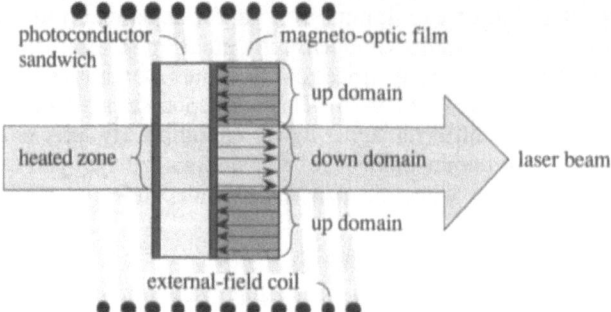

**Figure 9-18** *Principle of compensation point writing.* Local heating — assisted by a photoconductor sandwich — raises the magnetization and leads to domain flip according to the direction of the external magnetic field.

The technological use of iron garnets is favored by the possibility to substitute its metal ions partly or completely by other — paramagnetic or diamagnetic — ions of complementary sizes chosen to fit within the unit cell. In this way, the magnetic coupling between the different sublattices can be influenced. The wide range of possible substitutions enables to tailor the magneto-optic properties to specific needs. For example the compensation temperature can be adjusted to room temperature. Historically, the development of magnetic bubble memories was the incentive for investigating these materials.

### Observation of domain patterns

The inscribed domain patterns can be read-out using the Faraday rotation of linearly polarized light which depends on the magnetization of the domains (Figure 9-19).

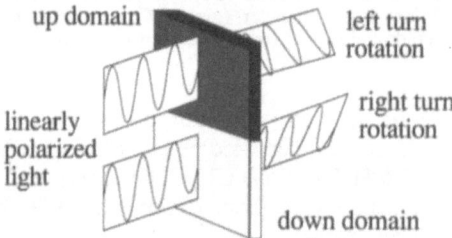

**Figure 9-19** Observation of magnetic domain patterns using the effect of Faraday rotation.

## 9.1.5    Generation of anisotropy

### Aligning magnetic domains along ion tracks

By the collective interaction of many ion tracks, new directionally dependent properties can be created, whereby the track angle controls the axis of anisotropy [1]. Oblique tracks tend to align along the track axis by magnetic domains by decreasing the alignment over the tracks (Figure 9-20). This is an example how ion tracks can be used to introduce an arbitrary axis of anisotropy in a solid.

[1]   Hansen, P., H. Heitmann, B. Strocka, R. Spohr: "Der Einfluß von Ionenstrahlen auf die magnetischen Eigenschaften von Ferrimagnetischen Schichten." BMFT Bericht NT 2007 3, FB 293/81, 1-179, (1981).

**Figure 9-20** *Generation of anisotropy along ion tracks.* Domain pattern of iron garnet irradiated by two mutually rectangular irradiations. The top half has been irradiated under 45° from left to right. The right half has been irradiated under 45° from bottom to top.

According to the described computer simulation, the alignment corresponds to the weakening of the interaction between neighboring domains on either side of the track (Figure 9-21).

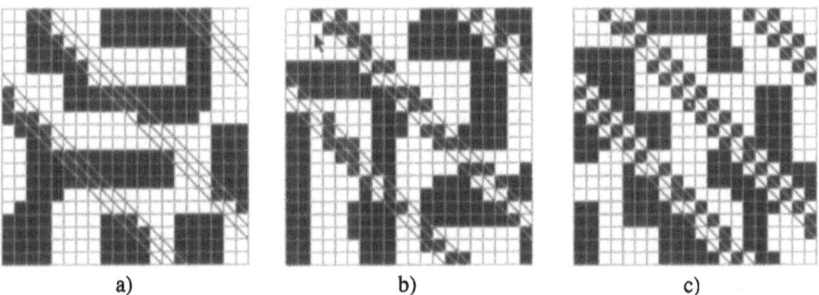

a)                                    b)                                    c)

**Figure 9-21** *Simulation of the influence of oblique ion tracks* (marked by 45° lines) on domain patterns in a matrix with a virgin-state alignment parameter $A = 1.3$. With increasing damage density — decreasing alignment parameter $A'$ of the damaged zones — the magnetic domains start to align along the tracks. In this process, the tracks become paramagnetic zones suppressing the mutual coupling between neighboring domains. (a) $A' = 1.2$. (b) $A' = 0.8$. (c) $A' = 0.4$. As settling temperature $T = 0.2$ was chosen.

# 10    Growth areas

## 10.1    Ion lithography

At high track densities, the ion tracks can be utilized for imprinting structures onto solids in the same way as conventional lithographies [1], [2]. This technique has several advantages in comparison with conventional techniques (Table 10-1).

**Table 10-1**   Technologically relevant parameters of the ion track technique.

1. **Single-particle tool**
   Each ion creates exactly one latent particle track and leads to exactly one characteristic hollow shape during the etching.

2. **Wide range of eligible materials**
   Besides light-sensitive photoresists, many radiation resistant polymers, glasses, and crystals can be used.

3. **High depth of resulting structures**
   Due to the small angular straggling of heavy ions in a light matrix very deep structures can be obtained.

4. **Controlled depth of tool**
   The depth of the resulting structure is determined by the energy of the incident ions. Range straggling is relatively small.

5. **High lateral resolution of tool**
   The finest structures obtained until now are channels of approximately 0.01 μm diameter.

6. **Generation of anisotropy realizable**
   New, directionally dependent properties can be created in solids by the collective interaction of many ion tracks.

7. **High quality of accessible radiation sources**
   Ion accelerators are able to provide very intense ion beams with very high beam parallelism.

While light particles deposit their energy rapidly in very few interactions, heavy ions release their energy very slowly and maintain simultaneously their direction with high accuracy. In light-particle lithographies the collective interaction of many particles is required to render the irradiated resist volume developable. However, already one single ion suffices to create a developable radiation damage in many radiation-resistant materials.

[1]    Yang, T.C., G. Welch, C.A. Tobias, H. Maccabee, T. Hayes, L. Craise, E.V. Benton, F. Abrams, Annals of the New York Academy of Sciences, **306**, pp. 3322-339, (1978)
[2]    Fischer, B.E., R. Spohr: "Heavy Ion Microlithography — a New Tool to Generate and Investigate Submicroscopic Structures." Nuclear Instruments and Methods, **168**, 241-246, (1980).

## 10.1.1    Basic techniques in lithography

### Basic techniques

Lithographic technologies are based on three principal processing steps:

1. *Irradiation* of the resist material with photons (visible light, ultraviolet light, x rays), electrons, or ions. The irradiation is used for transferring the pattern of a physical or nonphysical — computer-stored — mask onto the resist film.

2. *Development* (etching) in the liquid or gaseous phase, by plasma etching, or by ion sputtering. The development or etching removes preferentially either the irradiated or the unirradiated volume of the resist film, resulting in a stepped structure.

3. *Replication.* There exists a great wealth of available replica techniques — catalytic or galvanic deposition, chemical vapor deposition or ion sputter techniques — which can be used to transfer the created structure to another material.

**Positive and negative resist materials.** Resist materials are usually organic polymers consisting of long molecular chains. Depending on the prevalence of chain scission or chain bonding — in other words, depolymerization or polymerization — of the irradiated zone (Figure 10-1) the resist is termed a "positive" or a "negative" resist.

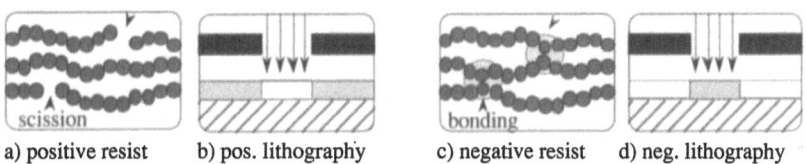

a) positive resist    b) pos. lithography    c) negative resist    d) neg. lithography

**Figure 10-1** *Definition of positive and negative resist material* and resulting lithography. The "positive" resist replicates the mask structure so that transparent zones are transferred into developable zones. The "negative" resist replicates the inverse of the mask structure so that transparent zones are transferred into undevelopable zones.

**Undercut problem.** Wet-chemical etching has the disadvantage that the etch process may lead to an isotropic or even worse to an interface-activated undercutting of the resist film. This problem can be circumvented by ion sputtering or by plasma-assisted sputtering (Figure 10-2).

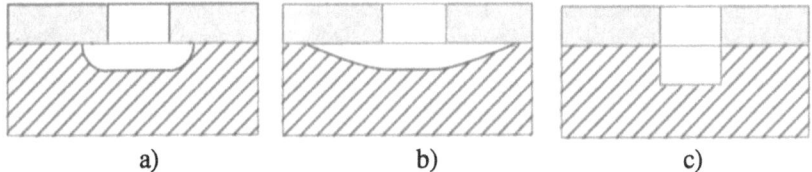

a)                        b)                        c)

**Figure 10-2** *Undercut problem.* (**a**) Isotropic undercut of resist film. The developent proceeds laterally at the same velocity as vertically. (**b**) Interface-enhanced undercut. The development proceeds frequently faster along the interface between the resist film and the substrate than in the substrate itself. **c**) Ion sputter techniques and reactive ion etching eliminate the undercut problem enabling aspect ratios (depth / width) up to about 10:1.

**Additive and subtractive technique.** In additive and subtractive lithographies always the resist is structured first, using irradiation and development. During these first processing steps, the resist film is partially removed. During further processing, material may be either deposited onto or removed from the bare areas of the substrate. The respective techniques are termed "additive" or "subtractive" (Figure 10-3).

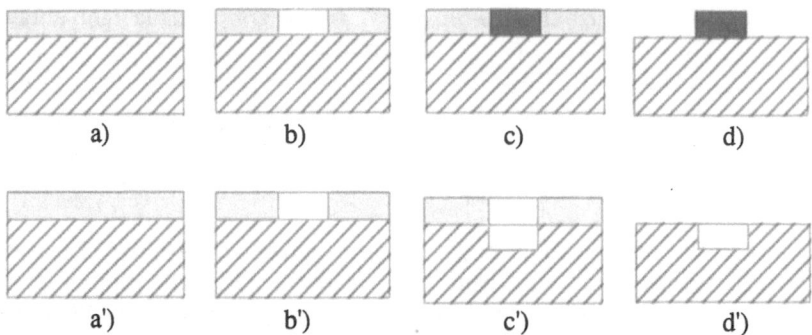

**Figure 10-3** *Principle of additive and subtractive lithography.* **Top:** Additive lithography. (**a**) Resist film on substrate. (**b**) Lithographically structured resist film. (**c**) Filling of trough — for example galvanically. (**d**) Removal of resist film. **Bottom:** Subtractive lithography. (**a'**) Resist film on substrate. (**b**) Lithographically structured resist film. (**c'**) Etch-attack of uncovered zone (**d'**) Removal of resist film.

Three additive techniques are shown in Figure 10 4.

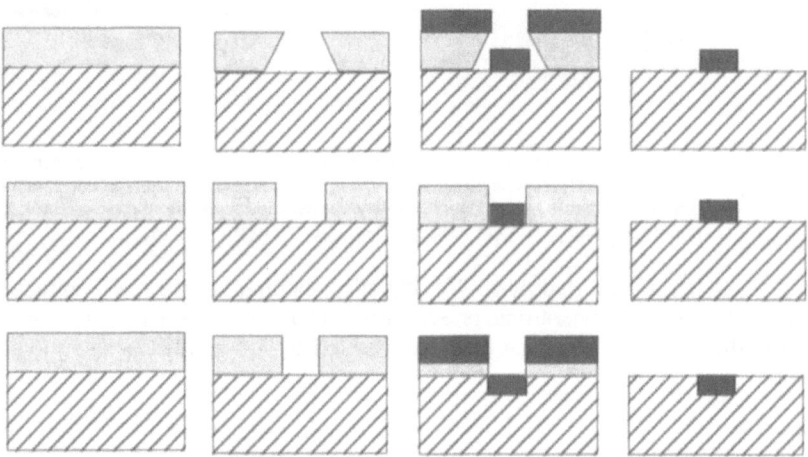

**Figure 10-4  Additive techniques.**
**Top:** *lift-off technique.* The used resist develops more rapidly with increasing depth. The overhanging resist prevents the establishment of contingent deposits and enables the selective removal of the resist together with its top deposit.
**Middle:** *Catalytic or galvanic deposition.* The catalytic technique uses the different chemical activity of the substrate with respect to the resist film. The galvanic technique requires a sufficiently conductive substrate for deposition and is mostly restricted to metal deposits.
**Bottom:** *Ion implantation* is mainly used for the controlled doping of semiconductors. It can also be used to generate buried insulating layers. Ion implanted material has to be thermally annealed to eliminate displacement damage in the crystal lattice.

### Contrast definition in conventional lithography

In conventional lithographies the contrast $C$ — or relative step height — is defined [1] as the ratio $D' / D$ of the step height and the maximally removed film thickness (Figure 10-5). It is related to the solubility $S(0)$ of the virgin material, and the solubility $S(\varepsilon)$ of the irradiated material exposed to the energy dose $\varepsilon$ in the developer:

$$C(\varepsilon)=\frac{D'}{D}=1-\frac{S(0)}{S(\varepsilon)} \text{ (pos. resist),} \quad C(\varepsilon)=\frac{D'}{D}=\frac{S(\varepsilon)}{S(0)}-1 \text{ (neg. resist).} \quad (10.1)$$

This definition leads to a positive contrast, $0 \le D'/D \le 1$, for positive resists and to a negative contrast, $0 \ge D'/D \ge -1$ for negative resists.

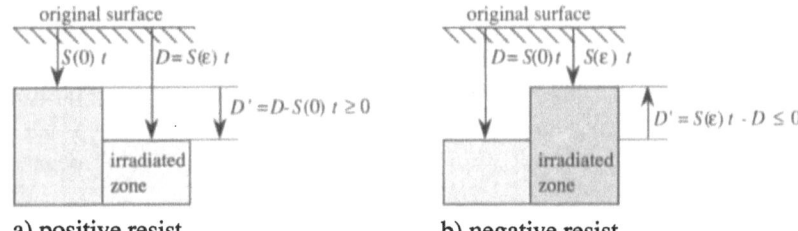

a) positive resist          b) negative resist

**Figure 10-5**  Definition of contrast in conventional lithographies.

### Solubility and deposited energy dose

The solubility $S(\varepsilon)$ of an irradiated resist film in the developer is a function of the energy dose $\varepsilon$ of the irradiation — the deposited energy per unit mass, measured in units of Gray = 1 Joule / kg = $10^4$ erg / g = $6.25 \cdot 10^{15}$ eV / g. The solubility of a polymer depends on the molecular weight $M$ of the polymer which in turn is related to the deposited energy dose $\varepsilon$. The semi-empiric relations are derived in the following.

**Positive resist.** First, the average number $P(\varepsilon)$ of chain scissions per molecule created by the irradiation is equal to the deposited energy dose $\Delta\varepsilon$ per molecule divided by the energy $\Delta E$ required per chain scission, or $P = \Delta\varepsilon / \Delta E$. Second, the deposited energy dose $\Delta\varepsilon$ per molecule is equal to the energy dose $\varepsilon \rho$ per unit volume — where $\varepsilon$ is the deposited energy per unit mass and $\rho$ is the mass density of the resist — divided by the number $n$ of molecules per unit volume, or $\Delta\varepsilon = \varepsilon \rho / n$. Third, the number $n$ of molecules per unit volume is equal to the Avogadro number $N_A$ — the number of molecules per mol — divided by the molar volume given by the ratio of molecular weight $M_0$ of the virgin resist and its mass density $\rho$, or $n = N_A / V_m = N_A \rho / M_0$. The three relations yield together:

$$P(\varepsilon) = \frac{\Delta\varepsilon}{\Delta E} = \frac{1}{\Delta E}\frac{\varepsilon \rho}{n} = \frac{\varepsilon \rho}{\Delta E}\frac{M_0}{N_A \rho} = \frac{\varepsilon M_0}{\Delta E N_A} . \quad (10.2)$$

Since every chain scission leads to one additional molecule, the scission process multiplies the total number of molecules by a factor $[1 + P(\varepsilon)]$ and thus reduces the original molecular weight $M_0$ to a molecular weight $M(\varepsilon)$, given by:

[1]     Wittels, N.D.: "Fundamentals of Electron and X-Ray Lithography." pp. 1-104, in "**Fine Line Lithography**.", Editor, R. Newman, North-Holland Publishing Company, Amsterdam, (1980).

$$M(\varepsilon) = \frac{M_0}{1 + P(\varepsilon)}, \quad \text{where } P(\varepsilon) = \frac{\varepsilon\, M_0}{\Delta E\; N_A}. \tag{10.3}$$

Assuming that the solubility $S(M)$ in the developer is proportional to $M^{-\alpha}$, where the empiric exponent $\alpha \le 1$, one obtains, using eq. (10.1) and eq. (10.3), (Figure 10-6):

$$C(\varepsilon) = \frac{D'}{D} = 1 - \frac{S(0)}{S(\varepsilon)} = 1 - \left(\frac{M(\varepsilon)}{M_0}\right)^{\alpha} = 1 - \frac{1}{\left(1 + P(\varepsilon)\right)^{\alpha}}. \tag{10.4}$$

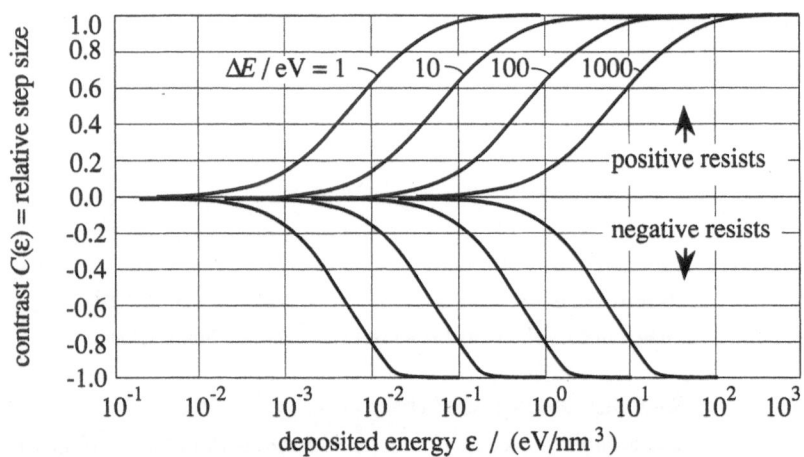

**Figure 10-6** *Contrast function of positive and negative resist,* assuming a molecular weight $M_0 = 10^5$ of the virgin resist material, assuming a solubility exponent $\alpha = 1$, assuming a mass density $\rho = 1$, for different values $\Delta E$ of the energy quantum required for a chain scission or an intermolecular bonding.

**Negative resist.** For negative resists the average number $P(\varepsilon)$ of created intermolecular bonds per molecule follows the same relation as for positive resists, given by eq. (10.2), but with a different energy $\Delta E$ required per created new bond.

Since every new intermolecular bond reduces the number of molecules by one, the intermolecular bonding process reduces the total number of molecules by a factor $[1 - P(\varepsilon)]$ and thus increases the original molecular weight $M_0$ to:

$$M(\varepsilon) = \frac{M_0}{1 - P(\varepsilon)}, \quad \text{where } P(\varepsilon) = \frac{\varepsilon\, M_0}{\Delta E\; N_A}. \tag{10.5}$$

This function has a singularity for $P(\varepsilon) \to 1$ which can be circumvented by replacing $P(\varepsilon)$ by the probability of creating at least one bond per molecule, or $w_{\ge 1} = 1 - w_0 = 1 - \exp(-P)$. This leads to $M(\varepsilon) = M_0 / \exp(-P)$ and, using eq. (10.1), yields a contrast function (Figure 10-6):

$$C(\varepsilon) = \frac{D'}{D} = \frac{S(\varepsilon)}{S(0)} - 1 = e^{-P\alpha} - 1. \tag{10.6}$$

**Resist characterization.** According to the above relations, resists can be characterized by the energy $\Delta E$ required for creating one chain scission or one intermolecular bond and by the solubility exponent $\alpha$. Another, experimentally directly accessible characterization is the maximum slope $\gamma_{max} = \{C(\varepsilon) / \log (\varepsilon / \varepsilon_0)\}_{max}$ of the contrast function taken from a semi-logarithmic plot such as shown in Figure 10-6, together with a sensitivity threshold $\varepsilon_0$ given by the corresponding value of the slope-line on crossing the $\varepsilon$-axis.

### 10.1.2  Ion lithographic techniques

Three possible ion lithographies are outlined in Figure 10-7.

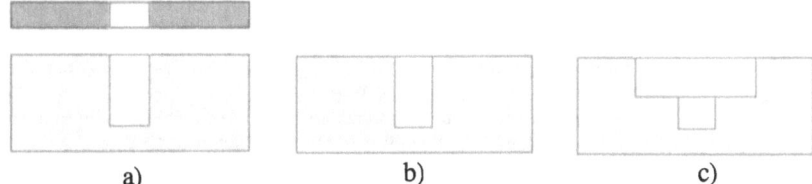

a)                                    b)                                    c)

**Figure 10-7** *Possible schemes of ion lithographies.* (**a**) Projection lithography, similar to the conventional lithographies using light particles such as photons. (**b**) Ion microscribing using a scanning ion microbeam. (**c**) Ion microscribing using two different ion energies in succession.

#### Depth definition in ion lithography

Due to the defined range of ions in matter, ion lithography can be considered as a cutting tool with well-defined depth (Figure 10-8).

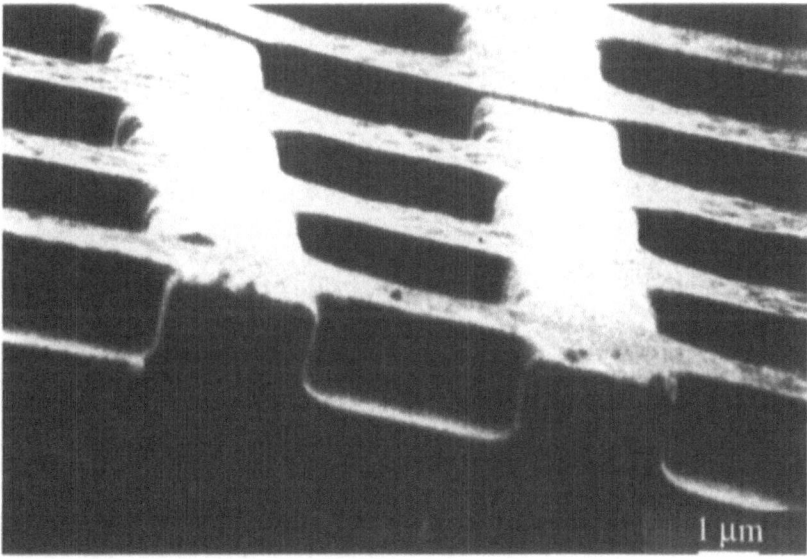

**Figure 10-8** *Principle of depth definition in ion lithography.* Troughs of precisely defined depth, about 4 μm wide, in PMMA obtained by ion-projection over a distance of about 1 mm, whereby the joint effect of many ions rendered the resist developable.

The depth definition is influenced on one hand by the natural range straggling of the ions in matter and on the other hand by the areal density and size of the etched tracks. Since for heavy ions in targets of low atomic number the range straggling is of the order of only a few percent, it can often be neglected. Therefore, in the following it is assumed that the depth definition depends exclusively on the areal density, shape and size of the etched ion tracks (Figure 10-9).

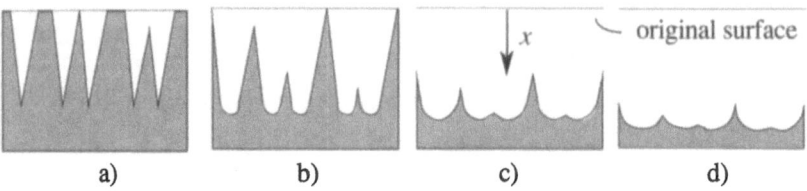

a)                          b)                          c)                          d)

**Figure 10-9**  *Depth definition in ion lithography*. The obtained structure becomes very rough after short etching times (**a**). With increasing track overlap the etched surface approaches asymptotically a smooth planar surface (**b** to **d**). Depth definition can be expressed as the effective track density at depth x below the original surface.

Depth definition can be expressed quantitatively by the effective density $\rho_{eff}(x, t)$ of the track recording material at the depth $x$ below its original surface and at the time $t$ after the etch process was started.Thereby stochastically distributed ion tracks are assumed.

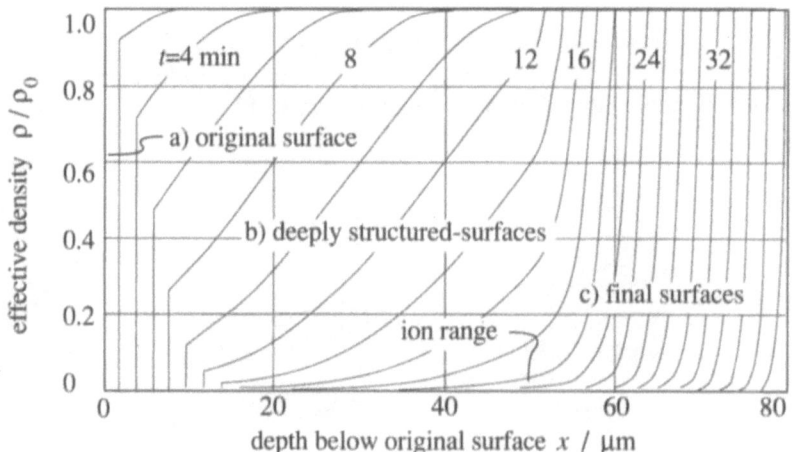

**Figure 10-10**  *Depth definition in ion lithography* for $N = 10^6$ cm$^2$, $g = 1$ μm / min, v = 5 μm / min, and $R_{ion} = 50$ μm. Three stages can be discerned. The originally flat surface (**a**) becomes deeply structured during etching (**b**) until the tracks are completely etched and the overlapping spherical sections yield increasingly smooth final surfaces (**c**).

In the following, $N$ is the areal density of the ion tracks, $g$ is the general etch rate of the virgin track recording material in the developing medium, v is the etch rate along the ion tracks, for simplicity assumed to be constant, and $R_{ion}$ is the ion range in the material. The ion tracks are assumed normal with respect to the surface plane of the track recorder.

The effective density $\rho_{eff}$ is given by the complement of the effective porosity $P_{eff}$ of the stochastic track pattern at the depth $x$ below the original surface of the track recorder. The effective porosity $P_{eff}$ in turn is given by the areal density $N$ of the tracks and by their cross-sectional area, $\pi r^2$. This yields the relation:

$$\rho_{eff} = \rho_0 (1 - P_{eff}) = \rho_0 \, e^{-P} = \rho_0 \, e^{-N \pi r^2}, \tag{10.7}$$

where $r = r(x, t)$ is the track shape according to eqs. (4.40) to (4.43). The factor $N \pi r^2$ of the exponential function determines the density at a given distance from the averaged final surface and is quantitatively related to the smoothness of the final surface. Very roughly speaking, the achievable smoothness increases with increasing areal density $N$ of the ion tracks and with increasing track etch ratio $v / g$, assuming sufficiently long etch times.

The evaluation of the analytically given $\rho_{eff}(x \, t)$ is shown in Figure 10-10, assuming $N = 10^6$ ions per cm², $g = 1$ µm / min, $v = 5$ µm / min, and $R_{ion} = 50$ µm.

### Projection lithography

Figure 10-11 shows the principle of a projection lithography in which the areal density distribution of an object or mask is projected along the axis of the well-collimated ion beam onto a track-recording material. During the development the latent ion tracks are etched sufficiently long, so that mutual track overlap is achieved, in order to remove the irradiated region completely. The resulting relief structure is directly related to the projected areal density of the object or mask. The technique can be used either diagnostically to gain information on the density distribution of an anknown object or as a lithographic tool transfering a mask pattern onto a resist material.

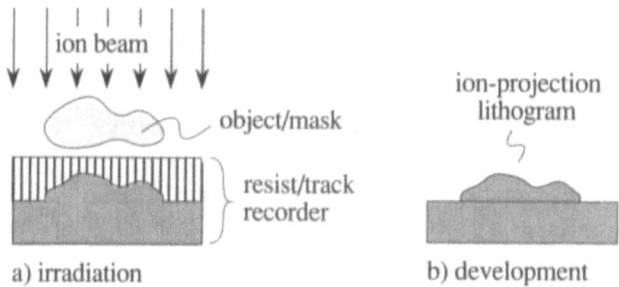

**Figure 10-11**  Principle of ion projection lithography.

Figure 10-12 shows an example in which a mask pattern was projected over about 1 mm distance onto a silicon dioxide film used here as the track sensitive resist material [1].

[1]   Fischer, B.E., R. Spohr: "Production and Use of Nuclear Tracks: Imprinting Structure on Solids." Reviews of Modern Physics, **55**, No. 4, pp. 907-948, (1983).

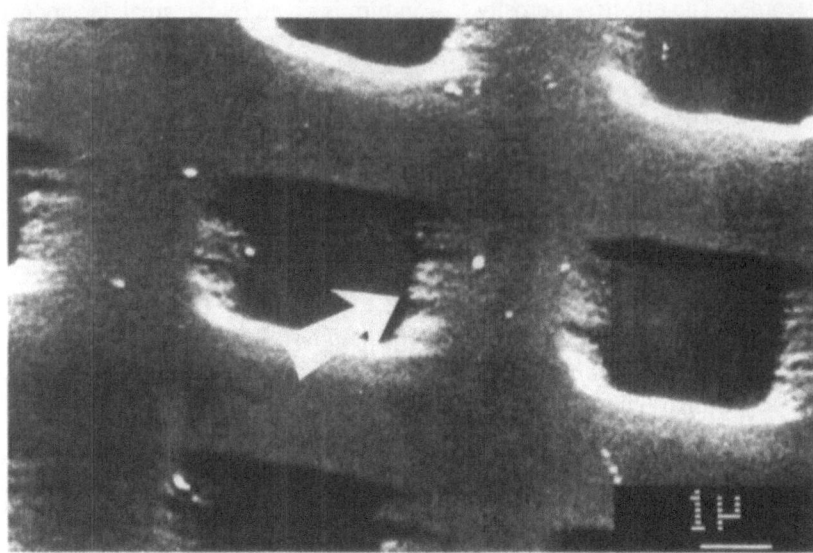

**Figure 10-12** *Silicon dioxide — a high-resolution resist for ion lithography.* Structures of about 0.1 μm size were transferred over a distance of about 1000 μm from the mask onto the quartz film using Argon ions of about 0.01 MeV/u.

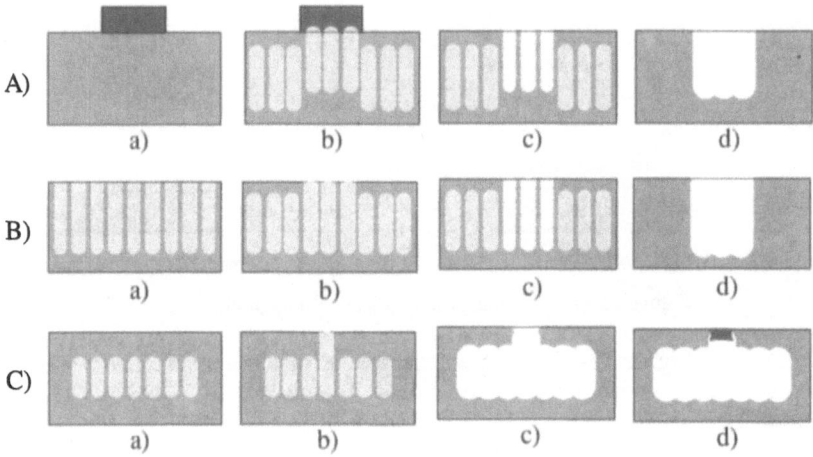

**Figure 10-13 Exploiting track etch threshold.** (**A**) *Range-reduction scheme* (**a**) Mask deposited on track-recording material. (**b**) Irradiation with energies above the energy-loss maximum, such that buried latent tracks are formed which are partially inaccessible to the etching process. (**c**) Short pre-etching of the accessible tracks after removal of the resist film (**d**) After annealing of the remaining latent tracks and a prolonged etching of the pre-etched tracks a deep trough results. (**B**) *Annealing scheme* (**a**) Pre-irradiated material. (**b**) Partial annealing of the top surface — for example by a scanning laser beam. (**c**) Short pre-etching of accessible tracks. (**d**) After annealing of the remaining latent tracks and prolonged etching of the pre-etched tracks. (**C**) *Double-irradiation scheme.* (**a**) First irradiation, using many (light) ions above their energy-loss maximum and leading to buried latent tracks. (**b**) Second irradiation, using few (heavy) ions of sufficient energy-loss. (**c**) After etching of latent tracks. (**d**) After closing top opening.

### Exploiting the track-etch threshold

Due to the existence a maximum for the electronic and the nuclear stopping function, the energy-loss of ions with sufficiently high energy can initially increase on its way through the solid, exceeding the track-etch threshold in the bulk material and ultimately falling to zero at the end of the ion range. There exist several possibilities for using the track-etch threshold along latent ion tracks (Figure 10-13).

### Track-sensitized materials

Another scheme for a lithography using track-sensitized material is suggested in Figure 10-14. It uses material pre-irradiated by a parallel ion beam and controls the access of the etchant to the latent tracks by a conventional lithographic masking technique.

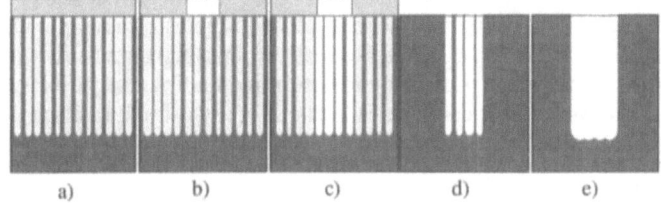

a)          b)          c)          d)          e)

**Figure 10-14** *Lithography using track-sensitized material.* (**a**) Conventional resist film on pre-irradiated material. (**b**) Conventionally structured resist film. (**c**) After short pre-etching of tracks in unprotected area. (**d**) After removal of resist film and annealing of remaining latent tracks. (**e**) After prolonged etching of pre-etched tracks.

### Scanning ion microbeam

Such devices enable a predefined track creation in solids (Figure 10-15) [1].

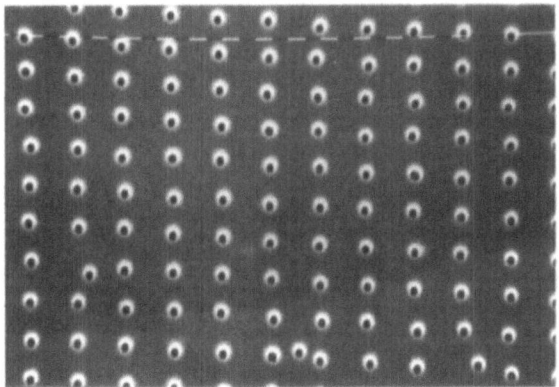

**Figure 10-15** *Regular track pattern of single-ion tracks* [2]. The conical tracks are inscribed with $^{52}Cr$ ions of 1.4 MeV/u specific energy. Normal soda-lime glass was used as a track recorder. The etched track cones have an opening angle of about 70°. The lattice has a characteristic distance of ca. 30 μm. Less than 1 % scattered particles are observed. Bar = 10 μm.

[1]     Fischer, B.E.: "The Scanning Heavy Ion Microscope at GSI." Nucl. Instr. Meth. in Physics Research, **B 10/11**, 693, (1985).

[2]     Fischer, B.E.: "The Heavy Ion Microprobe at GSI — Used for Single-Ion Micromechanics." Nucl. Instr. Meth. in Physics Research, **B 30**, pp. 284-288, (1988).

## 10.2    Surface texture

A very useful but little noticed aspect of the ion track technique is the possibility of generating microscopic surface textures with precisely defined parameters. A few potential applications are outlined in the following.

### 10.2.1    Light scattering devices

Etched tracks offer the possibility to generate ray-optical scattering devices with exactly predetermined properties (Figure 10-16) which for example can be used as reflectors with defined reflection-cone angle or as a focussing screens in optical systems. A special feature of tracks is the possibility to determine the fraction of the scattered beam by the density of ion tracks. This could be important if only a small fraction of the incident radiation shall be scattered.

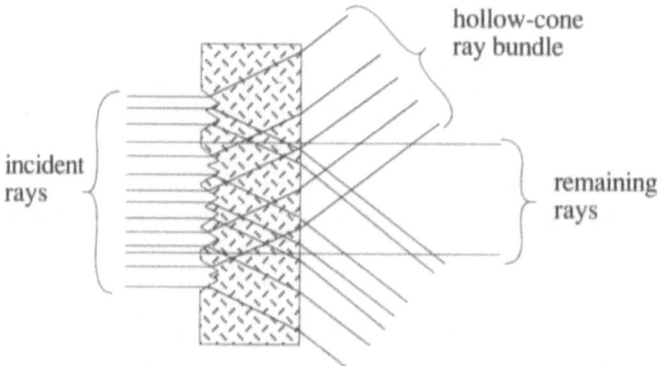

**Figure 10-16**  Scattering plate for scattering a defined fraction of an incident photon beam into a defined angle.

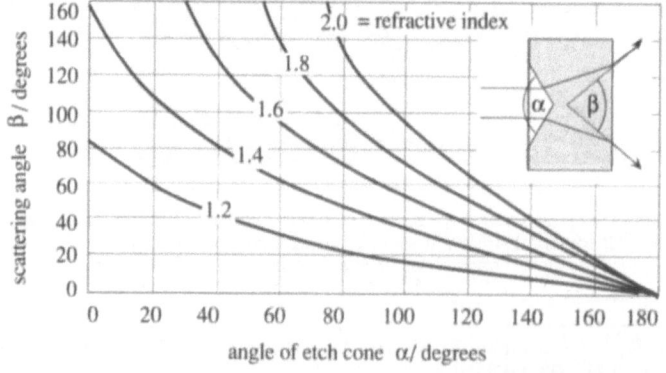

**Figure 10-17**  Scattering angle for track cones normal to the surface of the track recorder.

The scattering angle can be adapted to the acceptance angle of an optical system and is given by the cone angle of the individual etched ion tracks and the refractive index of the track recorder (Figure 10-17). The cone angle $\alpha$ of the etched ion tracks can be adapted to the specific requirements for example by partial annealing of the ion tracks. A necessary requirement for the use as a scattering plate is that the wave-length of the used radiation is smaller than the diameter of the individual tracks. Only in this case the contribution of the classical ray-optics is large in comparison with the diffraction of the waves.

## 10.2.2    Antireflection treatment

For etch cones with diameters and at distances small in comparison with the wave-length of the applied radiation, coherent scattering — wave optics — becomes dominant. At sufficient areal density of the ion tracks and sufficient depth of the etched track cones a drastic reduction of the reflected intensity is observed [1], [2]. In the following, the reflected intensity is determined according two different models [3]. First, the time-evolution of the reflectivity during etching is determined and second, an analytical upper limit for the reflectivity at sufficient long etch times is given .

In both models the areal density $N$ of tracks is assumed sufficiently high to ensure that the average distance between neighboring tracks is small in comparison with the wave length $\lambda$ of the incident radiation. The bulk etch rate $g$ is set to one and the track etch rate $v$ is assumed to be constant throughout the track recorder.

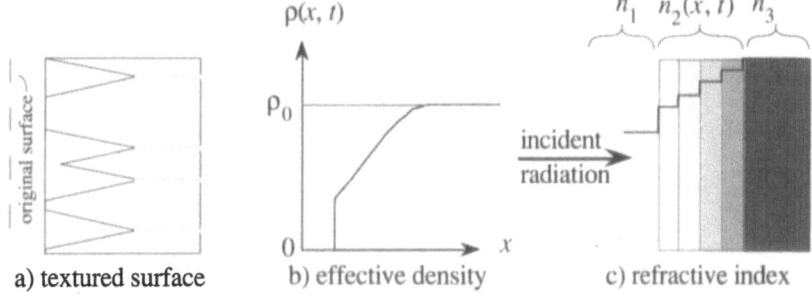

a) textured surface          b) effective density          c) refractive index

**Figure 10-18** *Principal steps for calculating energy reflection coefficient* — or reflectivity — $R_e$. The stochastic distribution of etched track cones (**a**) corresponds to an effective mass density increasing from a finite value at the surface to the density $\rho_0$ of the bulk material (**b**) which in turn corresponds to a refractive index increasing from the refractive index $n_1$ before the surface to the refractive index $n_3$ of the bulk material (**c**). The reflected amplitude is obtained by summing up the reflected amplitudes from individual layers of a staircase function.

### Time-evolution of reflectivity

According to the reasoning steps of Figure 10-18 the stochastic distribution of the etched tracks is replaced by the effective density $\rho_{eff}(x, t)$ — the complement of the effective porosity $P_{eff}$ — from which in turn the refractive index $n_2(x, t)$ is determined, and approximated by a sufficiently dense staircase function. The added-up reflection amplitudes from each step of this staircase function yield the total reflection amplitude and thus

[1]    Fischer, B.E.: "Graded-Index Antireflecting Surfaces." GSI 84-1, March 1984, ISSN 0174-0814. p. 216, (1984)
[2]    Fischer, B.E.: "Miscellaneous Applications of Nuclear Track Techniques." GSI 86-1, ISSN 0174-0814. p. 269, (1986)
[3]    Knittl, Z.: "**Optics of Thin Films**." John Wiley, London, pp. 455 - 458, (1967).

the reflected intensity by multiplication with its complex conjugate.

Using the track radius $r(x, t)$ from the track shape calculations according to eq. (4.38) one obtains the effective density:

$$\rho_{eff} = \rho_0 \, e^{-N \pi \, [(v \, t - x) \, \tan \, (\beta)]^2} \quad \text{for} \quad g \, t \leq x < v \, t \,. \tag{10.8}$$

This density function represents a step function, smoothing in time (Figure 10-19 left).

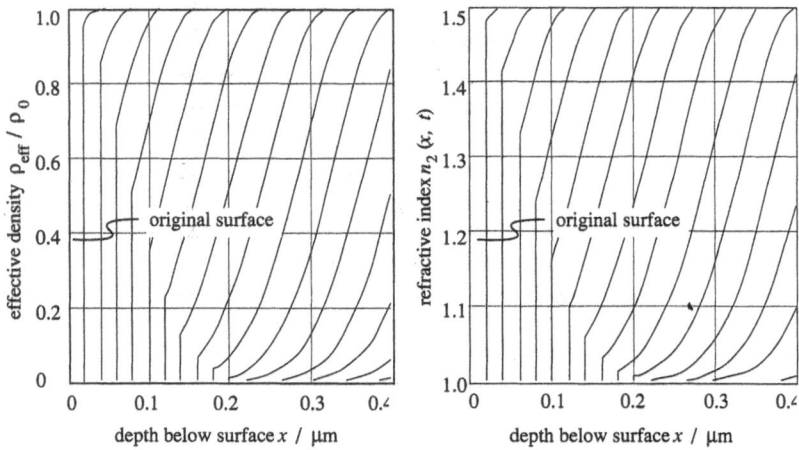

**Figure 10-19** (left) *Effective density* $\rho_{eff}(x, t)$ for $g = 1 \, \mu m \, / \, min$, $v = 2 \, \mu m \, / \, min$, $N = 10^{10} \, cm^{-2}$, and $t = \{0, 0.02, 0.04, ..., 0.28\}$ min. (right) Corresponding *refractive index* for same parameters with $n_1 = 1$, $n_3 = 1.5$.

According to the Lorentz-Lorenz formula, $(n^2 - 1)/(n^2 + 2) = \text{const} \, \rho$ [1], the effective density $\rho_{eff}$ corresponds to a refractive index

$$n = \sqrt{\frac{3}{1 - \text{const} \, \rho} - 2} \,, \quad \text{where} \quad \text{const} = \frac{n_3^2 - 1}{n_3^2 + 2} \, \rho_0 \tag{10.9}$$

is determined from the right-side boundary condition of the problem (Figure 10-19 right) and by its definition fits already to the left-side boundary condition, $n_1 = 1$.

Now the continuous refractive index $n_2$ is replaced by a step function whereby the steps size is chosen such that sufficiently small — here identical — phase shifts exist between successive reflections. The total phase shift $\Delta\phi_k$ — or $\Delta\phi(x)$ — of a wave originating at position 0, reflected at position $x_k$ — or $x$ — and returning to position 0 is

$$\Delta\phi_k = 2 \, \frac{2 \, \pi}{\lambda} \sum_{m=1}^{k} n(x_m) \, \Delta x_m \quad \text{or} \quad \Delta\phi(x) = \frac{4 \, \pi}{\lambda} \int_0^x n(\xi) \, d\xi \,. \tag{10.10}$$

The amplitude $A_k$ — or $A(x)$ — of the wave reflected at position $x_k$ — or $x$ —corresponds to the square-root of the Fresnel formula [2]:

[1]    Becker, R., F. Sauter: "**Theorie der Elektrizität.**" B.G. Teubner, D-7000 Stuttgart, Band. **3**, p. 152, (1969).
[2]    Flügge, S.: "**Lehrbuch der Theoretischen Physik.**" Springer-Verlag, Berlin, vol **3**, p. 211 ff., (1961).

$$A_k = \frac{n_{k+1} - n_k}{n_{k+1} + n_k} \quad \text{or} \quad A(x) = \lim_{dx \to 0} \frac{n(x+dx) - n(x)}{n(x+dx) + n(x)} = \frac{dn(x)}{2n(x)} \quad . \qquad (10.11)$$

The total reflected wave amplitude $A_r$ at position 0 is obtained by summation:

$$A_r = \sum_{k=1}^{k_{max}} A_k e^{-i\,\Delta\phi_k} \quad \text{or} \quad A_r = \int_0^{x_{max}} A(x) e^{-i\,\Delta\phi(x)}\,dx \ . $$

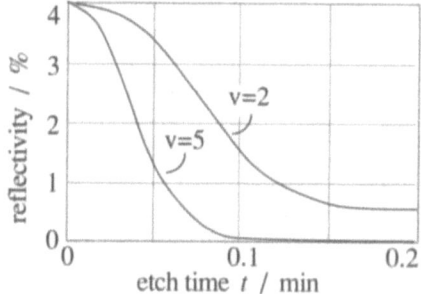

(10.12)

where $k_{max}$ — or $x_{max}$ — corresponds to the upper limit $n_3$ of the refraction index and $i = (-1)^{1/2}$. By multiplying the complex amplitude $A_r$ with its complex conjugate the energy reflection coefficient — or reflectivity — $R_e$ is obtained (Figure 10-20 ).

**Figure 10-20**   *Energy reflection coefficient — or reflectivity — as function of etch time.* Numeric calculation for $g = 1$ $\mu$m / min, $N = 10^{10}$ cm$^{-2}$, $n_1 = 1$, $n_3 = 1.5$, $\lambda = 0.5$ $\mu$m, and v = {2, 5} $\mu$m / min. With increasing v the smoothness of the density function increases for sufficiently long etch times and thus the reflectivity is reduced correspondingly.

### Stationary-state reflectivity

For sufficient long etch times the initial step of the effective density vanishes. The effective density assumes its stationary-state shape which shifts to higher $x$-values at the constant velocity v. An analytical approximation — actually an upper limit — for the energy reflection coefficient $R_e$ is obtained by replacing the stationary effective density by a straight line touching it in its inflection point (Figure 10-21):

$$x_{inflection} = -\frac{1}{\sqrt{2\,c}} + v\,t \ , \quad \text{where} \quad c = N\,\pi\,[\tan(\beta)]^2 \quad \text{and} \qquad (10.13)$$

$$\Delta x = \sqrt{\frac{e}{2\,c}} \ , \quad \text{where} \ e = 2.71828... \ . \qquad (10.14)$$

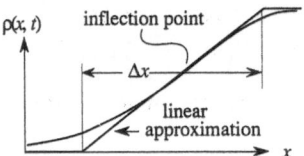

**Figure 10-21** Linearizing the effective density $\rho_{\text{eff}}(x, t)$.

According to reference [1] the reflected amplitude $A_r$ and the energy reflection coefficient $R_e$, due to a linearized density step, are given by:

$$A_r = \frac{n_1 - n_3}{n_1 + n_3} \, e^{-i \vartheta} \, \frac{\sin(\vartheta)}{\vartheta} \quad \text{and} \tag{10.15}$$

$$R_e = \left(\frac{n_1 - n_3}{n_1 + n_3}\right)^2 \left(\frac{\sin(\vartheta)}{\vartheta}\right)^2, \quad \text{where} \tag{10.16}$$

$$\vartheta = \frac{2\pi}{\lambda} \Delta s \quad \text{and} \quad \Delta s = \frac{\Delta x \, n_1 \, n_3}{n_1 - n_3} \ln\left(\frac{n_1}{n_3}\right) \tag{10.17}$$

is the optical path length — or optical thickness — for a wave starting at the beginning of the linearized step, reflected at the upper limit of the linearized step, and returning to its beginning.

The analytical eq. (10.16) is evaluated in Figure 10-22 and represents an upper limit for $R_e$ and enables to judge the influence of the various parameters $N$, $g$, v, $\lambda$ on the energy reflection coefficient directly. Due to the discontinuity of the linearized step at its beginning an at its end there exist, however, interferences which are not present in the numerical solution of the time-evolution of the reflectivity.

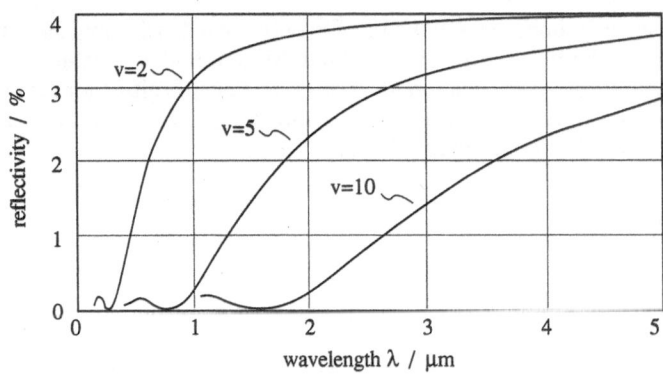

**Figure 10-22** *Reflectivity* according to linearized step as function of wave-length. Used parameters: Areal density $N = 10^{10}$ ions per cm$^2$, $g = 1$ µm / min, $n_1 = 1$, $n_3 = 1.5$, v = {2, 5, 10} µm / min.

[1]     Knittl, Z.: "**Optics of Thin Films.**" John Wiley, London, pp. 455 - 458, (1967).

### 10.2.3    Further possibilities

There exists an almost unlimited variety of ion track applications. Before closing this work just a few possibilities may be mentioned to underline this statement.

Deeply structured surface textures such as obtained by the ion track technique provide a high surface resistance [1] and are highly insensitive to metal deposition. Surface textures can be used to increase the mutual interlock between the track-structured surface and a deposited layer on top of it. Replicated track cones can be transformed into protruding very fine pins (Figure 10-23 left) [2], which could be used as finely distributed field emitters. Finally, the filling of etched track structures by other materials, for example metals, is a possibility to generate nicely oriented dipoles (Figure 10-23 right) and microcomposite materials (Figure 10-24).

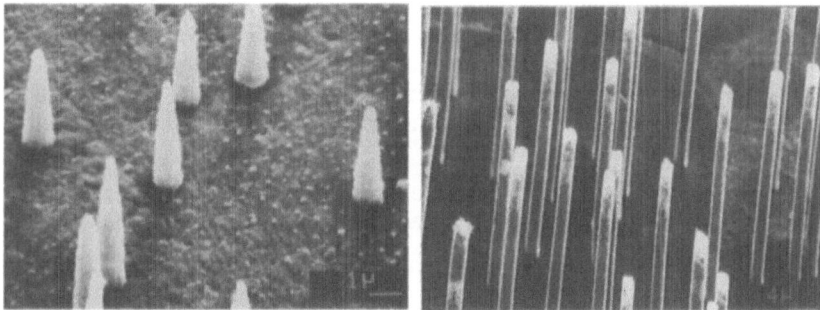

**Figure 10-23** (**left**) Copper pins obtained by replication of conical tracks. (**right**) Oriented metallic dipoles obtained by replication of etched ion tracks in mica.

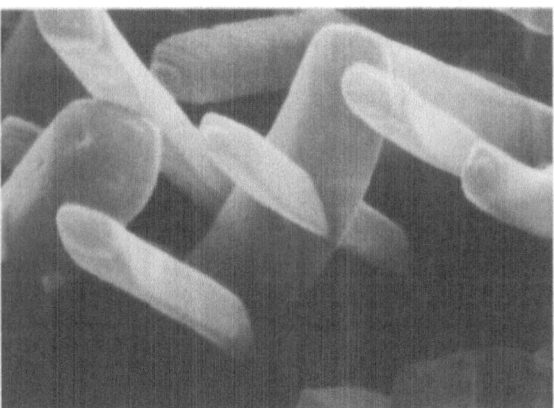

**Figure 10-24**  Interconnected fibers of a "microcomposite", obtained by filling criss-crossing etched ion tracks in mica by polystyrene.

[1]    B.E Fischer, D. Albrecht, R. Spohr: "Preparation of Superinsulating Surfaces by the Nuclear Track Technique." Radiation Effects, vol.65, pp. 143 - 144, (1982).
[2]    Fischer, B.E., R. Spohr: "Production and Use of Nuclear Tracks: Imprinting Structure on Solids." Reviews of Modern Physics, **55**, No. 4, pp. 907-948, (1983).

# Concluding remarks

The sweeping progress of ion source and accelerator technology is a challenge for engineers and scientists to promote as well their practical uses.

While low-energy ion implantation can be considered as a standard tool in semiconductor and surface technology, at considerably higher ion energies the astounding phenomenon of discrete and directed ion tracks appears distinctly. This opens the startling possibility to structure and modify materials with individual atomic particles. The unrivalled depth and sharpness of this "microknife" foreshadows new stimuli for an almost unlimited variety of fields.

The above work is devoted to the preliminary exploration of the technique, outlining tentatively its relation to materials and microstructure technology.

In realizing this task, "synergetically" several favorable circumstances conspired. The availability of heavy ions of sufficient energy (boosted by the needs of basic research), the possibility to amplify the engraved damage zone chemically by many orders of magnitude (a key to the practical applicability), and finally the responsive tolerance of the embedding environment (a necessary condition for a pleasant work).

Very much remains to be done to penetrate deeper, to understand better, to describe the transcription of energy into a lasting radiation effect and its development, and, finally, to derive useful applications. Some consolation about the incompleteness of this work can be found in the history of a closely related branch of science: Hundred and fifty years after the invention of photography, this field is still far from coherent and conclusive, but nevertheless sparkling with life and fruitfulness.

# Definitions and units

The following glossary is intended as a conveniently accessible source of technical terms used in context with ion tracks. It is compiled mainly on the basis of the references [1], [2], [3], [4], [5], [6].

---

[1] D.H. Menzel (editor): "**Fundamental Formulas of Physics.**" Dover Publications, New York, 1960
[2] "**Glossary of Terms used in Nuclear Science.**" British Standards Institution, B.S. 3455, 1962
[3] "**The Radiochemical Manual.**" Amersham, the Radiochemical Centre, 1966
[4] Cohen, E., P. Giacomo: "**Symbols, Units, Nomenclature and Fundamental Constants in Physics.**" International Union of Pure and Applied Physics (IUPAP), Revision 1987, 67 pp. (1987).
[5] Alan Isaacs (editor): "**Concise Dictionary of Physics.**" Oxford University Press, pp. 1-295, 1985.
[6] Hansen, J.W.: "Experimental Investigation of the Suitability of the Track Structure Theory in Describing the Relative Effectiveness of High-LET Irradiation of Physical Radiation Detectors." Risø National Laboratory, DK-4000 Roskilde, Denmark; Risø-R-507, (1984).

---

**absorbed dose**, absorbed energy per unit mass. SI unit is the Gray (Gy). $1 \text{ Gy} = 1 \text{ J} / \text{kg} = 10^4 \text{ erg} / \text{g} = 6.25 \cdot 10^{15} \text{ eV/g}$.

**actuator**, a device for inducing an action.

**alpha particle**, ($\alpha$), the nucleus of a $^4\text{He}$ atom consisting of two protons and two neutrons.

**areal density** of targets, ($\rho_s$), thickness-equivalent of targets, $\rho_s \equiv \rho\, d$, where $\rho$ is the mass density and $d$ the target thickness, usually given in $\text{mg} / \text{cm}^2$.

**areal dose** (fluence), number of accumulated (recorded) particle tracks per unit surface area.

**aspect ratio**, ratio between length and width of an ion track.

**atom**, a unit of matter consisting of a single nucleus surrounded by one or more orbital electrons. The number of electrons corresponds to the number of protons in the nucleus.

**atomic mass** ($A \cdot u$), mass of a nuclide measured in atomic mass units, where $A$ = atomic mass number, and $u$ = atomic mass unit.

**atomic mass number** ($A$), mass number of a nuclide, nucleon number, number of protons and neutrons in the nuclide. $A$ is the nearest integer to its atomic mass.

**atomic mass unit** ($u$), one-twelfth of the mass of a neutral atom of $^{12}\text{C}$, $u = 1.6606 \cdot 10^{-27} \text{ kg}$.

**atomic number** ($Z$) of an element, nuclear charge number, number of protons in the nucleus of the atoms of the element. The atomic number defines the chemical properties of the element.

**atomic radius** of hydrogen, $a_0 = h^2 / 4\pi^2\, m\, e^2 = 0.5292 \cdot 10^{-8} \text{ cm}$.

**atomic weight** ($W$), for a given specimen of an element, the mean weight of its atoms, expressed in either atomic mass units (physical scale) or atomic weight units (chemical scale).

**atomic weight unit** (awu), one-twelfth of the mean mass of the neutral atoms of naturally occurring carbon.

**Avogadro number**, $N_A = 6.0220 \cdot 10^{23} / \text{mole}$, number of molecules per mole.

**bar**, unit of pressure. $1 \text{ bar} = 10^5 \text{ Pa} = 750.19 \text{ torr}$.

**barn**, a unit of cross-section. $1 \text{ barn} = 10^{-24} \text{ cm}$.

**Becquerel** (Bq), unit of radioactivity, $1 \text{ Bq} = 1$ decay event / s. Official unit *before* 1986: 1 Curie (Ci) $= 3.7 \cdot 10^{10}$ decay events / s.

**beta particle, electron.**

**Bohr magneton**, $\mu_B = e\, h / (4\pi\, m_e\, c) = 9.2741 \cdot 10^{-24}$ Joule pro Tesla.

**Bohr radius**, $a_0 = \alpha / 4\,\pi\, R_\infty = 0.529177 \cdot 10^{-8} \text{ cm}$. $\alpha$ = fine structure constant, $\pi = 3.1415$, $R_\infty$ = Rydberg constant.

**Bohr velocity**, $v_0 = 2\pi\, e^2 / h = \alpha\, c = 2.1847 \cdot 10^8 \text{ cm} / \text{s}$, where $e$ = electron charge, $h$ = Planck constant, $\alpha$ = fine structure constant, $c$ = velocity of light.

**Boltzmann constant**, $k = R / N_A = 1.3806 \cdot 10^{-23} \text{ J}$

$/K = 1.3806 \cdot 10^{-16}$ erg / K. $R$ = molar gas constant, $N_A$ = Avogadro constant.

**Bragg peak:** region of maximum energy loss of a high energy ion traversing matter.

**bremsstrahlung**, the electromagnetic radiation resulting from the retardation of charged particles.

**calorie**, unit of thermal energy, 1 calorie = 4.18400 Joules.

**capacitance**, unit of capacitance is the Farad (F). 1 Farad = 1 Coulomb / Volt = $1\ m^{-2}\ kg^{-1}\ s^4\ A^2$.

**charge, electric** ~, given in units of Coulomb (C). $1\ C = 1\ A\ s = 3 \cdot 10^9$ esu.

**cross section** of a nucleus, atom, or grain for a given radiation, that area perpendicular to the direction of the radiation attributed geometrically for its interaction with the radiation; or, in other words, the number of interactions per unit time divided by the radiation flux and the number of scattering objects present. The cross section of a nucleus is roughly of the order of $10^{-24}$ cm, of an atom roughly of the order of $10^{-16}$ cm.

**Coulomb** (C), the unit of electric charge. The charge of 1 Coulomb corresponds to the charge of an electrical current of 1 Ampere flowing during 1 second. $1\ C = 1\ A\ s$. 1 Coulomb corresponds to the charge of $0.62414 \cdot 10^{19}$ electrons.

**Coulomb barrier** of nuclear reactions, minimum energy of projectile ions required for inducing nuclear reactions. The coulomb barrier corresponds to a specific energy in the range between 1 and 10 MeV/nucleon, depending on the projectile/target combination.

**Coulomb explosion**, origin of the atomic collision cascade in solids, caused by the passage of a rapidly moving ion and leading to displaced atoms and ultimately to a latent track.

**Coulomb per kilogramm (C / kg)**, "unit" of absorbed dose, defined as total electron charge generated during — preferentially x or γ — irradiation. Official unit *before* 1986: 1 Roentgen = $2.58 \cdot 10^{-4}$ C / kg. One Roentgen produces $3.7 \cdot 10^{10}$ ion pairs in air at standard pressure and temperature.

**Curie** (Ci), old unit of activity. One curie corresponds to $3.7 \cdot 10^{10}$ nuclear transformations per second. $1\ Ci = 3.7 \cdot 10^{10}$ Bq.

**De Broglie wavelength** of a particle, $\lambda = h / p$, where $h$ = Planck constant, and $p$ is the particle momentum. The magnitude of the associated wave vector is $k \equiv 2\pi / \lambda = 2\pi\ p / h$.

**delta ray:** secondary and higher order electrons generated by a primary radiation. This term is analogous to the terms alpha-ray and beta-ray.

**density, electron** ~ ($N_e$), number of electrons per volume element, $N_e = Z \rho / (A\ u)$, where $Z$ = nuclear charge number, $\rho$ = mass density, $A$ = atomic mass number, $u$ = atomic mass unit.

**density, mass** ~ ($\rho$), mass per volume element, $\rho = A\ u\ N_A / V_m$, where $A$ = atomic mass

number, $u$ = atomic mass unit, $N_A$ = Avogadro number, $V_m$ = molar volume.

**density, number** ~ ($N$), number of atoms per volume element, $N = \rho / (A\ u)$, where $\rho$ = mass density, $A$ = atomic mass number, $u$ = atomic mass unit.

**dose, energy** ~, absorbed energy per unit mass. SI unit is the Gray (Gy). 1 Gy = 1 J / kg = $10^4$ erg / g = $6.25 \cdot 10^{15}$ eV / g. Radiation energy deposited in a medium causes bond rupture, radical formation, and physico- and chemical changes, which are detectable in different ways.

**dose equivalent**, see Rem. A unit of biologically effective dose, defined as the absorbed dose in rad multiplied by the quality factor — or relative biological effectiveness (*RBE*) —. For all x rays, γ rays, β⁻ rays and β⁺ rays encountered from radioisotopes the *RBE* is 1. For α-particles and heavier ions the *RBE* can be different from 1, depending on the ion charge and its velocity.

**Dose-response:** relation between the deposited energy per volume, in other words the dose, and the observed effect, which may for example be optical absorption, survival of cells, etc.. For many physical radiation recorders, irradiated by low-*LET* radiation, the dose-response function is approximately linear at low doses and saturates exponentially at high doses and the response is a single-valued function of dose within a large range of initial photon and electron energies.

**dyne**, unit of force, 1 dyne = 1 g cm s⁻².

**electron**, the negatively charged particle which forms a constituent of all atoms. Electron charge $e = -1.60219 \cdot 10^{-19}$ Coulomb = $-4.8032 \cdot 10^{-10}$ esu. Electron mass $m_e = 9.1095 \cdot 10^{-28}$ g. Classical electron radius $r_e = e^2 / (m_e\ c^2) = 2.8179 \cdot 10^{-13}$ cm.

**electron volt** (eV), a unit of energy equal to the kinetic energy acquired by an electron when accelerated through a potential difference of 1 volt. 1 eV = $1.602 \cdot 10^{-12}$ erg = $1.602 \cdot 10^{-19}$ Joule.

**eV**, electron volt. 1 eV = $1.602 \cdot 10^{-12}$ erg.

**element**, matter consisting of atoms having the same atomic number $Z$.

**energy**, a measure of a system's ability to do work. *Potential energy* is the energy stored in a body or system as a consequence of its position, shape, or state. *Kinetic energy* is energy of motion and is usually defined as the work that will be done by the body when it is brought to rest. For a body of mass $m$ having a speed v, the kinetic energy is $m\ v^2/2$. The rotational kinetic energy of a body with an angular velocity ω is $I\ \omega^2 / 2$, where $I$ is its moment of inertia.

**energy units**, 1 erg = 1 g cm² s⁻². 1 J = $10^7$ erg = 0.239 calories. 1 erg = $6.25 \cdot 10^{11}$ eV. 1 eV = $1.6 \cdot 10^{-12}$ erg. 1 kJ / mole = $1.038 \cdot 10^{-2}$ eV / molecule.

**energy, free** ~, corresponds to the following two thermodynamic functions. *Gibbs free energy*,

$G=H-TS$, is the energy liberated or absorbed in a reversible process at constant pressure and constant temperature $(T)$, $H$ is the enthalpy and $S$ the entropy of the system. Changes in Gibbs free energy, $\Delta G$, are useful in indicating the conditions under which a chemical reaction will occur. If $\Delta G$ is positive the reaction will only occur if energy is supplied to force it away from the equilibrium position (when $\Delta G=0$). If $\Delta G$ is negative the reaction will proceed spontaneously to equilibrium. *Helmholtz free energy*, $F=U-TS$, where $U$ is the internal energy, is that portion of the energy of a system which is the maximum available energy for doing work. For a reversible isothermal process, $\Delta F$ represents the useful work available.

**energy, internal ~**, $U$, $U=Q+W$, where $Q$ and $W$ are heat and work transferred to the system, respectively. The relation $U=Q+W$ is referred to as the first law of thermodynamics. The internal energy of a body is the sum of the potential energy and the kinetic energy of its component atoms and molecules. It does not include the kinetic and potential energies of the system as a whole nor their nuclear energies or other intra-atomic energies. The value of the absolute internal energy of a system in any particular state cannot be measured; the significant quantity is the change in internal energy, $\Delta U$. For a closed system (a system that is not being replenished from outside its boundaries) the change in internal energy is equal to the heat absorbed by the system ($\Delta Q$) from its surroundings, plus the work done on the system by its surroundings ($\Delta W$), $\Delta U=\Delta Q+\Delta W$.

**energy loss** ($dE / dx$), linear energy transfer ($LET$), or stopping power ($S$), energy deposited by the incident ion per unit path length of the stopping medium.

**energy, specific ~**, ($T_s$) characteristic energy of a high frequency accelerator, corresponding to identical ion velocities, independent of the nucleon number. A given ion of nucleon number $A$ will be accelerated to an energy $T$ corresponding to the specific energy $T_s$ times the nucleon number $A$, $T = T_s A$. Unit of the specific energy is MeV/nucleon (MeV/u).

**enthalpy**, heat content, $H=U+pV$, where $U$=internal energy of system, $pV$=product of pressure and volume of the system. For chemical reactions at constant pressure one has $\Delta H = \Delta U+p \Delta V$. For an exothermic reaction $\Delta H$ is taken to be negative.

**enthalpy, free ~**, Gibbs free enthalpy $\equiv$ Gibbs free energy, $G=H-TS$.

**entropy**, a measure of the (isothermally) unavailable energy in a thermodynamic system. An increase in entropy is accompanied by a decrease in energy availability. When a system undergoes a reversible change the entropy S changes by an amount equal to the energy Q absorbed by the system divided by the thermodynamic temperature $T$ at which the energy is absorbed, $\Delta S=\Delta Q/T$. However, all real processes are to a certain extent irreversible changes and in any closed system an irreversible change is always accompanied by an increase in entropy. The increase in the entropy of a body during an infinitesimal stage of a reversible process is equal to the infinitesimal amount of heat absorbed divided by the absolute temperature of the body. Thus for a reversible process $dS=dQ/T$. $S(T)$ is the integral of $dQ/T = c_p\, dT/T$ from 0 to $T$, where $dQ$ = heat increment at temperature $T$, $c_p$ = specific heat at constant pressure. In a wider sense entropy can be interpreted as a measure of a system's disorder; the higher the entropy the greater the disorder.
*Boltzmann relation of entropy*, $S = k \ln W$, $W$ = probability, $k$ = Boltzmann constant.

**erg**, cgs unit of energy. 1 erg = 1 g cm s$^{-2}$. 1 erg = $10^{-7}$ Joule.

**excitation**, The addition of energy to a system, transforming it from its ground state to an excited state.

**Faraday constant**, $F = N_A\, e = 9.6484 \cdot 10^4$ C/mol.

**fine structure constant**, $\alpha = 2\pi\, e^2 / (h\, c) = 7.29735 \cdot 10^{-3} \approx 1 / 137$.

**Fermi level**, energy level in a solid at which the probability to find an electron is 1/2. At absolute zero all electrons would occupy energy levels up to the Fermi level and no higher levels would be occupied.

**fission**, a nuclear reaction in which a heavy nucleus splits into two (or rarely three) approximately equal parts. The specific energy of fission fragments is of the order of 1 MeV/nucleon.

**fluence of particles**, areal dose, number of accumulated (recorded) particle impacts (tracks) per unit surface area.

**flux of particles**, number of particles penetrating a unit surface area per unit time or product of the number of particles per unit volume and their average speed.

**gamma radiation**, gamma rays, electromagnetic radiation emitted by atomic nuclei.

**gas constant, molar ~**, $R = 8.314$ J / (mol K). $R = k\, N_A$; $k$ = Boltzmann constant, $N_A$ = Avogadro constant.

**grain**, sensitive element, sensitive volume, It is assumed that the medium consists of identical sensitive "elements", which may be either atoms, molecules or larger entities embedded in a more or less passive matrix acting as an energy transfer medium. Once the sensitive element has been activated it will in principle stay activated despite being hit several times. The amount of excitation and ionization energy deposited by secondary electrons is taken as a measure of the density of hits in the irradiated medium.

**Gray** (Gy), unit of absorbed energy dose, 1 Gy = 1 J / kg. Official unit *before* 1986: 1 rad = $10^{-2}$

$Gy = 10^{-2}$ J / kg = 100 erg / g.

**ground state**: the lowest-energy state of a system.

**half-life**, the time in which an ensemble of excited particles decays to half its initial number.

**hit**: a quantized interaction which implies that one event under consideration takes place in the sensitive element — grain — of the medium and initiates an effect. The activation may be brought about by a single electron passing through the sensitive element.

**hydrogen radius**, Bohr radius, $a_0 = 0.529177 \cdot 10^{-8}$ cm.

**inductance**, unit of inductance is the Henry (H). 1 Henry = 1 Weber / Ampere = $m^2$ kg $s^{-2}$ $A^{-2}$.

**ionization**, any process by which ions are formed; in particular, ionization of a gas by the passage of fast charged particles.

**ionization energy**, electron binding energy, energy required for removal of one electron. The hydrogen ionization energy is 13.6 eV.

**isotope effect**, differences that may be detectable in the chemical or physical behavior of two isotopes or their compounds.

**isotopes**, nuclides having the same atomic number Z but different mass numbers A.

**isotopic abundance**, the number of atoms of a particular isotope in a mixture of the isotopes of an element, expressed as a fraction of all the atoms of the element.

**Joule**, (J), SI unit of energy. 1 J = 1 kg $m^2$ $s^{-2}$. 1 J = $10^7$ erg.

**Josephson frequency/voltage ratio**, $2 e / h = 4.8359 \cdot 10^{14}$ Hz $V^{-1}$.

**keV**, thousand electron volts = $10^3$ eV.

*LET*, linear energy transfer, deposited energy per unit path length. More or less equivalent to energy-loss (d$E$ / d$x$) and stopping power $S$.

**linear energy transfer** (*LET*), energy loss (d$E$ / d$x$), or stopping power ($S$), energy deposited by the incident ion per unit path length of the stopping medium. The *LET* is roughly proportional to $Z_{eff}^2$ / $\beta^2$, where $Z_{eff}$ is the effective charge of the penetrating ion and $\beta$ = v / c its relative velocity, measured in units of the speed of light.

**Loschmidt constant**, $n_0 = N_A / V_m = 2.6868 \cdot 10^{19}$ $cm^{-3}$.

**magnetic flux**, unit of the magnetic flux is the Weber (Wb). 1 Weber = 1 V s = 1 = $m^2$ kg $s^{-2}$ $A^{-1}$.

**magnetic flux quantum**, $\Phi_0 = h / 2 e = 2.06785 \cdot 10^{-15}$ J s $C^{-1}$.

**mass number** of a nuclide, (A), number of protons and neutrons in the nuclide. A is the nearest integer to its atomic mass.

**metastable state**, long-lived state of an atom or nucleus with an energy above the ground state.

**MeV**, million electron volts = $10^6$ eV.

**MeV/nucleon**, MeV/$u$, unit of specific energy $T_s$ produced by a specific high frequency accelerator.

**milliatom**, one thousandth part of the atomic weight of the element in grams .

**millimole**, (mM), one thousandth part of a mole.

**mole** (M), molecular weight of a compound in grams .

**moment of inertia**, (*I*), $I = \int dm\ r^2$. For a mass point at distance r from the origin $I = m\ r^2$.

**momentum**, quantity of motion equal to the product of mass and velocity of a particle.

**multi-hit recorder**: A track-recording medium in which the sensitive grains consist either of several targets each of which has to be hit once or one target which has to be hit several times before the effect is observed. Accordingly, multi-hit recorders may either be described as single-hit multi-target systems or as multi-hit single target systems.

**neutron**, one of the particles of which nuclei consist; it is of zero charge and slightly heavier than a proton. $m_n = 1.67495 \cdot 10^{-24}$ g.

**Newton**, (Nt), unit of force, 1 Nt = 1 kg m $s^{-2}$.

**nuclear charge number** (Z), atomic number, number of protons contained within the nucleus.

**nucleon**, a particle constituting the atomic nucleus; proton or neutron.

**nucleon number**, (A), mass number, number of protons and neutrons in an atomic nucleus.

**nucleus**, the positively charged central portion of an atom, containing almost the whole mass of the atom, but only a minute part of its volume. Radius $r_A \approx 1.2 \cdot 10^{-13} \cdot A^{1/3}$, where A = atomic mass number.

**nuclide**, a species of atom characterized by its mass number A, atomic number Z, and nuclear energy state.

**Pasqual**, (Pa), unit of pressure. 1 Pa = 1 Nt / $m^2$ = $10^{-5}$ bar = $0.75 \cdot 10^{-2}$ torr.

**permeability of vacuum**, $\mu_0 = 1 / \varepsilon_0 c^2 = 4 \pi \cdot 10^{-7}$ H $m^{-1}$ = $1.256637 \cdot 10^{-6}$ H $m^{-1}$.

**permittivity of vacuum**, $\varepsilon_0 = 1 / (\mu_0 c^2) = 8.854188 \cdot 10^{-12}$ F $m^{-1}$.

**photon**, a quantum of electromagnetic radiation, possessing the energy $h\nu$ (h being Planck's constant and $\nu$ the frequency).

**Planck constant** (*h*), h = $6.6262 \cdot 10^{-34}$ J s = $6.6262 \cdot 10^{-27}$ erg s. Frequently used quantity: $h / 2\pi = 1.05459 \cdot 10^{-34}$ J s.

**pressure**, force per unit surface area. SI unit is Pasqual (Pa). 1 Pa = 1 Nt / $m^2$ = $10^{-5}$ bar = $0.75 \cdot 10^{-2}$ torr.

**proton**, a nuclear particle of mass number A = 1 having a charge equal and opposite to that of an electron and having a mass of $m_p = 1.67265 \cdot 10^{-24}$ g = 1836.16 $m_e$.

**rad**, the old unit of absorbed dose. 1 rad = $10^{-2}$ J $kg^{-1}$ = $10^{-2}$ Gy = 100 ergs $g^{-1}$. The unit rad expresses the energy absorbed in a particular material from a radiation flux. It makes no reference to the type or quantum-energy of the applied radiation.

**radiation, high-*LET* radiation:** neutrons or heavy charged particles (ions). Heavy charged particles essentially move in straight lines through the recording medium losing only a small fraction of their energy per collision. They distribute their energy very *inhomogeneously* in the medium through the ejected low-energy electrons, the delta-rays. The maximum energy and range of the delta-rays are only small fractions of that of the primary particle. High-*LET* radiation creates also primary and secondary ions and neutral atoms with a still smaller ejection range. The track of a heavy ion is constituted by a core of clusters, which consists mainly of very low-energy Auger electrons, excited atoms, and ions, and a penumbra of more energetic delta-rays clearly separated from the core. High-*LET* radiation is often arbitrarily defined as radiation having a *LET* well above that of electrons.

**radiation, low-*LET* radiation:** fast electrons, x-rays, and γ-rays. Low-*LET* radiation leads to a low ionization density and distributes its energy more or less *homogeneously* in the recording medium because of multiple scattering (Compton processes) and the relative long range of the formed secondary electrons. The secondary radiation of low-*LET* radiation contains no ions or neutral atoms.

**radiation biology**, the study of radiation effects in living matter.

**radiation chemistry**, the study of the chemical effects of radiation on matter.

**radioactivity**, the property of certain nuclides of emitting radiation by the spontaneous transformation of their nuclei.

**relative biological effectiveness**, *RBE*, biological effectiveness of a particular radiation with respect low *LET* radiation, such as γ radiation.

**recoil**, the motion acquired by a particle through ejecting another particle or photon.

**Rem** (Roentgen equivalent man), the old unit of dose equivalent, defined as the absorbed dose in rad, multiplied by the quality factor — or relative biological effectiveness (*RBE*) — of the particular type of radiation under consideration. The biological effects of equal absorbed doses, in rad (1 rad = 100 erg / gram), of different kinds of radiation are not the same, mainly due to the different local distribution of the deposited energy. For x rays, γ rays, β particles and positrons the *RBE* is 1. For α-particles and heavier ions the *RBE* can be different from 1, depending on the ion charge and its velocity.

**Roentgen**, the old unit of exposure dose of x or γ radiation. One Roentgen is the amount of x or gamma radiation that will produce ions in air containing a quantity of positive or negative charge equal to one electrostatic unit in 0.001239 grams of air, if fully absorbed. One electrostatic unit of charge corresponds to $1/4.8032 \cdot 10^{-10}$ = $2.0819 \cdot 10^9$ electrons.

**Rutherford scattering**, Coulomb scattering between two nuclei.

**Rydberg constant**, $R_\infty = \mu_0^2\, m_e\, e^4\, c^3\, /\, (8\, h^3)$ = $1.0973732 \cdot 10^5$ $cm^{-1}$.

**sensor**, a device for observing the environment.

**Sievert (Sv)**, unit of absorbed dose-equivalent, defined as energy dose in Gray, multiplied by a quality factor — the so-called relative biological effectiveness (*RBE*) of the particular type of radiation under consideration. 1 Sv = 1 J / kg. Official unit *before* 1986: 1 Rem (Roentgen equivalent man) = $10^{-2}$ J / kg.

**specific activity**, the activity per unit mass of an element or compound containing a radioactive nuclide.

**standard volume** of an ideal gas, $V_0 = 22.420 \cdot 10^3$ $cm^3$ $mole^{-1}$ at standard pressure (760 torr or $10^5$ Pa ) and temperature (0°C).

**stopping power** (*S*), energy loss (d*E* / d*x*), or linear energy transfer (*LET*), energy deposited by the incident ion per unit path length of the stopping medium.

**target**, object of an irradiation with rapidly moving particles for initiating atomic or nuclear processes.

**thickness equivalent** of targets, $(\rho_s)$, areal density of targets, $\rho_s \equiv \rho\, d$, where $\rho$ is the mass density of the target material and $d$ the target thickness.

**torr**, old unit of pressure, corresponding to a column of 1 mm Mercury (Hg). 1 torr = $1.333 \cdot 10^2$ Pa = $1.333 \cdot 10^{-3}$ bar.

**track, etched ~**, result of developing a latent track in an etching medium that selectively removes the material of the latent track.

**track, latent ~**, permanent ion track in a solid which may be enlarged by a later development process, for example by an etchant.

**track, ion ~**, approximately cylindrical damage zone, created along the path of a rapidly moving ion in matter.

**uv radiation**, electromagnetic radiation with wavelengths between roughly 0.02 and 0.2 μm.

**velocity of light ,** $c = 2.9979246 \cdot 10^{10}$ cm $s^{-1}$.

**volume, molar ~** ($V_m$) of ideal gas at $T_0 = 273.15$ K and $p_0 = 101\ 325$ Pa. $V_m = R\ T_0\, /\, p_0$ = $2.2414 \cdot 10^{-2}$ $m^3$ $mol^{-1}$.

**x rays**, electromagnetic radiation resulting from electronic transitions or from the acceleration of charged particles (typically electrons), having shorter wavelength than ultra-violet radiation. .

# List of symbols

In the following a list of the most frequently used symbols is given. In order to avoid multiply indexed symbols, the definitions depend on the context.

| | | | | | |
|---|---|---|---|---|---|
| $A$ | alignment parameter, amplitude, atomic mass number | $I$ | ment of inertia<br>electric current, heat flow, intensity, ionization potential | $\rho$ | dius, resistance<br>mass density |
| $\alpha$ | alpha particle, angle | $\vartheta$ | scattering angle | $\rho_\varepsilon$ | charge density |
| $a$ | distance, size | $J$ | linear ionization density | $R_c$ | condensation rate |
| $a_0$ | Bohr radius | $\varphi$ | scattering angle | $R_e$ | evaporation rate |
| $\boldsymbol{B}$ | magnetic induction | $\boldsymbol{j}$ | mass flow density, electric current density | $S, s$ | solubility constant, sorption constant, stopping power, surface area |
| $\beta$ | angle, relative velocity $v/c$ | $\varphi$ | phase | $\sigma$ | surface area, cross section |
| $b$ | distance of closest approach | $K$ | Kelvin | $T$ | kinetic energy, temperature, oscillation time interval |
| $C, c$ | capacitance, concentration, lithographic contrast, velocity of light | $k$ | Boltzmann constant | | |
| | | $\kappa$ | elastic compliance, thermal conductivity | $t$ | time |
| $c_v$ | specific heat | $k$ | wave number $2\pi/\lambda$ | $\tau$ | time interval, life time |
| $\chi$ | magnetic susceptibility | $\boldsymbol{k}$ | wave vector | $T_c$ | critical temperature |
| $D,d$ | diffusion constant, diameter, differentiation operator | $L, l$ | inductance, length | $T_s$ | specific energy |
| | | $\lambda$ | wave length | $U$ | potential energy, voltage |
| $\Delta$ | divergence operator | $\boldsymbol{M}$ | magnetization | $u$ | atomic mass unit |
| $\delta$ | displacement, variation operator | $M, m$ | mass, molecular weight | $V$ | potential energy, volume |
| | | $\mu$ | chemical potential | $v$ | track etch rate |
| $\boldsymbol{E}$ | electric field strength | $\mu$ | magnetic moment | $v$ | velocity |
| $E$ | energy | $m_0$ | rest mass at velocity $v = 0$ | $v_0$ | Bohr velocity |
| $e$ | Euler's number | | | $v_c$ | critical velocity |
| $e$ | electron charge | $\mu_B$ | Bohr magneton | $V_m$ | molar volume |
| $\varepsilon$ | energy density, dose | $M_s$ | saturation magnetization | $W$ | atomic weight, cumulative probability |
| $E_F$ | Fermi energy | $N, n$ | areal dose, fluence, integer number, number density, refractive index | | |
| $\boldsymbol{F}$ | force | | | $\Omega$ | solid angle |
| $F$ | particle flux (ions per $cm^2$ and per s) | $\nu$ | frequency | $\omega$ | angular frequency |
| | | $N_A$ | Avogadro number | $\omega_0$ | plasma frequency |
| $\gamma$ | lithographic contrast | $P, p$ | diffusion permeability, impact parameter, linear momentum, nominal porosity, static pressure, probability | $w$ | probability |
| $g$ | general etch rate, earth gravitational acceleration | | | $\xi$ | coherence length |
| | | | | $x$ | $x$-coordinate |
| $\gamma$ | relativistic mass ratio $m/m_0$, interface energy | $P_{eff}$ | effective porosity | $\psi$ | scattering angle, wave function |
| | | $Q, q$ | electric charge, thermal energy | $y$ | $y$-coordinate |
| $\boldsymbol{H}$ | magnetic field strength | | | $Z$ | nuclear charge number |
| $h$ | Planck constant | | | $z$ | $z$-coordinate |
| $\eta$ | viscosity constant | $R,r$ | ion range in matter, ra- | $Z_{eff}$ | effective charge number |
| $I$ | angular momentum, mo- | | | $\nabla$ | gradient operator |

# Index

# Probability and Heat

**Fundamentals of Thermostatistics**
by Friedrich Schlögl

*1989. XII, 249 pp. with 52 figs.*
*Hardcover DM 98,—*
*ISBN 3-528-06343-2*

Contents: General Statistics: Probability, Information Measures, Generalized Canonical Distributions – Thermodynamics of Equilibria: Thermal States, Statistical Foundations of the Macroscopic Scheme, The Phenomenological Framework, The Low Temperature Regime – Macroscopic Description of Special Systems: Gases and Solutions, Chemical Reactions, The Method of Cycle Processes – Microscopic Description of Special Systems: Thermal Equations of State, Specific Heat, Magnetism – Nonequilibria: Thermal Fluctuations, Nonequilibrium Dynamics, Linear Thermodynamics, A Model of Time Scale Separation.

*(. . .) The aim of the author, "to make things simpler and more transparent in statistical thermodynamics" is fully attained. (. . .)*

*(. . .) This textbook can not be only recommended to graduate students, but also to all those who do research work in physics, physical chemistry, and even in the modelling of technical processes.*

Prof. Dr. Schirmer,
Zeitschrift für physikalische Chemie 5/89

Dr. Dr. h. c. *Friedrich Schlögl* is Professor Emeritus for Theoretical Physics at the Technical University Aachen (Rheinisch-Westfälische Technische Hochschule Aachen), FRG.

Vieweg Verlag · Postfach 58 29 · D-6200 Wiesbaden 1

**vieweg**

# Festkörperprobleme
# Advances in Solid State Physics

Eine Buchreihe mit den Referaten des Fachausschusses Halbleiter der Deutschen Physikalischen Gesellschaft.

The aim of the "Festkörperprobleme" ist to stimulate the interest in modern problem in condensed matter physics covering fundamental and applied aspects, to show the state of the art of some examples, and to help the reader to find an approach to those topics and to the original literature.

**Festkörperprobleme 29**

Advances in Solid State Physics. Plenary Lectures of the Divisions Semiconductor Physics. Thin Films, Dynamics and Statistical Physics, of the German Physical Society (DPG), Münster, April 3 to 7, 1989.

*1989. VIII, 345 pp. with 204 figs.*
*Hardcover DM 168,–*
*ISBN 3-528-08035-3*

Die Frühjahrstagung des Arbeitskreises Festkörperphysik der Deutschen Physikalischen Gesellschaft fand 1989 in Münster statt. Der vorliegende Band 29 der *Festkörperprobleme* enthält eine Auswahl von Plenar- und Hauptvorträgen dieser Konferenz, darunter auch die Vorträge der diesjährigen Träger des Walter-Schottky-Preises zur Theorie quantenmechanisch kohärenter dissipativer Systeme. Bei den übrigen Beiträgen stehen Themen aus dem Bereich der Halbleiter- und der Oberflächenphysik im Vordergrund.

Vieweg Verlag · Postfach 58 29 · D-6200 Wiesbaden 1